The Harmony of the Spheres

The Harmony of the Spheres

A Sourcebook of the Pythagorean Tradition in Music

edited by
JOSCELYN GODWIN

Inner Traditions International
Rochester, Vermont

Inner Traditions International, Ltd
One Park Street
Rochester, Vermont 05767

Library of Congress Cataloging-in-Publication Data

Godwin, Joscelyn
        Harmony of the spheres: a sourcebook of the pythagorean tra-
dition in music/edited by Joscelyn Godwin.
        p.   cm.
    Includes bibliographical references.
    ISBN 0-89281-265-6
    1. Music—Philosophy and aesthetics. 2. Pythagoras and Pythagorean
school. I. Title.
ML3800.G573   1990
780'.1—dc20   89-24741
                                                            CIP
                                                            MN

Printed and bound in the United States

10   9   8   7   6   5   4   3   2   1

Distributed to the book trade in the United States by American International Distribution Corporation (AIDC)

Distributed to the book trade in Canada by Book Center, Inc., Montreal,
Quebec.

Contents

Preface and Acknowledgments   *xi*

I ✦ CLASSICAL

1 ✦ PLATO   *3*
Timaeus: Creation of the World-Soul

2 ✦ PLINY THE ELDER   7
Natural History: Pythagorean planet-tones

3 ✦ NICOMACHUS OF GERASA   *9*
Manual of Harmonics: Planet-tones and proportions

4 ✦ THEON OF SMYRNA   *16*
Mathematics Useful for Understanding Plato: Planet-tones

5 ✦ PTOLEMY   *21*
On Music: Harmonics and astronomy

6 ✦ CENSORINUS   *40*
On the Day of Birth: Harmonics, embryology, planets

7 ✦ THE HYMNS OF ORPHEUS   *46*
Orphic Hymn to Apollo

8 • SAINT ATHANASIUS   *48*
Against the Heathen: The cosmic lyre

9 • ARISTEIDES QUINTILIANUS   *51*
On Music: Music and the descent of the soul

10 • CALCIDIUS   *60*
Commentary on the Timaeus: World-Soul and universal harmony

11 • MACROBIUS   *64*
Commentary on the Dream of Scipio: Planetary music

12 • PROCLUS   *71*
Commentary on the Republic: Sirens and Muses
Commentary on the Timaeus: Division of the World-Soul

13 • BOETHIUS   *86*
The Principles of Music: Stars and Strings

II • MEDIEVAL

14 • HUNAYN   *91*
Maxims of the Philosophers: Aphorisms on music

15 • AURELIAN OF RÉÔME   *99*
Musica Disciplina: Modes and planets

16 • JOHN SCOTUS ERIUGENA   *104*
Commentary on Martianus Capella: Planet-tones

17 • REGINO OF PRÜM   *109*
Letter on Harmony: Planet-tones, psychology

18 • THE IKHWAN AL-SAFA' (BRETHREN OF PURITY)   *112*
Epistle on Music: Principles of music; Instrument symbolism;
Astronomy

19 • AL-HASAN AL-KATIB   *119*
The Perfection of Musical Knowledge: Resemblance of soul, music,
and the celestial sphere

20 • ANONYMOUS OF THE TWELFTH CENTURY *123*
"There is a concord of planets. . . ": Poem on planets and tones

21 • ISAAC BEN ABRAHAM IBN LATIF *126*
The King's Treasures: Music as a liberal art

22 • JACQUES DE LIÈGE *128*
Speculum Musicae: Musica mundana

23 • UGOLINO OF ORVIETO *141*
Declaratio Musicae Disciplinae: Musica mundana

24 • GIORGIO ANSELMI *145*
De Musica: Planet-tones, angels

25 • ISAAC BEN HAIM *152*
The Tree of Life: Music between heaven and earth

III • RENAISSANCE

26 • MARSILIO FICINO *163*
Letter on Music to Domenico Benivieni

27 • RAMIS DE PAREJA *170*
Musica Practica: Modes, planets, temperaments

28 • PICO DELLA MIRANDOLA *175*
Fourteen Pythagorean Conclusions: Ratios of parts of soul

29 • FRANCHINO GAFORI *177*
De Harmonia Musicorum Instrumentorum Opus: Muses, soul- and
sphere-music, number

30 • FRANCESCO GIORGI *185*
Harmonia Mundi: Timaeus numbers; macrocosm and microcosm

31 • HEINRICH GLAREAN *196*
Dodecachordon: Muses and tones

32 • GIOSEFFO ZARLINO   *205*
Institutioni Harmoniche: Musica mundana

33 • JEAN BODIN   *214*
Colloquium Heptaplomeres: Universal harmony

IV • BAROQUE

34 • JOHANNES KEPLER   *221*
Mysterium Cosmographicum: Planetary aspects and intervals
Harmonices Mundi: Planet songs

35 • ROBERT FLUDD   *236*
Utriusque Cosmi Historia: Music of spheres and elements; Harmony
of soul and body

36 • MARIN MERSENNE   *250*
Harmonie Universelle: Metaphysics of the unison

37 • ATHANASIUS KIRCHER   *263*
Musurgia Universalis: Musical experiments; Harmonies of the
planets and their satellites

38 • ANGELO BERARDI   *286*
Miscellanea Musica: Kabbalistic music

39 • ANDREAS WERCKMEISTER   *292*
Hodegus Curiosus: Allegory of Creation

V • ENLIGHTENMENT AND ROMANTICISM

40 • ISAAC NEWTON   *305*
Draft Scholia for the Principia: Ancient heliocentricity revealed

41 • JEAN-PHILIPPE RAMEAU   *309*
Traité de l'harmonie: Mathematical basis of music
Nouvelles réflexions: Universal principles of the arts

42 • GIUSEPPE TARTINI   *314*
Trattato di Musica: Music and geometry
Scienza Platonica: Rediscovery of ancient harmonic wisdom

43 · LOUIS-CLAUDE DE SAINT-MARTIN   *322*
Des erreurs et de la vérité: Metaphysics of the common chord

44 · JOHANN FRIEDRICH HUGO VON DALBERG   *335*
Polyhymnia: Music as a gift from Zeus

45 · ARTHUR SCHOPENHAUER   *338*
The World as Will and Idea: Nature and symbolism of music

46 · FABRE D'OLIVET   *344*
La Musique expliquée: Ancient musical cosmology
La Vraie maçonnerie et la céleste culture: Metaphysics of the
musical system

47 · ALPHONSE TOUSSENEL   *356*
L'Esprit des bêtes: Fourier's cosmology

48 · PETER SINGER   *362*
Metaphysische Blicke in die Tonwelt: Analogy of tone-world
with the Godhead

49 · ALBERT VON THIMUS   *370*
Die harmonikale Symbolik des Alterthums: Pythagorean table;
Ancient heliocentricity

50 · ISAAC RICE   *382*
What Is Music? Space and Time; Tone and Color

51 · SAINT-YVES D'ALVEYDRE   *395*
La Musique: Music's eternal principles

52 · AZBEL   *399*
Harmonie des mondes: Modern astronomy and harmonics

Notes   *403*
Bibliography   *477*
Index Nominum   *483*
Index Rerum   *489*

Preface and
Acknowledgments

✦

This publication completes, for the time being, a project in Speculative Music that began in 1978 with the systematic assembly, translation, and commentary on texts from Plato to the present day. Seeking material mainly in the British and Bodleian Libraries, I had by 1982 completed some thousand pages of typescript. For three years this unwieldy mass helped to clutter the office of a prominent London publisher, enthusiastic in principle but hesitant to launch a volume of this size. It was Eileen Campbell, then at Routledge & Kegan Paul, who urged me to cut the book in two and agreed to publish half of it. That half appeared in 1986 as *Music, Mysticism and Magic: A Sourcebook*, and the following year as an Arkana (now Penguin) paperback. (References in notes as well as parenthetical text references to this earlier work will use the abbreviation *MM&M*.)

Another visionary publisher, Ehud Sperling, has now brought the rest of this material out of oblivion, in a sourcebook that is in every way a sister-work to the earlier one. Whereas *Music, Mysticism and Magic*, true to its title, contained mainly writings on music as it affects the human being (what the ancients called "Musica Humana"), this one treats the loftier subject of "Musica Mundana," the music of the Worlds, or Spheres.

The Music (or *Harmony*) *of the Spheres* is a common enough phrase in poetic discourse, but how many know what is meant by it? This sourcebook exists to answer that question, as far as it can be answered in print. In one respect, the book is a many-faceted commentary on the passage in Plato's *Timaeus* that describes how the Demiurge fashioned the World-Soul. Our authors return again and again to this, each one adding to it from his store of knowledge — or prejudice. Several of the voices heard in *Music, Mysticism and Magic* return here to address the new level of their subject: they are Plato himself, of course, Censorinus, Boethius, the Brethren of Purity, Jacques de Liège, Ficino, Zarlino, Kepler, Fludd, Kircher, Werckmeister, von Dalberg, Schopenhauer, Fabre d'Olivet, and (interpreted by Toussenel) Charles Fourier. An additional thirty- eight authors, ranging from those with whose importance every educated person is familiar to others whose works have seldom seen the light of day, contribute to a picture of considerable richness.

I make no apology for including such recondite material, for it is my conviction that fame and a demonstrable influence on intellectual history are not the only standards by which a person should be judged. In this of all subjects, no one can do more than to chronicle his own eccentricities, and those of a Ptolemy or a Newton are not intrinsically superior to those of an Isaac ben Haim or an Azbel. Anyone who studies this subject at any length will conceive a special affection for his companions across the centuries, and will find a particular delight in the discovery of a new fellow-traveler on these little-trod roads of the intellect. My work during 1984–86, mostly in the Bibliothèque Nationale and the Bibliothèque de la Sorbonne in Paris, brought many such happy encounters, and enabled me to add the extracts from Azbel and Saint-Yves d'Alveydre. Much more on them, and on the French side in general, appears in my book *L'Esotérisme Musical en France, 1750–1950* (Paris: Albin Michel, 1991).

This sourcebook does not cover the twentieth century, as was attempted in *Music, Mysticism and Magic*. One reason is that the most important modern contributors to speculative music have been Hans Kayser, Rudolf Steiner, and their respective followers and pupils; and that these are represented, together with the solitary genius Marius Schneider, in my collection *Cosmic Music: Musical Keys to the Interpretation of Reality* (Rochester, Vt.: Inner Traditions International, 1989). Another reason is that modern work on the Harmony of the Spheres has become increasingly quantitative and technical, hence inaccessible to many a student, including myself; I cannot enjoy Clarence Hill's *Harmonia Harmonica*, for instance, or Dom Néroman's *La Leçon de Platon*. More gifted readers must discover these for themselves. I would only mention that the combined bibliographies of this work, *Music, Mysticism and Magic*, my book *Harmonies of Heaven and Earth* (Rochester, Vt.: Inner Traditions International, 1987), and K. S. Guthrie's *The Pythagorean Sourcebook* in the Phanes Press edition, contain as full a guide to the literature of speculative music as is currently available.

It is once again a pleasure to thank Colgate University for the several small grants that have enabled me to travel for my research, and for enabling the text to be typed onto disk by Patricia Ryan; the American Council of Learned Societies for a Grant-in-Aid during 1985-86; the National Endowment for the Humanities for Travel to Libraries Grants in 1986 and 1988. My colleagues Ross and Malva Ferlito helped me with some difficult passages in Tartini; Siegmund Levarie with information on Isaac Rice. Free use of their copyrighted translations was kindly granted by Professors Calvin Bower, Marion L. Kuntz, Clement A. Miller, Thomas E. Mathiesen, Amnon Shiloah, Mr. Arthur Farndell, and the editors of the Colorado College Music Press, the Scholars Press, the Wizard's Bookshelf, and the American Institute of Musicology. I have been unable to locate Mary Protase Le Roux to acknowledge my use of her work on Regino. It is only fair to add that these translations often contain valuable notes, which the restrictions of space and the nature of this sourcebook have compelled me to omit. Those notes that I have used are indicated by the writer's initials in brackets.

I

CLASSICAL

1

✦

Plato

c. 429–347 B.C.E.

Although Plato wrote his dialogue *Timaeus* some thirty years later than the *Republic*, the conversation it records is supposed to have taken place on the following day. The main speaker is Timaeus, a Pythagorean philosopher from Locri in southern Italy, and his subjects are the creation of the World-Soul, the structure of the elements, and the workings of the human body.

This passage concerning the making of the World-Soul by the Demiurge, or creator god, has rightly been called "the most perplexing and difficult of the whole dialogue,"[1] for it is an inextricable blend of mathematics and music, astronomy and metaphysics. Even in Plato's own century it was a cause of dispute among his successors at the Athenian Academy.

Among the modern guides, A. E. Taylor and Francis M. Cornford have approached the work from a basis of solid classical erudition. Cornford[2] sees this passage as primarily cosmological, and imagines Plato as having composed it with the visual aid of an armillary sphere or skeleton model of the cosmic orbits. Ernest McClain,[3] on the other hand, reads it as a musical myth, fitting naturally within the context of the ancient obsession with tuning systems that McClain has traced through Babylon, India, and China, as well as through all the works of Plato that deal in number. Plato would thus have used not the sphere but the monochord, an aural aid. Both these interpretations are elaborate and persuasive, being carried through consistently to the last detail.

Another independent researcher, John Michell,[4] finds that the *Timaeus* scale encodes and confirms the ancient canon of number, which Michell in his turn has unearthed in megalithic monuments and Gothic cathedrals, as well as in the Greek language itself. But for Proclus (see no. 12 of this sourcebook), whose commentary on the *Timaeus* is the most extensive to survive from the Academy, Plato's intentions were to be interpreted in the light of Orphic and Chaldaean theology, and his subject as the generation of the Gods.

Perhaps the *Timaeus* should be classed with those "revealed" scriptures, which seem to act as universal reflectors: each commentator projects onto such a work his own preoccupations and beliefs, and each receives from it perfect confirmation. One interpretation does not contradict another: each is valid in proportion to its interpreter's intellect and motives, and in the case of the five commentators named here, there is no question that these are of the highest. The notes are intended purely to make Plato's story more comprehensible.

The *Timaeus*, as it turned out, was the only Platonic dialogue that never sank into oblivion in the West. During the Dark and Middle Ages its first and most essential part was known in the Latin version of Calcidius (see no. 10). This freak of survival was probably the single most important fact in the transmission of the doctrines of cosmic and psychic harmony.

I use here the translation of Thomas Taylor (1758-1835), in homage to one who carried the torch of Platonism through another dark age, that of Regency England. Whatever the archaic qualities of his English style, one can be sure that he penetrated as deeply into the meaning of Plato's text as anyone in the modern age.

Source: The Works of Plato, translated by Thomas Taylor (London, 1804), vol. 2, pp. 484–488.

The Demiurge Fashions the World-Soul

[35a] From an essence impartible, and always subsisting according to sameness of being, and from a nature divisible about bodies, he mingled

from both a third form of essence, having a middle subsistence between the two.[5] And again, between that which is impartible and that which is divisible about bodies, he placed the nature of same and different. And taking these, now they are three, he mingled them all into one idea. But as the nature of different could not without difficulty be mingled in same, he harmonized them together by employing force in their conjunction.[6] [35b] But after he had mingled these two with essence, and had produced one from the three, he again divided this whole into becoming parts; at the same time mingling each part from same, different, and essence. But he began to divide as follows: –In the first place, he received one part from the whole. Then he separated a second part, double of the first; afterwards a third, sesquialters of the second, but triple of the first: then a fourth, double of the second; in the next place a fifth, triple of the third; a sixth, octuple of the first; [35c] and lastly a seventh, twenty-seven times more than the first.[7] After this, he filled up the double and triple intervals, [36a] again cutting off parts from the whole; and placed them so between the intervals, that there might be two mediums in every interval; and that one of these might by the same part exceed one of the extremes, and be exceeded by the other;[8] and that the other part might by an equal number surpass one of the extremes, and by an equal number be surpassed by the other.[9] But as from hence sesquialter, sesquitertian, and sesquioctave intervals were produced,[10] from those bonds in the first spaces, [36b] he filled with a sesquioctave interval all the sesquitertian parts, at the same time leaving a part of each of these.[11] And then again the interval of this part being assumed, a comparison is from thence obtained in terms of number to number, subsisting between 256 and 243.[12] But now the whole of that mixture from which these were separated was consumed by such a section of parts. Hence he then cut the whole of this composition according to length, and produced two from one; and adapted middle to middle, like the form of the letter X. Afterwards he bent them into a circle, [36c] connecting them, both with themselves and with each other, in such a manner that their extremities might be combined in one directly opposite to the point of their mutual intersection; and externally comprehended them in a motion revolving according to sameness, and in that which is perpetually the same.[13] And besides this, he made one of the circles external, but the other internal; and denominated the local motion of the exterior circle, the motion of that nature which subsists according to sameness; but that of the interior one, the motion of the nature subsisting according to difference. He likewise caused the circle partaking of sameness to revolve laterally towards the right hand; but that which partakes of difference diametrically towards the left. But he conferred dominion on the circulation of that which is same and similar: [36d] for he suffered this alone to remain undivided.

But as to the interior circle, when he had divided it six times, and had produced seven unequal circles, each according to the interval of the double and triple; as each of the intervals being three;[14] he ordered the circles to proceed in a course contrary to each other:—and three of the seven interior circles he commanded to revolve with a similar swiftness; but the remaining four with a motion dissimilar to each other,[15] and to the former three; yet so as not to desert order and proportion in their circulations.

After, therefore, the whole composition of the soul was completed according to the intention of its artificer, in the next place he fabricated within soul the whole of a corporeal nature; and, conciliating middle with middle, he aptly harmonized them together. [36e] But soul being every way extended from the middle to the very extremities of the universe, and investing it externally in a circle, at the same time herself revolving within herself, gave rise to the divine commencement of an unceasing and wise life, through the whole of time. [37a] And, indeed, the body of the universe was generated visible; but soul is invisible, participating of a rational energy and harmony,[16] and subsisting as the best of generated natures, through its artificer, who is the best of intelligible and perpetual beings.

2

$\blacklozenge$

Pliny the Elder
23 or 24–79 C.E.

On his retirement from military service, Gaius Plinius Secundus devoted himself to writing historical chronicles and making a collection of universal knowledge. The resulting work, a *Natural History* in thirty-seven books, is a stupendous achievement, and would supply scientists and philosophers with material until well into the seventeenth century. Pliny died accidentally while observing the interesting natural phenomenon of the eruption of Vesuvius.

In the first extract, Pliny expresses skepticism about the planetary music, while the second introduces us to the tangled skein of planet-tone theories. These are based not on their relative speeds, as in Cicero's conception (see Macrobius, no. 11), but on their distances, as if one were to stretch strings from the earth to each planet's sphere or orbit. In a later chapter, (II, xxi), Pliny denigrates all efforts to ascertain these distances, recommending "geometrical inference" as the only method likely to produce results, and then only conjectural ones. But curiously enough, he does not record any of these results, except those of Pythagoras, whom he dignifies as a "man of sagacious mind." Stranger still, he shows at the end of the book that he had access to the canon of exact measurements, at least of the earth's dimensions (see the mention of John Michell's work in note 3). Knowledge of this kind was a closely guarded secret, and it is impossible to penetrate the proverbial "Pythagorean silence" that veils the true opinions of that school. The symbolism of the Pythagorean theories given here, however, is quite plain: the sun is equidistant from the earth and the stars, whether the distances are reckoned as 378,000 stades or as a musical fifth.

Source: Pliny the Elder, *Natural History* (Cambridge, Mass.: Loeb edition, 1938), vol. I, pp. 172–174, 226–228. Translated by the Editor.

What Tones the Pythagoreans Assign to the Spheres

II, iii.   This form,[1] then, is eternal and unceasing in its revolution, turning with indescribable speed in the space of twenty-four hours, as the sun's rising and setting leave us in no doubt. Whether the sound of such a vast mass whirling in its ceaseless rotation is so loud as to exceed the capacity of the ear, I cannot easily say—no more, by Hercules, than whether there is at the same time a tinkling of the stars as they turn around with it, revolving in their orbits—nor whether it makes sweet music of incredible beauty. To us, who live within it, the world glides silently day and night.

xix. Many people have also attempted to discover the distances of the stars from the earth, and have proclaimed that the sun is nineteen times as far from the moon as the moon is from the earth.[2] Pythagoras, however, a man of knowing soul, made the distance from the earth to the moon 126,000 stades,[3] from the moon to the sun twice as much, and from the sun to the twelve signs (of the Zodiac) three times. Our countryman Gallus Sulpicius was also of the same opinion.
xx. But sometimes Pythagoras, using musical theory,[4] calls the distance from the earth to the moon a whole-tone, from the moon to Mercury half as much, the same from Mercury to Venus, a tone and a half from Venus to the Sun, a tone from the Sun to Mars (i.e., the same as from the earth to the moon), half a tone from Mars to Jupiter, half a tone from Jupiter to Saturn, and a tone and a half from Saturn to the Zodiac. Thus it makes seven whole-tones, which they call the harmony *diapason,* i.e., a universal harmony. Saturn moves therein at the Dorian pitch,[5] Jupiter at the Phrygian, and the others similarly—but this is more of a playful subtlety than a necessary one.

3

✦

Nicomachus of Gerasa
lived between 50 and 150 C.E.

Nicomachus was a native of Jerash, now in Jordan, and an important figure in the popular Neopythagorean movement of the first and second centuries. His *Introduction to Arithmetic* would later serve as a model for Boethius (see no. 13) and a standard textbook for the medieval Christian and Islamic worlds. Proclus (see no. 12) considered Nicomachus his own former incarnation, and the Neoplatonists in general adopted Nicomachean mathematics with its arithmology and distinctive classification of numbers.

The *Enchiridion harmonices* ("Textbook of Harmony") was an elementary work, a sketch written for an educated lady with the promise of a more detailed commentary to follow. (Probably the latter survives in Boethius' work.) This extract shows the breadth of Nicomachus' harmonic thought, embracing everything in the universal matrix of the musical Tetraktys (6:8:9:12 or the interval-sequence E' B A E) and the diatonic scale that fills it out: the movements of the planets, the acoustics of instruments, Pythagoras' discoveries, experiments, and innovations, and Plato's creation of the World-Soul. The harmony of tone with number became, for the Pythagoreans, a kind of Grand Unified Theory: an archetype of the harmony which permeates and unites both the greater and the lesser world.

For all the problems that Nicomachus poses when examined in detail (see the notes), his story of Pythagoras is the stuff not of mere history but of genuine myth. It records the semidivine origin of harmonic knowledge, replete with hints of deeper meaning. What were these four blacksmiths making, as they presided over the birth of Greek experimental science? Was Pythagoras' visit to their subterranean forge an initiation, or an initiatic vision? And again, in his experiment at home, do we not have an image of the *Lambdoma*, the diagram of the *Timaeus* numbers, in this "angle formed by two walls" with its single nail as the unitive point where all lines meet? (See von Thimus, no. 49.)

Source: Nicomacque de Gerase, Manuel d'harmonique, translated by Charles-Emile Ruelle in *Collection des auteurs grecs rélatifs à la musique* (Paris: Baur, 1881), pp. 13–26. Translated by the Editor.

The Scale and the Planets

Chapter III. 10. The names of sounds must have been borrowed from the seven stars which traverse the sky and circle around the earth. In fact they say that all bodies which whirl around in a very fluid and yielding motion must needs produce noises which differ from each other according to their size, the speed of their sound, and their position; that is to say, according to their own sounds, their own speeds, and the media in which the revolution of each body is accomplished, depending on whether these media are more fluid or more resistant.[1]   11. They point out clearly these same three differences in the case of the planets, which are distinguished from one another by size, speed, and position, and which circle without respite, whistling in the ethereal vapor.   12. It is precisely because of this that they are called by the name of *aster*, as one might say 'restless' or 'ever-running' (*aie theon*), whence are formed the words *theos* (God) and *aither* (ether).[2]

13. Now, because of the movement of Cronos, which is the planet situated the furthest above us, the lowest sound in the octave is called *hypate*, for *hypaton* signifies the most elevated.[3]   14. Because of the orbit of the Moon, placed the lowest in rank and the nearest to the Earth, we say *neate, nete*, for *neaton* signifies the lowest.   15. Of the two stars situated on either side, to that of Zeus, beneath Cronos, corresponds the *parhypate;* to the other, Aphrodite's, situated above the Moon, corresponds the *paranete*.   16. The most central star, which

is the Sun, in the fourth place counting from either end, becomes the origin of the *mese*, placed at the interval of a fourth from each end in the ancient heptachord, just as the Sun is fourth in rank among the seven planets counting from each extreme, for it occupies the middle point.   17. Of the two stars situated on either side of the Sun, to the one, Ares, to whom belongs the sphere placed between Zeus and the Sun, corresponds the *hypermese*, also called *lichanos;* to the other, Hermes, placed midway between Aphrodite and the Sun, corresponds the *paramese*.

18. We will corroborate these ideas for you with more precision and more fully by adding linear and numerical proofs in the commentaries which we have already promised you, O most wise and enlightened of women; and we will tell you the reason why we ourselves do not hear this cosmic symphony emitting sweet and utterly harmonious accents, as tradition[4] relates. But now, for lack of time, we must continue.

Acoustics

Chapter IV. 19. We say generally that sound is an uninterrupted percussion of the air which reaches as far as the auditive sense; that a note is a tension without dimension of the melodic voice; that tension is a state and an identity, like size, of a note without interval.   20. The interval is the path traversed from low to high and vice versa; the [scale-] system, a collection of several intervals. If several attacks or a strong wind happen to strike the surrounding air at several points, a loud note results; a weak one, if there are few attacks or little wind. If the attacks or the wind are regular and of constant force, the note will be consistent; if they are unequal, it will be uneven. If they strike slowly it will be low; if rapidly, it will be high. The effect produced will necessarily be inverse in wind instruments such as flutes, trumpets, the syrinx, hydraulus, etc., and otherwise in the stringed instruments: cithara, lyre, spadix, and the like. There also appear to be intermediate instruments, leaning to the one or the other or rendering similar effects: the monochords, commonly called pandoras and, by the Pythagoreans, canons; the trigons among stringed instruments, the plagiaules (transverse flutes), and the photinx, as will appear in the continuation of our discourse.

21. In stringed instruments, greater tension produces louder and higher sounds, and weaker tension slower and lower sounds. In fact, when the plectrum displaces the strings, these are drawn out of their normal position and, as they return, hit the surrounding air at many points with a very great rapidity and a strong vibratory movement, as if excited by the actual energy of their tension. Then they return to rest

and their vibration ceases, after the manner of a mason's plumb-line.   22. In wind instruments, on the contrary, the bigger bores and lengths produce a slow and relaxed note, because the wind escapes into the surrounding air after having exhausted its intensity on its long journey; it strikes and agitates it in an almost insensible fashion, and consequently the note produced is low.

23. We must consider here that 'more' and 'less' are dependent upon our own quantitative contribution, whether in blowing an aulos hard or soft, or in making the strings longer or shorter. It is evident that all this is numerically regulated, for it is agreed that quantity can properly be applied only to number.

The Discoveries of Pythagoras

Chapter V.[5] 24. Pythagoras was the first who sought to avoid the middle note of the conjunct tetrachords (being the same distance—a fourth—from the two extremes *hypate* and *nete*), in order to obtain a more varied system, as we may suppose, and also to make the extremes themselves produce the most satisfying consonance, the octave-ratio of 2:1.[6] This could not be the case with the existing tetrachords. He therefore inter-calated an eighth note between the *mese* and *paramese* and fixed it at the distance of a whole tone from the *mese*, a semitone from the *paramese*. In this way, the string which previously represented the *paramese* in the heptachordal lyre was still called *trite*—'third,' counting from the *nete*—and still occupies this position, while the intercalated string was fourth from the *nete* and sounds a fourth with it: a consonance which was originally sounded between the *mese* and *hypate*.

25. The note [B] placed between these two—the *mese* and the inter-calated string—received the name of the former *paramese*, and accord-ing to whether it was joined to one or the other tetrachord, was some-times more 'netoid' (if joined to the upper tetrachord), sometimes more 'hypatoid' (joined to the top of the lower one). It furnished the conso-nance of a fifth, marking the limits of a system formed by the tetrachord itself plus the added tone. Thus the sesquialtera ratio [3:2] of the fifth is acknowledged as the sum of the sesquitertia and the sesquioctave or whole-tone [4:3 times 9:8].

Chapter VI. 26. As for the numerical quantity which represents the distance of the strings sounding the fourth, fifth, and their sum the octave—in fact the additional note placed between the two tetrachords—here is how Pythagoras is said to have reported its discovery.[7]

27. One day he was out walking, lost in his reflections and in the thoughts which his schemes suggested to him, wondering whether he could invent an aid for the ear, secure and free from error, such as the

senses of sight and touch possess, the one in the compass, the rule, or even, we may say, in the dioptra;[8] the other in the scales or in the invention of measures. He happened by a providential coincidence to pass by a blacksmith's workshop, and heard there quite clearly the iron hammers striking on the anvil and giving forth confusedly intervals which, with the exception of one, were perfect consonances. He recognized among these sounds the consonances of the diapason (octave), diapente (fifth), and diatessaron (fourth). As for the interval between the fourth and the fifth, he noticed that it was in itself dissonant, but otherwise complementary to the greater of these two consonances.   28. Thrilled, he entered the shop as if a god were aiding his plans, and after various experiments discovered that it was the difference of weights that caused the differences of pitch, and not the effort of the blacksmiths, nor the shape of the hammers, nor the movement of the worked iron. With the greatest care he ascertained the weights of the hammers and their impulsive force, which he found perfectly identical, then returned home.

29. He fixed a single nail in the angle formed by two walls, in order to avoid even here the slightest difference, and lest a number of nails having each their own substance might invalidate the experiment. From this nail he hung four strings identical in substance, number of threads, thickness, and torsion, and suspended from the lower end of each of them a weight. He made the lengths of the strings, moreover, exactly the same, then, plucking them together two by two, he heard the above-mentioned consonances which varied with each pair of strings.   30. The string stretched by the greatest weight, compared to that which supported the smallest, sounded the interval of an octave. Now the former represented 12 units of the given weight, and the latter 6. He thus proved that the octave is in duple ratio, as the weights themselves had made him suspect. The greatest string compared to the next smallest, representing 8 units, sounded the fifth, and he proved that they were in sesquitertia ratio, that being the ratio of the weights. Then he compared it to the next one, with regard to the weight it supported. The larger of the two other strings, having 9 units, sounded the fourth; so he established that it was in the inverse sesquitertia ratio, and that this same string was in sesquialtera ratio to the smallest—for 9 to 6 is the same ratio, just as the second smallest string with 8 units is in sesquitertia ratio to the one of 6 units, and in sesquialtera ratio to the one of 12 units.

31. Consequently, the interval between the fifth and the fourth—the amount by which the fifth exceeds the fourth—was confirmed as being in the sesquioctave ratio, 9:8. The octave was the system formed by the union of one and the other, namely the fifth and the fourth placed side by side. So the double ratio is composed of the sesquialtera and

sesquitertia, 12:8:6; or, inversely, by the union of the fourth and fifth, so that the octave is composed of the sesquitertia and sesquialtera in this order, 12:9:6.

32. After having exercised his hand and his ear in the study of the suspended weights, and having established from these weights the ratios of the proportions stated, he ingeniously transferred the results obtained through strings hung on a nail placed in the angle of a wall to the soundboard of an instrument he called 'cordotone,' in which the tension, raised to a point proportional to that which the weights produced, passed to the movement of pegs placed in the upper part. Once installed on this terrain, and possessing, as it were, an infallible gnomon, he enlarged his experiment by making it on different instruments: for example, by striking vases, on flutes, syrinxes, monochords, trigons, and suchlike. He invariably found the numerical determination consonant and reliable.   33. He called the note corresponding to the number 6 *hypate*; the note of 8, a sesquitertia above, *mese*; the note of 9, a tone higher than the mean and consequently is sesquioctave, *paramese*; and finally 12 he called *nete*. Then he supplied intermediary points according to the diatonic genus, by means of proportional notes, and thus bound the octachord lyre to the consonant numbers, namely the double, the sesquialtera, the sesquitertia, and the difference between the latter pair, the sesquioctave.[9]

Chapter VII. 34. Pythagoras recognized in the following way, by virtue of natural necessity, the progression of sounds from the lowest to the highest, following this same diatonic genus; for the chromatic and enharmonic he described afterward, as we will explain one day. So this diatonic genus seems to have by nature certain degrees and certain progressions of which the details are as follows. A semitone, a tone, a tone, which form a system of a fourth, composed of two tones plus what is called a semitone;[10] then by the addition of another tone (the intercalated one), there results the system of a fifth composed of three tones plus a semitone. Next there follow a semitone, a tone, a tone: another system of a fourth, which is sesquitertia.   35. Thus in the heptachordal lyre, previous to this one, every fourth note starting from the lowest one was always a fourth away, the semitone occupying by turns, according to its placement, the first degree, the middle degree, and the third degree of the tetrachord. But in the Pythagorean or octachordal lyre, there is either—in the case of conjunction—a system composed of a tetrachord and a pentachord, or—in the case of disjunction—two tetrachords separated by the interval of a tone. Thus the progression from the lowest string will be such that every fifth note is consonant by the interval of a fifth, the semitone occupying by turns four different degrees: the first, second, third, and fourth.

The Means in Plato's Timaeus

Chapter VIII. 36. Having come thus far, it is time for us to comment on the passage of the *Psychogony* where Plato expresses himself in these words:[11]

> He placed them so between the intervals, that there might be two means in every interval; and that one of these might by the same fraction exceed one of the extremes, and be exceeded by the other; and that the other might by an equal number surpass one of the extremes, and by an equal number be surpassed by the other. He (the Demiurge) filled the distance which separates the sesquialtera and sesquitertia intervals with the interval of the sesquioctave.

37. These are in fact the double interval, that is the ratio of 12 to 6, and the two means, which are the number 9 and the number 8. The number 8, in harmonic proportion, is the mean between 6 and 12, being superior to 6 by a third and inferior to 12 by a third of that number 12. This is why (Plato) says that it is by the same fraction, considered in relation to the extremes themselves, that the mean 8 is respectively superior and inferior; for as the ratio of the greatest term is to the least, namely double, so the difference of the greatest to the mean, a difference which is 4, is to the difference of this mean to the smallest, a difference which is 2; and in effect their differences are in double ratio as 4 to 2.   38. The proper character of this mean is such that the sum of the extremes multiplied by the mean gives a product double the product of the extremes. In fact, 8 times the sum of the extremes which is 18 makes 144, double the product of the extremes, which is 72.

39. The other mean, 9, placed in the *paramese* position, is considered as the arithmetical mean between the extremes, being 3 less than 12 and 3 more than 6. Its proper character is such that the sum of the extremes is double this mean, and the square of the mean, which is 81, is superior to the product of the extremes by a quantity equal to the exact square of their mutual difference, namely to 9, the square of 3, which is their difference.   40. One can also demonstrate the third proportion, which is 'proportion' truly so called, in the two mean terms 9 and 8; for 12 is to 8 as 9 is to 6; these two ratios are sesquialtera, and the product of the extremes is equal to the product of the means, 6 times 12 being equal to 9 times 8.

4

Theon of Smyrna
fl. 115–140 C.E.

Theon's only surviving work is intended as a first step toward the studies recommended to philosophers by Plato in Book VII of his *Republic*: Arithmetic, Plane Geometry, Solid Geometry ("Stereometry"), Astronomy, and Harmonics. It is clear even from the elementary parts of Theon's treatise—which may have been all that he completed—that he viewed mathematics not as a self-sufficient science but as an entry to the world of real, divine, and almost personified numbers which, in the Pythagorean and Platonic worldview, embody the ultimate laws of the universe.

In the introduction to his second book, on music, Theon says that "after having finished our treatise on all mathematics, we will add to it a dissertation on the harmony of the world, and will not hesitate to relate what our predecessors have discovered, nor to make more widely known the Pythagorean traditions which we have inherited, without ourselves claiming to have discovered the least part of it." And at the end of Book III he promises a summary of his own work and that of Thrasyllus, the Emperor Tiberius' astrologer. Unfortunately we have no further material of this sort. Our extracts present only an outline of how an early Platonic commentator understood these sciences, and a summary of some planet-tone theories from the book on Astronomy. Theon's doubts with regard to these, and the poverty of his sources, make one the more curious as to what he would himself have written about the harmony of the world.

Source: Theon of Smyrna, *Mathematics Useful for Understanding Plato,* translated by Robert and Deborah Lawlor from the edition of J. Dupuis (San Diego: Wizard's Bookshelf, 1979), pp. 11–12, 91–94. Used by kind permission of the publisher.

On the Order in which Mathematics Must Be Studied

Book I, Chapter 2. We are going to begin with the arithmetic theorems which are very closely connected with the musical theorems which are transposed into numbers. We have no need for a musical instrument, as Plato himself explains, when he says that it is not necessary to agitate the strings of an instrument (with hand to ear) like curious folk trying to overhear something. [*Republic* 531a-b] What we desire is to understand harmony and the celestial music; we can only examine this harmony after having studied the numerical laws of sounds. When Plato says that music occupies the fifth rung (in the study of mathematics), he speaks of the celestial music which results from the movement, the order and the harmony of the stars which travel in space.[1] But we must give the mathematics of music second place after arithmetic, as Plato wished, since one can understand nothing of celestial music if one does not understand that which has its foundation in numbers and in reason. Then so that the numerical principles of music can be connected to the theory of abstract numbers, we will give them the second rung, in order to facilitate our study.

According to the natural order, the first science will be that of numbers, which is called arithmetic. The second is that whose object is surfaces and is called geometry.[2] The third, called stereometry, is the study of solid objects. The fourth treats of solids in movement and this is astronomy. As for music, whose object is to consider the mutual relations of the movements and intervals, whatever these relations be, it is not possible to understand it before having grasped what is based on numbers. Thus in our plan, the numerical laws of music will come immediately after arithmetic; but following the natural order, this music which consists in studying the harmony of the worlds will come in fifth place. Now, according to the doctrine of the Pythagoreans, the numbers are, so to say, the principle, the source and the root of all things.[3]

The Order of the Planets and the Celestial Concert

Book II, Chapter 15. Here are the opinions of certain Pythagoreans relative to the position and the order of the spheres or circles on which the planets are moving. The circle of the moon is closest to the earth, that of Hermes is second above, then comes that of Venus, that of the sun is fourth, next come those of Mars and Jupiter, and that of Saturn is last and closest to that of the distant stars. They determine, in fact, that the orbit of the sun occupies the middle place between the planets as being the heart of the universe and most able to command. Here is a declaration of Alexander of Aetolia:[4]

> The spheres rise higher and higher;
> the divine Moon is the nearest to the Earth;
> the second is Stilbon, 'the shining one,' star of Hermes, the
>   inventor of the lyre;
> next comes Phosphorus, brilliant star of the goddess of Cythera
>   (Venus);
> above is the Sun whose chariot is drawn by horses, occupying the
>   fourth rung,
> Pyrois, star of the deadly Mars of Thrace, is fifth;
> Phaeton, shining star of Jupiter, is sixth;
> and Phenon, star of Saturn, near the distant stars, is seventh.
> The seven spheres give the seven sounds of the lyre and produce a
>   harmony (that is to say, an octave), because of the intervals
>   which separate them from one another.

According to the doctrine of Pythagoras, the world being indeed harmoniously ordained, the celestial bodies which are distant from one another according to the proportions of consonant sounds, create, by the movement and speed of their revolutions, the corresponding harmonic sounds. It is for this reason that Alexander thus expresses himself in the following verse:[5]

> The Earth at the center gives the low sound of the hypate;
> the starry sphere gives the conjunct nete;
> the Sun placed in the middle of the errant stars gives the mese;
> the crystal sphere gives the fourth in relation to it;
> Saturn is lowest by a half-tone;
> Jupiter diverges as much from Saturn as from the terrible Mars;

the Sun, joy of mortals, is one tone below;
Venus differs from the dazzling sun by a trihemitone;
Hermes continues with a half-tone lower than Venus;
then comes the Moon which gives to nature such varying hue;
and finally, the Earth at the center gives the fifth with respect to
    the Sun; and this position has five regions, from wintry to
    torrid,
accommodating itself to the most intense heat, as to the most
    glacial cold.
The heavens, which contain six tones, complete the octave.
The son of Jupiter, Hermes, represents a Siren to us,
having a seven-stringed lyre, the image of this divine world.

In these verses Alexander has indicated the order for the spheres that he has determined. It is evident that he arbitrarily imagined the intervals which separate them, and nearly all the rest. Indeed, he says that the seven-stringed lyre, the image of the universe, was constructed by Hermes, and that it gives the consonances of the octave; then he established the harmony of the world with nine sounds which, however, include only six tones.

It is true that he attributes the sound of the *hypate*, as being lower than the others, to the earth; but being immobile at the center, it renders absolutely no sound. Then he gives the sound of the conjunct *nete* to the sphere of the stars, and places the seven sounds of the planets between the two. He attributes the sound of the *mese* to the sun. The *hypate* does not give the sound of the fifth with the *mese*, but that of the fourth, and it is not with the *nete* of the conjuncts that it gives the consonance of the octave, but with the *nete* of disjuncts.

The system does not conform to the diatonic type, since the melody of that genus allows for neither a complete trihemitone interval, nor two half-tones one after the other. Neither is it chromatic, for in the chromatic genus, the melody does not include the unbroken tone. If it be said that the system is formed of the two genera I would answer that it is not melodious to have more than two half-tones following one another. But all of this is unclear to those who are not initiated into music.

Eratosthenes, in a similar manner, exposes the harmony produced by the revolution of the stars, but he does not assign the same order to them.[6] After the moon which is above the earth, he gives the second place to the sun. He says that in fact Mercury, still young, having invented the lyre, first rose up to the sky, and passing near the stars called "errant," he was astonished that the harmonies produced by the speeds of their revolutions were the same as those of the lyre which he had constructed. In the epic verses, this author appears to leave the earth immobile and determines that there are eight sounds produced by the starry sphere and by the seven spheres of the planets which he makes circle around the earth. It is for this reason that he made an eight-stringed lyre including the consonances of the octave. This explanation fares better than that of Alexander.

The mathematicians establish neither this nor that order among the planets. After the moon, they place the sun, and some put Hermes beyond it, then Venus, and others put Venus, then Hermes. They arrange the other planets in the order we have mentioned.

5

Ptolemy
fl. 127–148 C.E.

Claudius Ptolemaeus, the great Alexandrian astronomer, was a scientist of universal scope. His works on astronomy (the *Almagest*), astrology (the *Tetrabiblos*), and geography became standard texts for the Arabic, Byzantine, and Western Middle Ages. After Aristotle, he was the major scientific authority for the medieval period.

The first and second books of Ptolemy's *Harmonics* are an ordinary treatise on scales and intervals. With the third chapter of Book III, the tone suddenly changes to that of a fantasia on psychic and cosmic harmony. If, as is suggested, he was prevented by death from completing the work, what are we to make of this as the swan song of Antiquity's greatest astronomer? Commentators from Kepler to the present have been embarrassed by it, calling it ingenious nonsense. But Kepler himself suffered the same posthumous verdict on what he considered the crown of his work, his explanation of the eliptical planetary orbits through musical harmony. Does each age simply have its own distinctive nonsense?

At the beginning of this section, Ptolemy is quite eloquent about the wonder and the awe which the study of harmonics arouses. As our whole collection demonstrates, whose who have devoted themselves intensively to it often believe that they have discovered the key to the universe. In some respects it is a different key for each one—and it is this personal dimension that may later be ridiculed—while in other respects the key is always the same, unlocking the doors of insight into a numbered and harmonious universe, in which microcosm and macrocosm reflect in time and space endless variations on a central theme. The conviction through intuitive perception that this is so is the core of the matter. After that, the theorist hears harmonies everywhere, just as some mystics see God in everything.

As a single example, consider Ptolemy's sevens. At the end of Book III, Chapter 5, he correlates the seven faculties of the soul and the seven virtues of reason with the seven notes of the scale. To one who has had a direct intuition of the meaning of Seven, every sevenfold grouping will become transparent, and the soul's divisions will reveal it no less than the planets and the notes of the scale. One is then no longer playing number games, but pointing to the basic truth that whenever sevens appear, the archetypal Seven is manifest in them.

This mode of thinking and experience is unfamiliar today, but has to be taken seriously if a man of Ptolemy's stature is not simply to be judged more gullible than ourselves. If this imaginative leap can be made, his chapters on cosmic correspondences will be read as the report of a man for whom every number struck an inner tone, every proportion a chord; every movement in the pure circular form proper to the souls of men and stars resonated as the two-octave matrix of audible harmony.

Source: Ptolemy, *Harmonics,* German translation by Ingemar Düring in his *Ptolemaios und Porphyrios über die Musik* (Göteborg: Elanders, 1934), pp. 114–136. Translated by the Editor, with reference also to the Latin translation of John Wallis, *Claudii Ptolemaei Harmonicorum Libri Tres* (Oxford, 1682; repr. Broude).

———◆———

How Should We Relate the Power of Harmony to Its Doctrine?

Book III, Chapter 3. I think I have sufficiently demonstrated that the harmonic intervals up to the *emmeleis*[1] are intrinsically defined by certain fundamental ratios, and have also answered the question of which ratio belongs to each of them. Anyone who has concerned himself deeply with the perceptual cause of our reckonings, as well as with their practical investigations—i.e., with the methods I have discussed for using the monochord—can no longer doubt that in all tunings the corroboration of the ear holds good. The natural consequence is that anyone who has practiced these calculations, if he retains any feeling for beauty, must be amazed at the power and beauty dwelling within harmonies; yet this is something which also agrees completely with the calculations of the intellect, and with the greatest precision discovers and produces the tunings in practical use. He will also be seized as it were by a holy yearning to behold and understand the true relationship in which this power stands to other phenomena in our world. Therefore we will try to treat this last part of our scientific task in the broadest possible way, so as to give expression to the sublimity of this wonderful power.

The principles of all beings are Matter, Movement, and Form: Matter as the material of their origin, Movement as their cause, and Form as their purpose. The power of harmony cannot be regarded as an object: it is something active, not receiving impressions from without—nor as a goal, because it already possesses something: harmonic and rhythmic rightness, or order and beauty according to rules—but as a cause which orders material and gives it natural form.

Causes, too, are divided into three main categories: the first refers to Nature and to mere being, the second to Reason and a good condition of being, the third to Divinity and to the eternally good being. The power of harmony is not a cause relating to Nature, for it creates no being from formless material; nor one which relates to Divinity, for it is not in the first place a cause of eternal being. We have instead to do with Reason, the middle one of the three causes, which does good in both directions. For it belongs eternally to the gods, who remain forever the same, while Nature on the contrary is nowhere and never so. Now the cause belonging to Reason divides into Intellect, which refers to divine Form; Art, which refers to Reason in the actual sense of the word; and Practice, which refers to Nature.

As we can see, the power of harmony seeks in all this its special task. Reason, regarded simply and generally, is the creator of order and conformity; the laws of harmony are valid in the field of acoustics, just as those of sight are for visible things and those of judgment for all that is graspable by intellect. The power of harmony establishes in acoustics the rules which we call harmonic consonance; and this occurs partly through theoretical discovery of the correct ratios with the aid of intellect, and partly through successful cultivation of these discoveries with the aid of practice.[2] Reason generally finds the Beautiful through theoretical speculation, then confirms it through its means of expression (the activity of the hands), then through practice molds the formless material into conformity with itself. Hence it goes without saying that the science which is common to all developments of reason, namely Mathematics, is concerned not only with the theoretical investigation of the Beautiful, as some would believe, but also with the exhibition and practical employment of those things whose evaluation falls within its realm.

The power of harmony uses as a kind of tool or servant the two highest and most wonderful of our senses, Seeing and Hearing, which hearken most of all to the ruling part of the Soul, and which do not judge objects so much by desire as by beauty. One can find for each of the senses particular differences in perception, e.g., for sight, white and black; for hearing, high and low; for smell, fragrant and stinking; for taste, sweet and sour; for touch, soft and hard, and, by Zeus! for every sense pleasure and displeasure. But no one would use the words *beautiful* or *ugly* in connection with the senses of touch, taste, or smell: only of visible and audible things, as for instance a form and a melody, or the movement of the stars and the acts of men. Therefore these senses are also the only ones which support the reasoning part of the soul with their respective perceptions, as if they were truly sisters. That which is only visible manifests also to hearing with the help of the latter's medium of expression [speech]; that which is only audible manifests to sight with the help of writing; and thus each of these senses often achieves more than if it had only conveyed its own impressions. For example, that which is to be communicated through speech is made easier for us to learn and remember by figures and letters, and that which we grasp through the sight is made more lively through poetic form: such as the view of a sea, the aspect of a place, a battle, the outward symptoms of mental states; so that by this means one's whole soul is brought into a certain mood, just as if one had seen it all with one's own eyes.

The fact that one need not grasp the object of a sense with that sense alone, but that each may rival the other in order to find out and seize everything which fits its own function, is particularly valid in the case of beauty and utility. It is so in the most rational of the relevant sciences, namely Astronomy, which is concerned with the sight and the movement of heavenly bodies which are only visible, and Harmonics, which is concerned with hearing and the movement of things which are only audible, namely sounds. Both employ as uncontradictable helpmeets Arithmetic and Geometry, in order to measure the quantity and quality of the principal movements. They are, so to speak, cousins, for they stem from the sister senses, sight and hearing, and have as their foster mothers Arithmetic and Geometry, to which they are related.[3]

Harmonic Power Dwells within Everything which is Complete by Nature, and Appears Most Clearly in the Human Soul and in the Movements of the Stars

Chapter 4. From these discussions it follows that the power of harmony belongs to the causes which are founded upon reason, in that it produces equality of movements and—since its theoretical science is a form of mathematics—is concerned with the mathematical ratios perceptible by the ear. Its goal, therefore, is to regulate practical activity with an orderly influence consequent upon its theories. Here we must also mention that, like all similar powers [of the soul], this one necessarily dwells within everything which has any movement in itself, however insignificant. It is especially found, in its highest degree, in those phenomena which are by nature complete and rational, which then possess this quality inherently. Only in such phenomena can it confirm, beyond doubt and at the highest possible level, the universal agreement of the ratios which instill conformity and order in different formations.

Everything that is governed by natural law partakes of some rational order with regard to its movements and the object of its movements. To the degree that this [rational order] is perceptible in its orderly rhythm, so it becomes for these phenomena their mother, nurse, and healer, and everywhere causes what one may call beauty. But if anything is deprived of this indwelling power, then everything is reversed, so far as this may be, and its course becomes worse and worse. This power is imperceptible in those movements which change the object itself, in which because of its indefinite character neither its quantity nor its quality can be defined; but it is seen clearly in those [movements] which come forth of its own fashioning. The latter, as we have said, applies to those [movements] which are most complete and rational, like the movements of the stars in the divine realm and those of the soul in the human realm. For only these two partake of the original and noblest movement—the spatial one—and are at the same time rational. [Harmonic power] reveals and informs us, as far as man can grasp, that tones are ruled by certain harmonic relationships. We will understand this by investigating separately each of these forms; and so we turn first to the movements of the human soul.

How Do the Consonant Intervals Correspond to the Original Diversity of the Soul's Powers, and to Their Natural Divisions?

Chapter 5. The original powers of the soul are three: the power of Thought, the power of Feeling, and the power of Life. The original identical and consonant intervals are also three: the octave identity and the consonances of fifth and fourth. One can therefore compare the octave to the power of thought—for in both there prevails simplicity, equality, and equivalence—, the fifth to the power of feeling, and the fourth to the power of life. The fifth is closer to the octave than the fourth, and sounds better because its surplus [$\frac{3}{2}$] is closer to unity. Even so, the power of feeling is closer to thought than is the power of life, since it partakes to a certain extent of consciousness. Some things have being but no feeling; others have feeling but no thought. On the other hand, all things that feel also have being, and all that have thought also possess feeling and being. So in harmony, where the fourth is present there is not necessarily a fifth, nor where the fifth is, an octave; but a fifth always contains a fourth, and an octave both fifth and fourth. The point is that the powers of life and feeling correspond to the incomplete *emmeleis* intervals and their combination, and the power of thought to the complete one.

One can divide the soul's power of life further into three, just as the fourth breaks down into three intervals. They are the three stages of life: Growth, Maturity, and Decline. The power of feeling similarly divides into four, as does the fifth into four intervals; they are the powers of Seeing, Hearing, Smelling, and Tasting (but we regard the sense of Touch as common to them all, since all perception takes place with some assistance from touch).[4] Lastly, the power of thought has a sevenfold division,[5] corresponding to the seven intervals of the octave. These are the faculties of Imagination, for the analysis of what is perceived; Understanding, for its first formation; Reflection, for retaining and inculcating what is formed; Meditation, for recalling and investigating it; Opinion, for superficial conjecture; Reason, for critical evaluation, and finally Knowledge, for grasping it in truth and clarity.

We could also divide our soul's powers in another way: into Reason, Emotion, and Desire. We compare reason, on account of the above-named correspondences, to the octave; emotion, which stands quite close to it, to the fifth; and desire, which is somewhat lower, to the fourth.[6] Everything else that arises from the respective values and compasses of these intervals could also be derived from them: we would then find that the clearest distinctions within the original unities (virtues) would coincide with the number of those within each of the first symphonious intervals. Among tones, too, consonance is a virtue and dissonance an evil; and, *mutatis mutandis*, the same applies to the human soul: virtue is a kind of consonance of the soul, evil a dissonance. What is common to both [tones and soul] is that a harmonically regulated ratio of parts is natural, and an unregulated one against Nature.

One might also say that Desire has three virtues, comparable to the three intervals of the fourth: they are Temperance, when it is a question of spurning pleasure; Self-control, when one has to bear privations; and Modesty, when one must protect oneself from dishonor. Feeling, moreover, has four virtues comparable to the four intervals of the fifth, and these are Mildness, when one must not be carried away by anger; Courage, when one has to look imminent evil fearlessly in the eye; Boldness, when one must spurn dangers; and Tolerance, when one has to bear hardship. The seven virtues of Reason may be given as follows: Perspicacity, for quick understanding; Subtlety, whereby one detects truth; Presence of mind, to see through things; Insight, to judge rightly; Wisdom, to know all; Prudence, to act practically; Experience, to live simply.[7]

In measuring the harmonic ratios, furthermore, it is necessary to establish first the identical ones, next the consonant and the *emmeleis;* for an insignificant fault in the small intervals harms the melody less than one in the larger and more important ones. It is even so in the soul: those parts which are ruled by intellect and reason have by nature a ruling influence on the lower parts. They require greater precision in the use of the reason, because they contain all or most of the faults [of the lower parts]. The highest expression of the soul, Justice, is like a whole, sounding in unison; and the relations between its parts are the same as determine the predominant ones. So between the parts which represent Prudence and Intelligence there is as it were the unison; between correct Comprehension and Skill, or Boldness and Temperance, the consonant intervals; and between Activity and Harmonious Fulfillment, the *emmeleis.* Finally, one can compare a philosopher's entire state of soul with the tunings of the whole *teleion* system.

Now we have completed in detail the comparison of the consonant intervals with the virtues, thereby fulfilling our claim that every harmony contains all harmonies and virtues: which includes as it were, all the chords which sound both in music and in the human soul.

Comparison between the Genera in Music and the Principal Virtues

Chapter 6. The two fundamental powers of the soul, Theory and Practice, each have three subdivisions: the theoretical power embraces Physics, Mathematics, and Theology (concerned respectively with Nature, Science, and Religion); the practical embraces Ethics, Economics, and Politics (concerned with Morals, Management, and Public Life). These subdivisions of the soul's powers are relatively undifferentiated, for the virtues of all three are common and interdependent; but they do differ in their meaning, value, and the scope of their structure. Hence one could compare each of these threefold groups in an appropriate way with the three types of harmony, called the Enharmonic, Chromatic, and Diatonic genera. These, too, are distinguished by the size and range of the portions which are removed or added. For a distinction of this kind is made in them, through absolute pitch and relative movement, between *pyknon* and *apyknon.*[8]

We would compare the Enharmonic genus to Physics and Morals, because both deal in very small spatial relationships; the Diatonic to

Theology and Politics, on account of its equal ordering and nobility; the Chromatic to Mathematics and Economics, because it strikes a mean between the extremes. Mathematics is often involved in Physics and Theology; Economics sometimes has its part in Morals (when it is a matter of the individual and the subservient), sometimes in Politics (when it concerns the community and rulership). The Chromatic is in a way a bridge between the Enharmonic and the Diatonic, inclining to the former through its limpness and softness, and to the latter through its seriousness and strength, and showing in comparison to each the contrary qualities. Just so, the *mese* lies an octave higher in relation to the *proslambanomenos*, and an octave lower in relation to the *nete hyperbolaion*.[9]

The Changes within the Genera Resemble Those of the Human Soul in the Vicissitudes of Life

Chapter 7. In the same way we could now compare the changes [i.e., modulations and transpositions] of the tonal system with the changing states of the soul during life's vicissitudes. For in each, the melody alters—even when the genus is retained—so that the place [on the instrument] which imparts a certain character to the melody is silent, or at least not used in the same way. And so it is with changes in the life of men: a certain attitude of soul, unaltered in itself, will be forced into a new course and compelled to adapt to the existing way of life of a community, even to make this its own way of life. Something akin takes place in lawgiving, where laws are adapted to different forms of justice according to the ruling trend. Peaceful circumstances thus convert the souls of citizens to steadfastness and equity; warfare, on the other hand, awakens courage and self-awareness; danger and hunger call forth self-sufficiency and thrift, but wealth and surplus lead to licentiousness and gluttony, and so forth. A similar effect is shown in melodic modulations. One and the same compass calls forth an enlivening expression in the higher modes, but a dejected one in the lower, because a high range causes the soul to tense, while low tones make it relax. Hence the middle modes, in the region of the Dorian, are compared with orderly and steadfast conditions of soul; higher ones, toward the Mixolydian, with restless and active ones; and deep ones, toward the Hypodorian, to limp and dull ones.

For this very reason, an emotion in the soul can immediately occur in the actual life of a melody; the soul, so to speak, recognizes the affinity between the harmonic relationships and its own situation; it is molded by the movements peculiar to certain melodic expressions, so that it plunges betimes into pleasure and diversion, and at others feels sympathy and humility.[10] It may be lulled into repose, then again spurred into wakefulness. Sometimes it sinks down into ease and relaxation, then flames forth in passion and enthusiasm. All this the melody may do by modulating in one direction or another, while the soul is simultaneously shifted to the appropriate conditions because of the inner resonance of the two.

I believe Pythagoras had this in mind when he decreed that one should busy oneself with music and pleasant melody right upon waking, before proceeding to one's work, so that the disturbance of the soul's repose consequent upon awakening should pass over into a pure and tranquil mood, preparing one for one's daily tasks with a well-tuned and harmonious attitude.[11] The fact that the gods are also summoned by music and song, such as hymns and the music of auloi and Egyptian harps, seems to me to indicate that we beseech them to hear our prayers with kindly benevolence.

The Correspondence between the Perfect System and the Ecliptic

Chapter 8. Thus we have explained the correspondences between the human soul and the nature of harmony. To summarize our findings briefly: we found that the identical and consonant intervals were constructed in correspondence with the original parts of the soul, the divisions of the *emmeleis* agreed with the divisions of the virtues, the generic differences of the tetrachords with those of the virtues according to value and size, and the modulations with the changing moods in the course of life. It remains to compare with these the movements of the stars, which are accomplished according to harmonic laws; and therefore we will make first a general comparison of all or most [phenomena], then undertake a more particular comparison of individual ones according to their divisions. Thus we begin now with the first, general [comparison].

Let the following discussion therefore begin immediately on the assumption that both tones and the stars' movements only take place through intervallic motion, without the actual and fundamental modifications consequent upon this.[12] The rotating movements of the stars, moreover, are all circular and regular, and the movements within the tone-system are also arranged similarly. One can imagine the order of tones [on an instrument] and their pitches arranged as it were along a straight line, while their relative meanings [in the tone-system] which characterize them can be completed on one and the same closed circuit. This is how one grasps the actual nature of the tones, without any beginning point; but if one only sees their placement [on an instrument] one can shift the beginning of the series to one point or another.

Let us now intersect the ecliptic at one of the equinoctial points, then bend them round, as it were, and place them at equal intervals (so that the parts match each other) on the two-octave Perfect System.[13] Thus we lay the uncut equinoctial point on the *mese*, the two ends of the divided one on *proslambanomenos* and *nete hyperbolaion*:

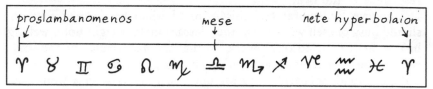

FIGURE 1: *The Greek Scale on a Linear Zodiac*

On the other hand, one could begin with the harmonic meanings of the single tones and bend the double-octave into a circle, joining *nete hyperbolaion* and *proslambanomenos* and uniting the two tones. This juncture will then lie diametrically opposite the *mese*, and be in octave relationship to it.

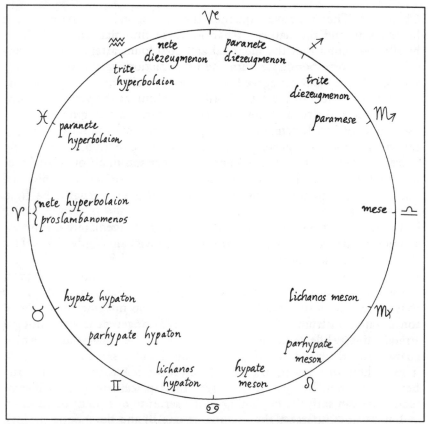

FIGURE 2: *The Greek Scale on a Circular Zodiac*

What was established in the aforementioned comparison is thereby confirmed, namely that the same relationships hold for the position of the diameter of the circle as for what was said of the octave. For the ratio 2:1 denotes in this comparison the ratio of the circle to the semicircle and the greatest mystery of all: for only the diameter must necessarily cut the center of the circle, that being the cause of this figure's symmetry. Other lines may cut the circumference of the circle into so many parts, but they never divide the whole area of the circle in this way [i.e., into the ratio 2:1]; but the diameter divides both the area and the circumference into two equal parts.

For this very reason the effect of the planets is at its strongest in opposition, when they occupy diametrically opposed positions in the Zodiac, and a similar relationship obtains among tones which are an octave apart from one another.[14]

Comparison of the Consonant and Dissonant Intervals in the Tone-System with the Aspects in the Zodiac

Chapter 9. If we now proceed further, we find that the consonant intervals in music can always be divided into four parts; for the largest one, the double octave, is expressed by a fraction whose numerator is four times the denominator [4:1], and the smallest, the fourth, by a fraction whose numerator exceeds the denominator by a quarter of itself [4:3]. In the same way, the observed harmonic and powerful aspects in the Zodiac are determined by the division of the circle into four.

Let us draw a circle AB, and from the same point A divide it by AB in two equal parts, by AC in three, by AD in four, and by CB in six. (See figure 3.) Then AB will produce the opposition, AD the square, AC the trine, CB the sextile. The relationships of the segments cut off from the point A are the same as those of the identical and consonant intervals and the wholetone, which is illuminating, if we divide the circle into 12 equal segments (since this is the smallest number that can be the common denominator of 1:2, 1:3, and 1:4). ABD will contain 9 of these segments, ABC 8, the semicircle AB 6, ADC 4, AD 3. The segments produce the ratio 2:1 of the first identical interval, the octave, thrice; the 12 of the whole circle to the 6 of the semicircle, the 8 of ABC to the 4 of AC, and the 6 of ACB to the 3 of AD. The ratio 3:2 of the largest consonant interval, the fifth, is also thrice present: the 12 of the whole circle to the 8 of ABC, the 9 of ABD to the 6 of AB, and the 6 of AB to the 4 of AC. The ratio 4:3 of the smallest consonant interval, the fourth, is also contained thrice: the 12 of the whole circle to the 9 of ABD, the 8 of ABC to the 6 of AB, and the 4 of AC to the

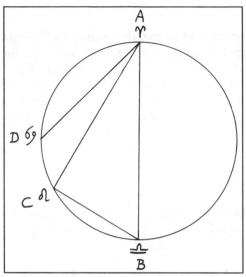

FIGURE 3: *Proportionate Division of a Circle*

3 of AD. Moreover, the 3:1 ratio of the twelfth is here twice: the 12 of the whole circle to the 4 of AC, and the 9 of ABD to the 3 of AD. The ratio 4:1 of the double octave identity is here once, in the 12 of the whole circle to the 3 of AD; the ratio 8:3 of the eleventh also once, in the 8 of ABC to the 3 of AD. Finally, the 9:8 ratio is present once, in the 9 of ABD to the 8 of ABC. These are also the ratios of the numbers beneath our figure.[15]

Following this table one could compare the consonant intervals: the fifth with the trine, the fourth with the square, the whole-tone with the sextile. For the ratio of the circle to the semicircle AB yields 2:1; the ratio of the semicircle to the third of the circumference AC 3:2; that of AC to the quarter of the circumference 4:3; and the difference between these two, which yields a whole-tone, the segment CD, is $1/12$ of the circle. In a similar way, the Zodiac is divided by nature into twelve houses, so that in the double-octave Perfect System, which contains approximately 12 whole-tones, the whole-tone interval can be compared with the twelfth part of the Zodiac.[16] Remarkably enough, of the Zodiacal aspects only the ones which would be expressed as 1:12 are not reckoned consonant, but are counted in the category of *emmeleis*.[17] On the other hand, such as are expressed by 5:12 are counted in the *emmeleis* category, but their nature is dissociate [i.e., two such houses are not in effective aspect] and they are called such [by the astrologers]. If one cuts with a straight line a twelfth of the circle, the whole circle will stand in the ratio of 12:1 or 12:11 to the cut-off segment: ratios excluded from the consonances but not from the *emmeleis*. If we cut off with a straight line five twelfths of the circle, the whole circle will stand in the ratio 12:5 or 5:12 to the cut-off segment: intervals foreign both to consonance and to the *emmeleis*, since they are neither superpartial nor multiple, nor compounded of intervals which can constitute any consonant interval. Through all points of the circle which is divided into twelfths one can draw only three squares—the same as the number of species of fourth; only four triangles—the same as the number of species of fifth; for only these intervals occur uncompounded among the consonances.

The Longitudinal Motion of the Stars Resembles the Continual Movement in the Tone-System

Chapter 10. We must now conclude our discussion of the theories of the actual circular motion of the stars and the tones, and of what are called concord and discord among aspects [and tones]. Next we must consider the most significant differences in the celestial motions. These are threefold: [first], longitudinal motion, forward and backward, by which alterations are made from east to west or vice versa; [second], the upward or downward motion found in the stars which approach the Earth or depart from it; [third], the latitudinal motion in which position changes toward the North or the South.[18] It is justifiable simply to compare the first, East-West motion with the movement from higher pitch to lower or vice versa, for both are ruled by uninterrupted connection; we compare the region of rising and setting with the lowest pitch, the region of culmination with the highest pitch. Rising and setting comprise the beginning and end of perception: in the former, something emerges from invisibility, and in the latter disappears into it. The lowest tones comprise the beginning and end of vocal motion: they come out of silence and fall silent again, so that the deepest pitch is the nearest to complete disappearance of the voice, the highest pitch the furthest therefrom. Therefore singing students begin their practice with the lowest notes and also end there.

We can compare the culminations, which are furthest removed from the invisible, with the highest pitches, since these are furthest from silence. As the lowest parts produce the deepest notes, so the uppermost parts produce the highest ones; we say, moreover, that the lowest notes come from the vitals, the highest from the temples. At the lowest point are rising and setting: at the highest, culmination. So the former may well be compared to the lowest notes, the latter to the highest. Hence the motions of the stars up to their culminations show many resemblances to the movements of tones from deeper to higher pitch, and their declinations to the falling of pitch.

Comparison of the Rising and Falling Motion of Stars with the Genera in Music[19]

Chapter 11. We will find that the second motion, rising and falling, relates to the movement in music within the so-called genera. For as [music] contains three of these, the enharmonic, chromatic, and diatonic genera, distinguished by the size of the steps in the tetrachord, so this [motion comprises] three types of distance: smallest, medium, and largest, measured according to its distance from the orbits.[20] The rotations that take place at a medium distance and always encircle the middle orbits could best be compared to the chromatic genus, since the lichanos notes here also halve the tetrachord.[21] Those orbits which proceed with the smallest rotations, whether in [orbital] apogee or perigee, are like the enharmonic, since here two intervals together (the so-called pyknon) are smaller than the third; while those orbits which proceed with the largest rotations, whether in apogee or perigee, are like the diatonic, because two intervals there are never smaller than the third (the so-called apyknon). Moreover, a compression occurs both in the enharmonic genus and the small orbits, respectively in intervals and speed; in the diatonic genus and the largest orbits, an expansion occurs; while the chromatic genus and the medium orbits adopt a middle course between the extremes.

The North-South Motions of the Stars Accord with Modulation

Chapter 12. The third and last of the celestial motions, namely the North-South one, can be likened to modulation. In the latter there is no change in regard to the genus when we move to another mode, just as in the former there is no perceptible inequality with respect to the orbit during changes in elevation relative to the celestial equator.[22]

Here, too, we should compare the Dorian mode, the middle one of all, with the medium of the North-South motions, i.e., to those which proceed along the celestial equator in both hemispheres; the Mixolydian and Hypodorian, as the highest and lowest modes, with the most northerly and southerly which one must imagine positioned at the tropics; and the remaining four modes, which lie between those named, with the ones that take place in the parallels intervening between the tropics and the celestial equator. These are in fact four, in conformity with the

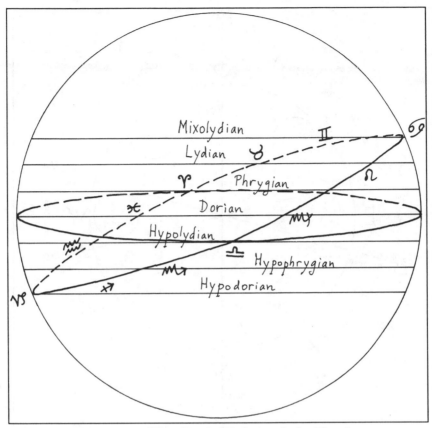

FIGURE 4: *The Greek Modes and the Tropics*

division of the planetary orbits into twelve parts, according to the respective twelve houses of the Zodiac. (See figure 4.)

For each of the tropics produces, in a way, one parallel, but each pair
of points which are the same distance from these on either side produces one and the same parallel. So five points of intersection arise in
accordance with the distances of the houses, and hence five parallels,
which together with the tropics make seven: the same number as that
of the modulations among the modes. The three above the Dorian can
be compared, on account of their higher position, to the parallels in the
direction of the visible pole and the summer solstice point, which in
northerly latitudes is the North Pole and in southerly ones the reverse.
The modes deeper than the Dorian can be compared, on account of
their lower position, with the parallels in the direction of the invisible
pole, that of the winter solstice, which in southerly latitudes is the
North Pole and in northerly ones the reverse.

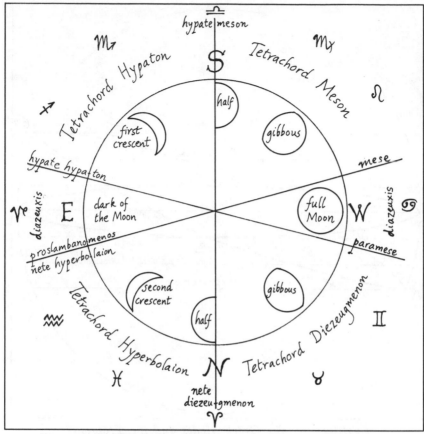

FIGURE 5: *The Tetrachords and the Phases of the Moon*

The Similarity of the Tetrachords and the Aspects to the Sun

Chapter 13. Finally, as we shall show, the arrangement of the tetrachords in the Perfect System agrees with the arrangement of the aspects with regard to the sun.[23] We will first place one of the disjunctive whole-tones on the position between the heliacal rising and setting, and the other at the point of opposition of the planets and moon. Thus those tones which conjoin tetrachords, namely *hypate meson* and *nete diezeugmenon*, come on the points at right angles to each of these, at the half-moon points. (See figure 5.)

The aspect from the rising point of any planet up to the first octant may be compared to the tetrachord *hypaton*, since both rising and deep pitch betoken a beginning; the next aspect, up to the second octant, can be compared to the tetrachord *meson*. Furthermore, that aspect which prevails at the third octant, when Mercury or Venus or the other three planets rise in opposition,[24] or at the opposition of the waning moon, is comparable to the tetrachord *diezeugmenon*, and in regard to the first octant and the tetrachord *hypaton* acts as opposition and as homophonous octave. Lastly, the next aspect up to the heliacal setting at the fourth octant is comparable to the tetrachord *hyperbolaion*, and this in turn acts as opposition and homophonous octave to the tetrachord *meson*.

The spaces between heliacal rising and setting, and those in opposition between late setting and early rising or lunar opposition, comprise about a twelfth [of the Zodiac], just like the disjunct whole-tone [as a proportion of the double octave]. Moreover, the four aspects each cover about 2½ twelfths, so that the four tetrachords each comprise about 2½ whole-tones. Finally, those aspects of the moon's phases which are in opposition make a unity together and reciprocally complete each other, just as all the tones of the double octave struck two by two seem to give the impression of single tones.

Calculation of the Smallest Numbers by which the Fixed Tones of the Perfect System May Be Compared to the Major Orbits of the Planetary System

Chapter 14. We can best understand, by means of such comparisons, what the various forms of *emmeleis* and the movements of the stars have, broadly speaking, in common. It now remains to observe what can be discovered in each particular case on the grounds of the foregoing. [lacuna][25] . . . of the numbers discussed and the ratios expressed through them. Let us divide the whole circle into 360 degrees, and place the sun diametrically opposite the moon, or any of the planets; then the distance between them will be 180°, reckoning along the circumference. Twice this distance gives the 360° of the entire circle. If the planets are in trine, we say that they are 120° apart, and if we triple this we reach 360°, the number of the whole circle. If they are square to one another, we say that they are 90° apart, and four times 90° make 360°. If they are sextile, they are 60° apart, for six times 60° make 360°. Now if we liken this to the Perfect System in music, we can compare the fixed tones to the position of just these numbers, as follows: *proslambanomenos* with the 180° position, *hypate meson* with 120°, *nete diezeugmenon* with 90°, and *nete hyperbolaion* with 60°; the two fixed tones which comprise the disjunctive whole-tone correspond with the position from which one reckons the distances, i.e., with the places where one locates the sun or one of the other planets. From here the distances are reckoned along both sides of the circle.

How Can the Relations of the Original Movements Be Numerically Expressed?

Chapter 15. Under these conditions the number of the square is 90, the mean between 120, the number of the trine, and 60, the number of the sextile; and it produces two ratios, 3:2 and 4:3, in accordance with the first consonant intervals, the fifth and fourth. As in music these first consonant intervals, the fifth and fourth, together make the octave identity, so also the two ratios mentioned, 3:2 and 4:3, produce the ratio 2:1 in conformity with the octave identity. If we use the number of the whole circle, 360, it will stand in 4:1 ratio with 90, in conformity with the double octave of the musical Perfect System. In another way, too, one could derive a similar agreement from the twelve houses of the Zodiac: 120° represents the distance of four houses, 90° that of three, 60° that of two, of which 3 is the mean and makes with 4 the ratio 4:3; with two the ratio 3:2. From these two ratios together we can build the ratio 2:1 (for the product of these two numbers is 2:1), i.e. 4:2. To this end we use the number of the whole Zodiac, 12, which stands in 4:1 ratio with 3 and duly produces the agreement with the double octave in the musical Perfect System. If now we recall the polygons, namely the triangle, square, and hexagon, the corresponding ratios in harmony can also be deduced, partly from the number of their angles, partly from other features which may present themselves.[26] The reckonings with which we have expanded [Ptolemy's text] supply, we believe, the most necessary requirements. The rest we have deferred until a more opportune moment.

Comparison Between the Qualities of the Planets and of the Tones [27]

Chapter 16. No one should be surprised that the note of Jupiter is consonant with the two lights, sun and moon, while that of Venus is only [consonant] with the moon (for the whole-tone is no consonance). [Venus' note] belongs to the lunar realm [the planets ruled by the moon], while Jupiter's note belongs to the solar realm [the planets ruled by the sun], and for the following reason: since the note of each of the two malefics [Saturn and Mars] makes a consonant fourth with the note of each of the two benefics [Jupiter and Venus]—the *nete hyperbolaion* of Saturn to the *nete diezeugmenon* of Jupiter, and the *nete synemmenon* of Mars to the *mese* of Venus—it follows that Saturn belongs more to the solar, Mars to the lunar realm. As for the aspects, those of Saturn to Jupiter are all good; of those of Saturn to the Sun, only the trine is good, which makes a more harmonious ratio than the others. In the same way, the aspects of Mars to Venus and to the moon are not all good, only the trine; all aspects of Saturn to the moon and to Venus are bad; those of Mars to the Sun and Jupiter are all treacherous.

Lunar realm				Solar realm
		Saturn	a'	*nete hyperbolaion*
		Jupiter	e'	*nete diezeugmenon*
nete synemmenon	d'	Mars		
		Sun	b	*paramese*
mese	a	Venus		
hypate hypaton[28]	e	Moon		

6

✦

Censorinus
fl. early third century C.E.

Censorinus' one surviving work, *On the Day of Birth*, was written as a birthday present for Quintus Caerullius in the year 238. It is a compilation of anecdotes and beliefs about birth, among them the superstition about the months of pregnancy in which a successful birth can take place. Censorinus offers two explanations for this, one drawn from the astrological aspects, the other from musical intervals. Chapters XI and XII, which relate musical numbers to embryology, were given in *Music, Mysticism and Magic*. Here are the surrounding chapters, which furnish the cosmic dimension of the question.

Censorinus' idea is, of course, laughable today, but the way in which he draws together three different disciplines—astronomy or astrology (the distinction makes little sense in his time), embryology, and music—is an epitome of the philosophy behind this collection of texts. The triple link between music, mankind, and the stars serves several integrative purposes. First, it puts number in its rightful place as the ultimate constituent of the universe. Next, it applies number and calculation to the stars, random and incalculable as they are to the casual eye. Through the agency of the stars, number controls the soul's entry into this world, expressed in the precise patterns of the natal horoscope. But what makes some numbers better than others? The answer lies in music, where certain numbers make harmony, others discord. Therefore music furnishes the qualitative control to an otherwise too quantitative view of the universe.

This is perhaps more than Censorinus himself would have understood, simple grammarian that he was, but it provides us with a context for understanding how, and why, he produced this curious birthday present.

Source: Censorinus, *De die natali*, VIII-X, XIII, ed. Otto Jahn (Berlin: G. Reimer, 1845). Translated by the Editor.

The Relation of Astrological Aspects to Gestation

Chapter VIII, 1. Now[1] we will treat briefly the arguments of the Chaldaeans, explaining why they believe that man can be born only in the seventh, ninth, and tenth months.   2. First, they say that our activity and life are subject to the stars, both wandering and fixed, and that the human race is governed by their manifold and various courses. But their motion, patterns, and effects are continually altered by the sun. For the setting of some stars and the standing of others, all affecting us according to their different temperatures, occur through the sun's power.[2]   3. Thus the one who moves the very stars by which we are moved, and gives us the soul by which we are governed, is the most powerful thing controlling us when after conception we emerge into the light.

The sun does this according to three aspects. I will explain briefly what an aspect is and what kinds of them there are, so that it may be clearly understood.   4. There is a circle supposed to carry figures, which the Greeks call the Zodiac, and through which are borne the sun, the moon, and all the other planets. This is divided equally into twelve parts corresponding to as many signs. The sun traverses it in the course of a year, remaining in each sign for about a month. Now each sign has mutual aspects with certain other ones, not all uniform but some stronger, some weaker. So at whatever time conception occurs, the sun must be in a certain sign and a certain degree [*particula*], and this is called the place of conception.   5. There are thirty of these degrees in each sign, making 360 in the entire Zodiac. The Greeks call them *moiras*, whence one can see why they name the Fates *moiras*, just as we call these degrees *fata*: for the one that is rising when we are born is of the utmost importance.

6. When the sun moves into the next sign, it forms either a weak aspect or no aspect at all with the place of conception, for most adjacent zodiacal signs will mutually negate each other. When it comes to the third sign, i.e., with one sign interposed, then it is said to aspect (*videre*) the place from which it set out, albeit very obliquely and with a feeble light; and this aspect is called a sextile because it subtends the sixth part of the circle. (If you draw a line from the first sign to the third, from thence to the fifth, to the seventh, and so on, you will inscribe an equilateral hexagon in the circle.)   7. This aspect is not accepted anywhere, because it would bring about birth at too early a stage.

8. When the sun comes to the fourth sign, leaving two in between, the aspect is called a square because the line which joins the aspected signs cuts off a fourth part of the circle.   9. When it is in the fifth sign with three intermediate, then it is in the aspect of a trine, measuring out a third part of the signs. These two aspects, square and trine, are very efficacious and make for a far greater chance of birth.[3]   10. The next aspect, from the sixth sign, lacks any efficacy, for its line does not form the side of any polygon.

But the seventh zodiacal sign, opposite to the first, makes the most pregnant and powerful aspect, capable sometimes of producing mature infants who, born in the seventh month, are called *septemmestres*.   11. But should the womb not be able to ripen in this period, it will not bring forth in the eighth month (the aspect from the eighth sign being inefficacious, like that from the sixth), but in the ninth or tenth month.   12. Placed in the ninth sign of the Zodiac, the Sun aspects the place of conception by a trine, and in the tenth by a square: the most efficacious aspects, as we have said above.

13. For the rest, they do not believe birth to take place in the eleventh sign, which sends forth a ray of faint light in a sextile; much less in the twelfth, which has no aspect at all. So according to this explanation *heptameroi* are born when the sun is in opposition, *enneameroi* when in trine, and *dekameroi* when in square aspect to its position at conception.

IX, 1. From this account of the Chaldaeans' beliefs I turn now to the Pythagorean opinion treated by Varro[4] in his book concerning human origins called *Tubero*, which seems to give an explanation closest to the truth.   2. Since all the members [of the embryo] do not come to maturity at the same time, many other authorities give the various times at which they reach it. Diogenes Apolloniates[5] says that the male body is formed in four months, the female in five. Hippon,[6] on the other hand, writes that the infant is formed in 60 days; that in the fourth month the flesh solidifies, in the fifth the nails and hair grow, and in the seventh the being is complete. Pythagoras says,[7] more credibly, that there are two kinds of birth: one in the seventh and the other in the tenth month, corresponding to their respective numbers of days. He holds that the periods during which semen is converted into blood, blood into flesh, and flesh into the human form, are proportionate to one another in the same way as those musical tones called *symphonoi*.

An Explanation of Embryology through Music

Chapter X, 1. To make this more understandable, it is necessary first to say something about the rules of music, the more so since I shall speak of things unknown even to musicians.   2. For to treat sounds scientifically and reduce them to order, to find out their motion, their mode, and their measure, is more the work of geometry than of music.   3. Music is the science of good modulation:[8] but this lies in the tone which is sometimes higher, sometimes lower. Single tones of whatever pitch are called notes *(phthongoi)*; the distinction of higher notes from lower ones is called an interval *(diastema)*.   4. Between the lowest and the highest notes there are many possible intervals *(diastemata)*, greater and smaller, such as the one they call the tone, the smaller semitone, and the other intervals of two, three, or any succeeding number of tones.

But combining all the notes promiscuously with each other, according to one's fancy, does not make for harmonious results in song.   5. Just as our letters mixed up together at random will not combine into words or coherent syllables, so in music there are only certain intervals productive of consonance *(symphonias)*.   6. A consonance is the pleasant agreement between two disparate tones in combination. The simple or primary consonances are three, to which the others correspond: first the interval of two tones and a semitone, called a fourth, then that of three tones and a semitone, called a fifth, thirdly the octave, whose interval contains the other two.   7. It comprises either six tones, as Aristoxenus and the musicians[9] aver, or five tones and two 'semitones,' according to Pythagoras and the geometers who demonstrate that two semitones cannot fill out a tone. On this account Plato wrongly calls this interval a semitone,[10] whereas it should be called a *dialeimma*.

8. In order to show plainly how the tones cannot be measured by sight nor by touch, I refer to the excellent experiment of Pythagoras,[11] who by observing the secrets of nature discovered that the notes of musicians correspond to numerical ratios. He stretched strings equal in

weight and in length by means of different weights, and plucking them repeatedly while changing the weights found at first that they did not form any consonance. Finally, however, after frequent attempts, he noticed the two strings sounding a fourth, while their weights were proportionate as three to four. This the Greek arithmeticians call *epitriton*, the Latins *supertertium*.   9. Then he found the consonance of a fifth when the difference between the weights was in sesquialtera proportion: that of two to three, called *hemiolion*. And when one string was stretched by a weight double that of the other, in the proportion *displasion logos*, it sounded the octave.   10. He also investigated whether the same was true of pipes, and found it no different. He made four pipes of equal bore and unequal length, the first as it were six inches long, the second with 1/3 added, i.e., 8", the third 9", half as long again as the first, the fourth 12", doubling the first. He blew these pipes,   11. bound together in pairs, and demonstrated to the ears of all musicians that the first and second (in supertertiary proportion) sounded the consonance of a fourth; the first and third pipe (in sescuple proportion) a fifth; the first and fourth (in duple proportion) an octave.   12. But there is this difference in the nature of pipes and strings: whereas pipes increased in length sound lower, the addition of weight to a string makes it higher; yet the proportion is the same in both cases.[12]

The Intervals from the Earth to the Planets

Chapter XIII, 1. In support of this there is Pythagoras' assertion that the whole world is made according to a musical plan *(ratione)*, and that the seven stars wandering between heaven and earth, which affect the birth of mortals, move rhythmically and in positions corresponding to musical intervals, and give forth various sounds consonant with their altitude which together make the most exquisite melody. But this is inaudible to us because of the grandeur of the sound, which our limited ears are incapable of grasping.

2. Now as Eratosthenes calculated by geometrical method *(ratione)* that the greatest circumference of the earth is 252,000 stades, so Pythagoras indicated how many stades there are between the earth and the individual stars. (The stade most probably to be understood in this measuring of the world is the Italian one, so called, of 625 feet, but there are other ones different in length, such as the Olympic one of 600 feet—that is, Pythian feet.) Pythagoras believed the distance from the Earth to the Moon to be about 126,000 stades, and this to be the interval of a tone.   3. Then from the Moon to the planet *(stellam)* Mercury, called Stilbon, is the half of that distance, or a semitone.

From Mercury to Phosphor or the planet Venus is about the same, i.e., another semitone; thence to the Sun is thrice as far, making a tone and a half.   4. The star (*astrum*) of the Sun is thus three and a half tones distant from the Earth, making a fifth, and two and a half tones from the Moon, making a fourth.

From the Sun to the planet Mars, whose name is Pyrois, the interval is the same as from the Earth to the Moon, i.e., a tone; from Mars to the planet Jupiter, called Phaethon, it is the half of that, a semitone. From Jupiter to the planet Saturn, called Phaenon, the distance is another semitone, and thence to the highest heaven where the signs of the Zodiac are, a semitone again.   5. So from the highest heaven to the Sun the interval is a fourth (two tones and a half), and from the Earth's highest point to the same heaven is six tones, which make the consonance of an octave. Much else of which musicians treat refers to the other stars,[13] and demonstrates that this whole universe is a harmony. This is why Dorylaus wrote that the world is God's instrument (*organum*); others added that it is a *heptachordon*, because there are seven wandering stars which are the most moved.   6. But this is no place for the detailed treatment of all these matters: yet if I wanted to assemble them in a separate book, I would be beset by difficulties. Since the allure of music has already distracted me too long, I had better return to my subject.

7

The Hymns of Orpheus
between first and fourth
centuries C.E.

Although the Orphic religion was one of the oldest in Greece, the poems known as the Orphic Hymns date from the period of religious syncretism and revival of ancient cults and mysteries in the early centuries of our era. Very little is known, or written, about them, and Athanassakis' suggestion is the most plausible one: that "newly founded *thiasoi*, or cult associations, commissioned one or more able men to provide them with religious poetry which should be as authentic and comprehensive as possible" (p. ix of his edition).

The Hymns would have been used in theurgic ceremonies, for each of which an appropriate incense is prescribed, the intention being to attract the virtues of the various gods and goddesses into the initiates. Hymn 34 invokes Apollo as a god of cosmic and natural order and harmony, who stamps the manifested world with the seal of the divine laws. His role is thus analogous to that of Plato's Demiurge.

Source: The Orphic Hymns, translated by Apostolos N. Athanassakis (Missoula, Mont.: Scholars Press, 1977), pp. 47–49. Used by kind permission of the publisher.

Hymn 34: To Apollo

(Incense: powdered frankincense)

Come, O blessed Paian, O slayer of Tityos, O Phoibos, O Lykoreus;
A giver of riches are you and an illustrious dweller of Memphis, O god
     to whom one cries "IE."
To you, O Titan and Pythian god, belong the lyre, and seeds and plows.
Grynean, Sminthian, slayer of Pytho, Delphic diviner,
you are a wild, light-bringing and lovable god, O glorious youth.

You shoot your arrows from afar, you lead the Muses into dance,
and, O holy one, you are Bacchos, Didymeus, and Loxias, too.
Lord of Delos, eye that sees all and brings light to mortals,
golden is your hair, and clear your oracular utterance.
Hear me with kindly heart as I pray for people.
You gaze upon all the ethereal vastness,
and upon the rich earth you look through the twilight.
In the quiet darkness of a night lit with stars
you see earth's roots below, and you hold the bounds
of the whole world. Yours, too, are the beginning and the end to come.
You make everything bloom, and with your versatile lyre
you harmonize the poles,[1] now reaching the highest pitch,
now the lowest, and now again with the Doric mode
balancing the poles harmoniously, as you keep the living races distinct.[2]
You have infused harmony into all men's lot,
giving them an equal measure of summer and winter.
The lowest notes you strike in the winter, the highest in the summer,
and your mode is Doric for spring's lovely and blooming season.
Wherefore mortals call you lord, and Pan,[3]
the two-horned god who sends the whistling winds.
For this, too, you have the master seal of the entire cosmos.
O blessed one, hear the suppliant voice of the initiates and save them.

8

✦

Saint Athanasius
c. 298–373

This great Alexandrian theologian, bishop, and politician was a principal actor in the period in which Catholic Christianity achieved its status as the established religion of the Roman Empire. The main task of his long life was the defense of the unity of God and the divinity of Christ against the doctrines of Arius. But the present work was written before the Arian controversy broke out in 318, hence when Athanasius was quite young. It was in fact his first work, and bears the stamp of his Greek education—nowhere more than in this passage, which uses the venerable idea of universal harmony to prove the existence and unity of God.

Like many of our writers, Athanasius assumes that the harmony and order of the macrocosm are reflected both in the microcosm of the human being and in the intermediate "mesocosm" of society. Coming shortly after a period in which the worship of the Sun had almost become the sole official religion of the Roman Empire, Athanasius' simile of the cosmic lyre-player would surely have evoked the image of Apollo with his all-harmonizing lyre, as we meet him in the Orphic Hymn (no. 7). But while they were anxious to appropriate the pagan god's attributes for their own, the Christian Fathers were not at all interested in how Apollo's lyre worked or sounded. In accordance with the Christian revelation of the supreme Deity incarnate in man, they considered it their privilege to bypass the ladder of ascent through gradual knowledge of the arts and sciences, and to embrace God in his immediacy and unity. In medieval scholasticism we will find another view (see Ugolino of Orvieto, no. 23).

Source: Saint Athanasius, *Select Works and Letters,* translated by Archibald Robertson (New York: Christian Literature Co., 1892), p. 24.

The Harmony of All Things

Chapter 38, 1.[1] Since then, there is everywhere not disorder, but order, proportion and not disproportion, not disarray but arrangement, and that in an order perfectly harmonious, we needs must infer and be led to perceive the Master that put together and compacted all things, and produced harmony in them. For though He be not seen with the eyes, yet from the order and harmony of things contrary it is possible to perceive their Ruler, Arranger, and King.   2. For in like manner as if we saw a city, consisting of many and diverse people, great and small, rich and poor, old and young, male and female, in an orderly condition, and its inhabitants, while different from one another, yet at unity among themselves, and not the rich set against the poor, the great against the small, nor the young against the old, but all at peace in the enjoyment of equal rights,—if we saw this, the inference surely follows that the presence of a ruler enforces concord, even if we do not see him; (for disorder is a sign of absence of rule, while order shows the governing authority: for when we see the mutual harmony of the members in the body, that the eye does not strive with the hearing, nor is the hand at variance with the foot, but that each accomplishes its service without variance, we perceive from this that certainly there is a soul in the body that governs these members, though we see it not); so in the order and harmony of the Universe, we needs must perceive God the governor of it all, and that He is one and not many.

3. So then this order of its arrangement, and the concordant harmony of all things, shews that the Word, its Ruler and Governor, is not many, but One. For if there were more than one Ruler of Creation, such an universal order would not be maintained, but all things would fall into confusion because of their plurality, each one biasing the whole

to his own will, and striving with the other. For just as we said that polytheism was atheism, so it follows that the rule of more than one is the rule of none. For each one would cancel the rule of the other, and none would appear ruler, but there would be anarchy everywhere. But where no ruler is, there disorder follows of course.   4. And conversely, the single order and concord of the many and diverse shews that the ruler too is one. For just as though one were to hear from a distance a lyre, composed of many diverse strings, and marvel at the concord of its symphony, in that its sound is composed neither of low notes exclusively, nor high nor intermediate only, but all combine their sounds in equal balance, —and would not fail to perceive from this that the lyre was not playing itself, nor even being struck by more persons than one, but that there was one musician, even if he did not see him, who by his skill combined the sound of each string into the tuneful symphony;[2] so, the order of the whole universe being perfectly harmonious, and there being no strife of the higher against the lower or the lower against the higher, and all things making up one order, it is consistent to think that the Ruler and King of all Creation is one and not many,[3] Who by His own light illumines and gives movement to all.

Aristeides Quintilianus
fl. probably c. 300

Aristeides, the author of one of the three major Greek treatises on music—the others being by Ptolemy (see no. 5) and Aristoxenus—remains a misty figure, his time and place unknown. What is abundantly clear, however, is that he was a fervent Platonist who saw music in its broadest context of educating the soul. This did not prevent him from devoting much of his treatise to valuable documentation of the Greek musical system and its notation, but, as his translator Thomas Mathiesen observes, *De Musica* "is a highly systematic work of philosophy in which even the technical materials are introduced for the purpose of philosophical demonstration." (ed. cit., p. 11)

Our interest is accordingly in the passage in Book III which comments on Plato's *Timaeus*, and in the section of Book II which offers a cosmic explanation for the musical sensitivity of soul and body.[1] Aristeides has begun his second book with a discussion of the role of music in education, not only in childhood but throughout life. He draws his arguments largely from Plato, with the addition of a section on Cicero and Rome. A brief study of musical instruments leads him to pose the question: Why do these have so great a power over the soul? Our passage follows directly.

There are resonances here with the Neoplatonists Plotinus and Porphyry, to which A. J. Festugière has devoted an important article (see Bibliography). Especially evocative is Aristeides' description of the formation of the subtle body as the soul descends through the planetary spheres on its way into incarnation. The latticework or netlike forms which it assumes recall Porphyry's commentary on Homer's Cave of the Nymphs, where bodies are woven from red threads on beams of stone. Likewise Aristeides indulges in a piece of Homeric exegesis, giving a delightful Neoplatonic interpretation to the famous myth of Aphrodite and Ares caught in Hephaistus' net. Then, in a masterly stroke such as is only possible—and thinkable—to those versed in the Hermetic doctrine of correspondences, the author applies the image to musical instruments, seeing in their structure a similarity to the subtle reticulations of the etheric vehicle. With this, the question is answered.

The chapter from Book III is not apparently musical, but is so important as a commentary on the prime text of this sourcebook that it is included, with the suggestion that the meanings attributed to the *Timaeus* numbers be applied also to their musical equivalents. One would then have the idea of the "body" of octave-spaces becoming "ensouled" by the entry of the circle of fifths and the creation of scales.

Source: Aristeides Quintilianus, *On Music, in Three Books*, translated with introduction, commentary, and annotations by Thomas J. Mathiesen (New Haven and London: Yale University Press, 1983), pp. 151–157. Used by kind permission of Professor Mathiesen and the publishers.

How the Soul Descends to Earth

Book II, 17. Is there not more strongly born in those hearing these things a yearning to seek after the reason and to learn what coerces the soul to be so readily conquered by melody played on instruments? I shall refer to an argument that is ancient, certainly, but since it comes from wise men, it is not untrustworthy. Even if it should be unconvincing in other ways, at least as far as what appears to be is concerned, it undoubtedly proves true. For that the soul is moved naturally by music through instruments, everyone knows. So in respect to the argument set out below, if it is possible to find another reason and if a better one, then we must reject what will be said. But if we are not able to do so, how shall we fail to trust from palpable things even what follows of necessity?

There is the one argument that the soul is a certain harmonia and harmonia exists through numbers. Of course, since harmonia in music is composed through the same proportions, when the similar proportions are moved the similar passions are also moved at the same time. This we shall thoroughly examine in the future.[2] Another argument says the following: because of the first constitution of the soul, through which it made conjunction with the body here,[3] the material and nature of instruments are analogous to each other. For the soul—while it is seated in the purer region of the universe, being unmixed with bodies—is unadulterated and immaculate and goes around together with the sovereign of this present universe.[4] Whenever from a declension to the things here it takes on certain appearances from the things around the earthly region, [87] then little by little the soul is forgetful of the beautiful things of that place and sinks down. And by however much it is divorced from the things above, by so much the soul, approaching the things in this world, is filled full of greater folly and is turned to bodily darkness, on the one hand because of a diminution of its earlier worth, since it is no longer able to be mentally coextensive with the universe, and on the other hand on account of forgetfulness of the beautiful things of that place and consternation with earthly things, as it descends into the more solid things and those familiar with matter.

Therefore, desiring a body, the soul, they say, takes and draws along from each of the higher regions some portions of bodily combination. Going through the ethereal orbits,[5] the soul partakes of everything so far as it is luminous and adapted for warming and naturally enclosing the body, plaiting certain bonds from these orbits for itself as a sort of latticework by irregular movement of the reciprocal lines among the movements themselves. And being carried through the regions around the moon, which shared in an aeroid and thus resistant breath, making much and vehement whistling because of its natural motion, the soul begins to be filled with the underlying breath, and extending the surfaces and lines of its orbits—partly being dragged down by the masses of the breath, partly holding on naturally to the far side—it loses its spheroid form and is changed in a manly form. It exchanges its surfaces, which are in accord with luminous and ethereal matter, for a membranaceous figure; and it turns its lines, which are reduced around the empyrean[6] and tinged by the yellowness of the fire, into the semblance of sinews; [88] and then it adds wet breath from the things of earth, so that this, for the first time, is a certain natural body for the soul, welded together from some membranaceous surfaces, sinuouslike lines, and breath.[7] This is the root of the body, and this they named "harmonia." In this way, they say that this oysterlike instrument here on earth is both congealed and enclosed.

The Poet, too, shows the following constitution of the soul; he says:

The sinews no longer hold the flesh and bones together.
[*Odyssey* 11, 219]

Elsewhere calling the soul Aphrodite and the bodily nature Ares—
because he has substance in blood, the Poet says that by certain such
bonds the soul was fastened together by the Demiurge, whom he names
Hephaistus. He speaks in this way:

and spun his fastenings around the posts from every direction,
while many more were suspended overhead, from the roof beams,
thin, like spider webs.
[*Od.* 8, 278-280]

That the bedposts, which happen to be named after eloquent Hermes,
are ratios and proportions through which it happened that the soul was
entangled with the body, we would argue not absurdly;[8] and that the
spider webs are the surfaces and figures by which the human form is
defined; and that the roof beam itself is, I presume, the dwelling place
fabricated by the soul. That for Homer this is a discourse about the
soul, the following things show. Reciting the separation of Ares and
Aphrodite into their kindred regions, he dispatches the one into his
cognate region of irrationality, to the barbarians and Paions, adding
nothing further by way of explanation; and he dispatches the other into
the ancestral region of her creation and blessed pastime, to Cyprus:

[89] where lies her sacred precinct and her smoky altar;
[*Od.* 8, 363]

and he purges and consecrates her as she withdraws from the more
inferior things; he says:

and there the Graces bathed her and anointed her with ambrosia.[9]
[*Od.* 8, 364]

Wise Heracleitus also somewhere speaks of something, not at variance
with the preceding. Showing the soul enjoying itself in the ether, he
says: "a dry, dessicated soul is wisest." [Fragment 118] And he reveals
the soul becoming turbid from the storm of air and rising vapor, saying:
"for souls, it is death to become water."[10] [Fragment 36]

Physicians also prove this. For they argue that the elements, those
things analogous to the natural masses, and the most essential parts of

the body (and when these have suffered even a little, they say the animal is in danger) are tissues and arteries, which are nothing but some sinewlike, cobweblike, pipelike membranes containing in the middle the breath, through which the soul is moved but certainly does not extend together with the body when the parts are growing and collapse when they decay.[11] They also show this from the pulse beats, the regular motion of which they explicitly state as the healthy condition of the animal, the irregular and agitated they foretell as a threat of death, and the strictest absence of motion they stoutly maintain as the complete retirement of the soul.

18. What is the marvel if the soul, after taking by nature a body similar to the things moving the instruments—strands and breath, is moved at the same time as these are moved; and when the breath sounds harmoniously and rhythmically, the soul is affected at the same time as the breath beside it; and when a strand is struck harmoniously, [90] the soul sounds and intensifies at the same time as the specific strands, since indeed such a correspondence is observed in the kithara?[12] For if anyone should place one of two unison strings inside a small and light straw and should strike the other stretched out at a distance, he will most palpably see the straw-bearing string moved at the same time.[13] It is strange how the divine art seems to operate and to do something even through inanimate things. Indeed, by how much more in the case of things moved by the soul is it necessary for the cause of similarity to operate?

Of instruments, those fitted together of strands closely resemble the ethereal, dry, simple region of the cosmos and part of spiritual nature, being more without passion, immutable, and hostile to wetness, and displaced from their proper setting by damp air; the wind instruments closely resemble the windy, wetter, changeable region, making the hearing overly feminine, being adapted for changing from the straightforward, and taking their constitution and power by wetness. The better instruments are similar to the better things, and the lesser instruments are the others. These things also demonstrate the legend, they say, that esteemed the instruments and mele [songs] of Apollo over those of Marsyas. The Phrygian, having been hung over the river in Celaenae after the manner of a wineskin, happens to be in the aerial, full-windy, and dark-colored region, since he is on the one hand above the water and on the other hand suspended from the ether; but Apollo and his instruments happen to be in the purer and ethereal essence, and he is the leader of this essence.[14]

19. In their discourse about the use of instruments, the ancients gradually reveal the following things to us. Harmful melody and that to be avoided as gradually leading to evil and destruction, they bestowed

as a figure on the brutal and mortal women, the Sirens, whom the Muses conquer and wise Odysseus with headlong speed avoids.[15] Since useful musical creation is twofold (one type is useful for the benefit of serious men, the other for the harmless indulgence of the common herd, even [91] if some of these are quite lowly), the educational type on the kithara, which happens to be manly, they dedicated to Apollo; and the type necessarily pursuing delight because it aims at the multitude, they bestowed on the feminine of the gods, on one of the Muses, on Polyhymnia.[16] And of musical creation in accord with the lyre, the type useful for paideia, as suitable for men, they appointed to Hermes;[17] the other type adapted to softening, as often appeasing the feminine and epithymetic part of the soul, they fastened upon Erato.

Again, in the case of the auloi, the melody flattering the quantity of men and part of the soul desiring pleasure, they assigned to the one advising that the sweet vie with the beautiful, in accord with her appellation, to Euterpe; the other type, able occasionally to benefit through much science and discretion but not however departing absolutely from its natural femininity, they allot no longer to the masculine of the gods, but rather to the one feminine in genus, discrete and warlike in ethos, to Athena.[18] So then, in displaying that the benefit through auletic melody is short and advising the wise to avoid for the most part the facility of it, they say that the goddess threw away the auloi as not adding suitable pleasure for those desiring wisdom; but this type of melody is utilized for those of mankind worn out and exhausted because of continuous work and labor.

They also brought up how a penalty pursued Marsyas, who dignified his music beyond its worth: whose instruments fell behind those of Apollo by so much as handicraftsmen and ignorant men fall behind wise men and Marsyas himself behind Apollo. That is why Pythagoras, too, counselled his disciples who heard the aulos to cleanse their hearing as defiled by breath and to thoroughly purify the irrational impulses of the soul with righteous mele to the accompaniment of the small lyre.[19] For the aulos [92] cultivates what is the leader of the worse portions, but the lyre—because it takes care of the rational nature—is friendly and welcome.

The learned in every group of mankind confirm my view that not only our souls but also that of the universe used such a constitution: those persons cultivating the region under the moon, which is full windy and of a wet constitution but which procures its actuality from ethereal life, are soothed by both kinds of instruments—wind and stranded; and those persons cultivating the pure and ethereal region on the one hand deprecated every wind instrument as defiling the soul and dragging it down to the things here and on the other hand hymned and held in honor the kithara and lyre alone of the instruments as the purer. Of this latter region, surely, wise men are imitators and emulators, on the one hand divorced from the disorder and variety of the things in this world—at least in inclination, even if present in body—and on the other hand holding on to the unbroken simplicity and reciprocal concord of the beautiful things of that place through a likeness to virtue.

How the World-Soul Was Organized

Book III, Chapter 24. That not only the body of the universe but also the soul was organized and is considered through consonant numbers, the ancient and wise men affirmed confidently. [126] Even the divine Plato speaks in a certain way in *Timaeus* about the following things. After taking an essence between the separable and inseparable essence, and in the case of the separable and inseparable nature of the Same and the Other, after constructing means with the mean of the essences and after making a blend of these three, the Demiurge of the soul in turn divided the entire compound by the following numbers, even and odd.[20] Then he augmented the even numbers up to eight by the duple ratio, the odd numbers until twenty-seven by the triple ratio. These numbers, some say, have been so specified because the soul operates by numbers—the soul of each person by numbers corresponding to the arts and the soul of the universe by numbers corresponding to nature; but the more precise say that these numbers present the property of the soul's power and essence.[21] The displays in numbers[22]—the nature of which is external to bodies[23]—demonstrate the incorporeal beginning of the soul,[24] and the increments through ratios and proportions demonstrate movement in depth: that by two (for the cube on two is eight) demonstrating bodily depth, which we call natural (for this is destructible and divisible); that by three demonstrating—like three—the incorporeal, the indivisible, and the active (for the cube on three is twenty-seven).[25] The soul uses bodily depth, becoming three-dimensional with the body, and is often in a contrary condition to the spiritual (for this is stronger). Sometimes the soul is turned to better things, which—it was resolved—are like the odd because this is inseparable, which incorporeal things are like;[26] and sometimes it is turned to the opposite, the nature of which is even and divisible like that of bodies.[27]

The wise men called the greatest of the good things of the soul what closely resemble the terms posited, for the four virtues do not happen to be otherwise than similarities to numbers: [127] judgment analogous to the monad (for monadic is even the simple knowledge of each person); manly spirit to two and akin to the second position, revealing an impulse and transition from one to another; discretion to three, blending a symmetry in between deficiency and excess; and righteousness to four (for it is the first virtue to demonstrate equality, the first number composed of equals multipled equally).[28] Of bodily good fortune,[29] strength is analogous to manly spirit (therefore also to two), beauty to discretion because of the symmetries of the parts and colors (therefore also to three), and a healthy condition to righteousness because of the mental concord[30] of these one to another. Although we see no similarity to judgment in bodies,[31] we do see that the wise man with reason contrives his discourse on the soul through the number seven in two ways. He assigns the measure by the duple and even positions to affective depth and the measure by triple and odd to the rational and incorporeal,[32] both of them suspended from the monad itself as the sole beginning and cause. The diagram will make this clear by accepting all the geometric and musical proportions in the double tetraktys and having the odd numbers in a straight line and the even around the circumference.[33]

In the body, the circular is more valuable, for this is the lightest and purer, while the straighter is lower and more material. In the soul, the straight and uninclined is the good, for this one by equality and identity is senior, while the Other is subordinate, hinting at the fickleness of the affective through the crookedness of the line as surely as this one in accord with the Same is expressive of concavity and convexity together.[34] [128] Therefore the learned of the Greeks not absurdly fastened upon Pan the shepherd's crook, for the one named after the animation of the universe not unreasonably used and adorned the instrument demonstrative of this animation.[35]

Of the proportions of the posited diagram, the arithmetic, considered in terms of the identity of the excesses, exhibit bodily size; and the harmonic, being symmetries of both proportions, display the constitution of the animal from the soul and the body. The theory pertaining to this was previously stated in respect to the 256 to 243.[36] Plato contrives the consonance of the parts of the soul through the first consonant ratio, the sequitertian.[37] His argument on this signifies by length and breadth, which in any relative distance is able to complete depth.[38]

Since the two circles show the circuit relative to the authoritative force of the appearances analogously with the ascents and descents of the voice,[39] the one circle derived from the even-times tetraktys shows the practical part of the soul of the universe, which was also fastened together with the body,[40] and the other derived from the tetraktys by odd numbers shows the theoretical, most divine, and external part, which associates with the better.[41] So, Plato surely ordains that the circle of identity reveals the unchangeableness of the essence corresponding to judgment,[42] and he names the circle of the Other what defines the unreliability of affective and irrational nature.

10

◆

Calcidius
fl. c. 400

From what little is known about Calcidius, he seems to have been a Christian scholar active in Cordova at the start of the fifth century. His only work is this Latin translation, with commentary, of the first part of Plato's *Timaeus* (up to 53c). Calcidius is not original. His commentary on the harmonic intervals of the World-Soul is a laborious explanation of the obvious, not even accurate in rendering Plato's text. However, for the Middle Ages this was the only *Timaeus* known, and so Calcidius' reflections on number and the soul were taken very seriously. This and a handful of other secondary sources (Macrobius, fragments of Proclus, Apuleius' *The Golden Ass*, the Hermetic *Asclepius*, Boethius' *Consolation*, and Martianus Capella) had to serve as the foundation for the whole of medieval Platonism.

This extract gives a fundamental statement of macrocosmic-microcosmic resemblance and harmony, which leads not only to a particular analysis of the human soul but also to a theory of perception, on the basis that "like is known only by like." Only through recognizing this connection, or rather reflection, of the cosmos in the soul can the concept of the Harmony of the Spheres become meaningful; for the spheres are then contained as possibilities of realization within the human being, besides surrounding him physically.

Calcidius' words on number are also fundamental to these doctrines, embodying the original dichotomy within the soul on which Proclus (no. 12) will have much to say. Finally, in christianizing the Platonic creation myth, Calcidius touches on the very points at which early Christianity and paganism meet: the compound nature of the human being, as a triad of divine Spirit, celestial Soul, and elemental Body.

Source: Timaeus a Calcidio translatus commentarioque instructus, ed. J. H. Waszink (London and Leiden: Warburg Institute and E. J. Brill, 1962; vol. 4 of *Plato Latinus*), pp. 100–103. Translated by the Editor.

Creation, Number and Soul

51. Now that we have explained the powers from which the World-Soul is made, the parts and members out of which it is assembled, and how its nature consists of the same powers and numbers and modes, it remains to be shown what Plato has said about this in explaining the matter, so that we may reach the appointed end of our task guided by reason, rationally and methodically. It is his opinion that the soul of the sensible world is born (as license has it) a witness to all things, both intelligible and sensible. It is moreover the Pythagorean doctrine that "like are known only by like"; and Empedocles has said the following in his verses:

> By the earthly we know earth, by flames the aether,
> By liquid, moisture, by our breath the spirituous,
> By the tranquil, peace, and by the striving, strife.
>   [Fragment 109]

In fact he called these the elements and origins of the universe, from which he held that the substance of the soul also consists; and on that account he said that it contains full knowledge of all things, comprehending by its resemblance those things which have a likeness to it.

52. Plato, too, asserting the same thing, conflates the soul from all these origins so that it might have knowledge and discrimination of these same origins and of what follows them, and indeed of all existing

things. Then, after he had mixed it together out of these powers from which he believed it to be compounded, and divided the mixture into parts according to certain geometrical and arithmetical and musical ratios,[1] he proceeded thus:

> Then he cut the whole of this fabric along its length, and produced two from one; and adapted middle to middle, like the form of the Greek letter X, and bent them into a circle so that their extremes might be combined. [. . .] And he inserted one circle into the other, so that one should revolve laterally and the other obliquely.[2]

He [Plato] augmented this with some additional matter:

> Seeing that it is compounded from the Same and the Different mixed with Existence, lacking movement but revolving upon itself with a circular motion, whenever it meets with any substance, either dissoluble or indivisible, it can easily recognize this indivisible or diverse and dissoluble nature, and see the causes of all that will come to pass, and judge from those that happen the ones which are to come.[3]

53. It is clear from this that the primal essences or substances of things are the oldest of all—and these twofold, the one kind indivisible, the other divisible—and the twofold diversity of nature most ancient; and that therefore the soul has been established from the beginnings, conflated from two substances, whose natures are those of the Same and the Different. Therefore its nature is most in agreement with that of the numbers which, it is said, are anterior even to the geometrical forms themselves. Geometrical forms are necessarily to be found in any number, such as the figures of three, four and more sides, and also those which are called hexahedra and octahedra. But they cannot subsist without numbers, whereas nothing prevents numbers from existing without these forms. Thus the oldest species of numbers is found to exist in all calculations. The first and foremost among these numbers are unity and duality, said to be the origins of the rest of the numbers.[4] For this reason it is concluded that the soul, made from a twofold substance and from the double nature of the power of numbers, builds and vivifies the celestial bodies and creatures in which there is reason and discipline; for it has knowledge of all the things from whose powers it is itself composed.

54. This is the rational Soul of the World, with its doubly venerable character, which through its better nature furnishes support to inferior

things following the divine commandments, and imparts providence to things created. It is blessed through its knowledge by similitude of things eternal, and is a helper and protectress of dissoluble things. Its powers of counsel and forethought are evidenced by the customs of man, the preeminent worshiper of God, who bestows his care on tame animals.[5] These virtues of the soul, indeed, by which the sensible world is fostered, are added to the other vital energies common to man, beasts, and even to plants, viz. those of growth, automation, desire and imagination, supplemented by reason which completes the soul peculiar to man, that he might not merely possess life, but also not lack the desire to live well. Thus the life of man is mingled in due proportion from both the natural and the rational soul in consort.[6]

55. That this is indeed true is attested by a certain lofty doctrine of a sacred school, wise in their understanding of the divine. This asserts that God founded the human race after completing and adorning the sensible world; that he actually built and formed its body after this pattern by taking a piece of soil, derived the life for it from the celestial vaults, and then introduced the breath of life into its inmost parts by his own breath, this latter signifying the intentions of God and the wisdom of the soul. And God's wisdom is a god who cares for human affairs, being the cause of men's living happily and well, so long as they do not neglect the gift that God Almighty has bestowed upon them. And now sufficient has been said of the Soul of the World, assembled by the compacting of two essences whose natures are those of the Same and the Different; of its division made according to the harmonic, arithmetical and geometrical ratios, and of the way in which its nature is put together from numbers and sounds.

11

◆

Macrobius
fl. c. 400

Macrobius' life is so obscure that scholars are still undecided as to whether he came from Greece or Africa, whether his native tongue was Latin or Greek, and whether or not he was a Christian. He was a typical popularizer of his time, filling a demand for a broad liberal education at a modest level. This commentary on the "Dream of Scipio" with which Cicero concludes his *De Republica* gave Macrobius an excuse to write a miniature encyclopedia on such topics as dreams, arithmology, life-cycles, virtue, suicide, astronomy, geography, music theory, and the Neoplatonic doctrine of the soul. He was evidently most interested in the latter, and his summary is a very adequate one.

Macrobius' work was treasured in the Middle Ages, along with the compilations of Boethius, Martianus Capella, Calcidius, Cassiodorus, and Isidore of Seville, far beyond its intrinsic merits. These works were copied and recopied as authoritative sources of ancient wisdom. Macrobius' book was tailor-made for an epoch that loved anecdote and trusted the Ancients implicitly, yet preferred to give its more serious efforts to the intricacies of Church doctrine. The treatment of music, given here in its entirety, encompasses skillfully all the main points of the Pythagorean-Platonic musical worldview.

Source: Macrobius, *Commentary on the Dream of Scipio,* translated by William Harris Stahl (New York: Columbia University Press, 1952), pp. 189–200. Copyright ©1952 by Columbia University Press. Used by permission.

Plato's Numbers and Their Significance

Book II, Chapter II [14] Plato's Timaeus, in disclosing the divine plan in the creation of the World-Soul, said that the Soul was interwoven with those numbers, odd or even, which produce the cube or solid, not meaning by this that the Soul was at all corporeal; rather, in order to be able to penetrate the whole world with its animating power and fill the solid body of the universe, the Soul was constructed from the numbers denoting solidity.[1]

[17] The fabrication of the World-Soul, as we may easily see, proceeded alternately: after the monad, which is both even and uneven, an even number was introduced, namely, two; then followed the first uneven number, three; fourth in order came the second even number, four; in the fifth place came the second uneven number, nine; in the sixth place the third even number, eight; and in the seventh place the third uneven number, twenty-seven.

Since the uneven numbers are considered masculine and the even feminine, God willed that the Soul which was to give birth to the universe should be born from the even and uneven, that is from the male and female; and that, since the Soul was destined to penetrate the solid universe, it should attain to those numbers representing solidity in either series.

[18] And then the Soul had to be a combination of those numbers that alone possess mutual attraction since the Soul itself was to instill harmonious agreement in the whole world. Now two is double one and, as we have already explained, the octave arises from the double; three is one and one-half times greater than two, and this combination produces the fifth; four is one and one-third times greater than three, and this combination produces the fourth; four is also four times as great as one, and from the quadruple ratio the double octave arises. [19] Thus the World-Soul, which stirred the body of the universe to the motion that we now witness, must have been interwoven with those numbers which produce musical harmony in order to make harmonious the sounds which it instilled by its quickening impulse. It discovered

the source of these sounds in the fabric of its own composition.

[20] Plato reports, as we have previously stated, that the divine Creator of the Soul, after weaving it from unequal numbers, filled the intervals between them with sesquialters, sesquitertians, superoctaves, and semitones.   [21] In the following passage Cicero shows very skillfully the profundity of Plato's doctrine: "What is this great and pleasing sound that fills my ears?" "That is a concord of tones separated by unequal but nevertheless carefully proportioned intervals, caused by the rapid motion of the spheres themselves."   [22] You see here how he makes mention of the intervals and states that they are unequal and affirms that they are separated proportionately; for in Plato's *Timaeus* the intervals of the unequal numbers are interspersed with numbers proportional to them, that is, sesquialters, sesquitertians, superoctaves, and half-tones, in which all harmony is embraced.

[23] Hence we clearly see that these words of Cicero's would never have been comprehensible if we had not included a discussion of the sesquialters, sesquitertians, and superoctaves inserted in the intervals, and of the numbers with which Plato constructed the World-Soul, together with the reason why the Soul was interwoven with numbers producing harmony.   [24] In so doing we have not only explained the revolutions in the heavens, for which the Soul alone is responsible; we have also shown that the sounds which arose from these had to be harmonious, for they were innate in the Soul which impelled the universe to motion.

Chapter III [1] In a discussion in the *Republic* about the whirling motion of the heavenly spheres, Plato says that a Siren sits upon each of the spheres, thus indicating that by the motions of the spheres divinities were provided with song; for a singing Siren is equivalent to a god in the Greek acceptance of the word. Moreover, cosmogonists have chosen to consider the nine Muses as the tuneful song of the eight spheres and the one predominant harmony that comes from all of them. [2] In the *Theogony* [lines 78-79], Hesiod calls the eighth Muse Urania because the eighth sphere, the starbearer, situated above the seven errant spheres, is correctly referred to as the sky; and to show that the ninth was the greatest, resulting from the harmony of all sounds together, he added: "Calliope, too, who is preeminent among all." The very name shows that the ninth Muse was noted for the sweetness of her voice, for Calliope means "best voice."[2] In order to indicate more plainly that her song was the one coming from all the others, he applied to her a word suggesting totality in calling her "preeminent among all."   [3] Then, too, they call Apollo, god of the sun, the "leader of the Muses," as if to say that he is the leader and chief of the other spheres, just as Cicero, in referring to the sun, called it "leader, chief, and regulator of the other planets, mind and moderator of the universe." [4] The Etruscans also recognize that the Muses are the song of the universe, for their name for them is *Camenae*, a form of *Canenae*, derived from the verb *canere*.

That the priests acknowledged that the heavens sing is indicated by their use of music at sacrificial ceremonies, some nations preferring the lyre or kithara, and some pipes or other musical instruments. [5] In the hymn to the gods, too, the verses of the strophe and antistrophe used to be set to music, so that the strophe might represent the forward motion of the celestial sphere and the antistrophe the reverse motion of the planetary spheres; these two motions produced nature's first hymn in honor of the Supreme God. [6] In funeral processions, too, the practices of diverse peoples have ordained that it was proper to have musical accompaniment, owing to the belief that souls after death return to the source of sweet music, that is, to the sky.

[7] Every soul in this world is allured by musical sounds so that not only those who are more refined in their habits, but all the barbarous peoples as well, have adopted songs by which they are inflamed with courage or wooed to pleasure; for the soul carries with it into the body a memory of the music which it knew in the sky, and is so captivated by its charm that there is no breast so cruel or savage as not to be gripped by the spell of such an appeal. [8] This, I believe, was the origin of the stories of Orpheus and Amphion, one of whom was said to have enticed the dumb beasts by his song, the other the rocks. They were perchance the first to attract in their song men lacking any refinement and stolid as rocks, and to instill in them a feeling of joy. [9] Thus every disposition of the soul is controlled by song. For instance, the signal for marching into battle and for leaving off battle is in one case a tone that arouses the martial spirit and in the other one that quiets it. "It gives or takes away sleep,"[3] it releases or recalls cares, it excites wrath and counsels mercy, it even heals the ills of the body, whence the statement that gifted men "sing out remedies" for the ailing. [10] Is it at all strange if music has such power over men when birds like the nightingale and swan and others of that species practice song as if it were an art with them, when creatures of land, sea, and air willingly fall into nets under the spell of music, and when shepherd's pipes bring rest to the flocks in pasture?

[11] We have just explained that the causes of harmony are traced to the World-Soul, having been interwoven in it; the World-Soul, moreover, provides all creatures with life: "Thence the race of man and beast, the life of winged things, and the strange shapes ocean bears beneath his glassy floor."[4] Consequently it is natural for everything that breathes to be captivated by music since the heavenly Soul that ani-

mates the universe sprang from music.   [12] In quickening the spheres to motion it produces "tones separated by unequal but nevertheless carefully proportioned intervals," in accordance with its primeval fabric.

Now we must ask ourselves whether these intervals, which in the incorporeal Soul are apprehended only in the mind and not by the senses, govern the distances between the planets poised in the corporeal universe.   [13] Archimedes, indeed, believed that he had calculated in stades the distances between the earth's surface and the moon, between the moon and Mercury, Mercury and Venus, Venus and the sun, the sun and Mars, Mars and Jupiter, Jupiter and Saturn, and that he had also estimated the distance from Saturn's orbit to the celestial sphere. [14] But Archimedes' figures were rejected by the Platonists for not keeping the intervals in the progressions of the numbers two and three. They decided that there could be but one opinion, that the distance from the earth to the sun was twice as great as from the earth to the moon, that the distance from the earth to Venus was thrice as great as from the earth to the sun, that the distance from the earth to Mercury was four times as great as from the earth to Venus, that the distance from the earth to Mars was nine times as great as the distance from the earth to Mercury, that the distance from the earth to Jupiter was eight times as great as from the earth to Mars, and that the distance from the earth to Saturn was twenty-seven times as great as from the earth to Jupiter.   [15] Porphyry' includes this conviction of the Platonists in his books which pour light upon the obscurities of the *Timaeus*; he says they believed that the intervals in the corporeal universe, which were filled with sesquitertians, sesquialters, superoctaves, half-tones, and a *leimma*, followed the pattern of the Soul's fabric, and that harmony was thus forthcoming, the proportional intervals of which were interwoven into the fabric of the Soul and were also injected into the corporeal universe which is quickened by the Soul.   [16] Hence Cicero's statement, that the heavenly harmony is a "concord of tones separated by unequal but nevertheless carefully proportioned intervals," is in all respects wise and true.

Chapter IV [1] At this point we are reminded that we must discuss the differences between high and low notes of which Cicero speaks. "Nature requires that the spheres at one extreme produce the low tones and at the other extreme the high tones. Consequently the outermost sphere, the star-bearer, with its swifter motion gives forth a higher-pitched tone whereas the lunar sphere, the lowest, has the deepest tone."

[2] We have stated that sound is produced only by the percussion of air. The blow regulates the pitch of the sound: a stout blow swiftly dealt produces a high note, a weak one lightly dealt produces a low

note.[6]   [3] For example, if one lashes the air with a staff, a swift movement produces a high note, a slower movement a lower tone. We see
the same phenomenon in the case of the lyre: strings stretched tight
give high-pitched notes, but when loosened produce deep notes.   [4]
Accordingly the outer spheres, revolving at high speeds on account of
their great size and constrained by a breath that is more powerful because it is near its source, as Cicero puts it, "with their swifter motion
give forth a higher-pitched tone, whereas the lunar sphere, the lowest,
has the deepest tone"; the latter is motivated by a breath which at that
great distance is weak, and revolves at a slow speed because of the small
space in which it, the sphere last but one, is confined.   [5] We have
further proof in pipes, which emit shrill notes from the holes near the
mouthpiece but deep notes from those near the other end, and high
notes through the wide holes but low notes through the narrow holes.
A single explanation underlies both circumstances: the breath at the
start is vigorous but weakens as it is spent, and it rushes with greater
force through broad openings; but when the openings are narrow and
far removed from the source the opposite is true.[7]   [6] Therefore the
outermost sphere, being of immense proportions and constrained by a
breath that is more vigorous as it is near its source, emits in its motion
high-pitched sounds, whereas the lowest sphere, because of its narrow
confines and remoteness, has a feeble sound.

[7] Here we also have a clear demonstration that the breath, as it
draws downward away from its source, becomes slower and slower;
consequently it gathers about the earth, the last of the spheres, in such
a dense and sluggish mass that it is responsible for the earth's remaining in one place and never moving. That the lowest portion in a sphere
is the center has been proven in a preceding passage.   [8] Now there
are nine spheres in the whole body of the universe. The first is the
starbearer, properly called the celestial sphere, and *aplanes* by the Greeks,
"confining and containing all the others." It always revolves from east
to west, whereas the seven lower spheres, the so-called errant spheres,
revolve from west to east, and the earth, the ninth, is without motion.   [9] Thus there are eight moving spheres but only seven tones
producing harmony from the motion of the spheres, since Mercury and
Venus accompany the sun at the same rate of speed and follow its
course like satellites; they are thought by some students of astronomy
to possess the same force, whence Cicero's statement: "The other eight
spheres, two of which move at the same speed, produce seven different
tones, this number being, one might almost say, the key to the universe."   [10] That seven is the key to the universe we clearly showed
in an earlier discussion about numbers.

I believe that this discussion, extremely condensed, will suffice to clarify the obscurity in Cicero's words on music.   [11] To refer even cursorily to the *nete* and *hypate* and the other strings and to discuss the subtle points about tones and semitones and to tell what corresponds in music to the letter, the syllable, and the whole word is the part of one showing off his knowledge rather than of one teaching.   [12] The fact that Cicero made mention of music in this passage is no excuse for going through all the treatises on the subject, a mass of literature that, it seems to me, is without end; but we should follow up those points which can clarify the words we have undertaken to explain, for in a matter that is naturally obscure the man who in his explanation adds more than is necessary does not remove the difficulty but aggravates it.

[13] Accordingly, we shall conclude this chapter in our treatise with one addition which seems worthy of cognizance: although there are three types of musical harmony, the enharmonic, the diatonic, and the chromatic, the first is no longer used because of its extreme difficulty, and the third is frowned upon because it induces voluptuousness; hence the second, the diatonic, is the one assigned to celestial harmony in Plato's discourse.[8]

[14] Then, too, we must not overlook the fact that we do not catch the sound of the music arising in the constant swirl of the spheres because it is too full to be taken into the narrow range of our ears. Indeed, if the Great Cataract of the Nile withholds the ominous thunder of its falls from the ears of the inhabitants, why is it surprising that the sound coming from the vastness of the universe surpasses our hearing?   [15] No idle words were these: "What is this great and pleasing sound that fills my ears?" Cicero would have us understand that if the ears of a man who deserved to participate in the heavenly secrets were filled with the vastness of the sound, surely the hearing of other mortals would not catch the sound of celestial harmony.

12

◆

Proclus
410–485

Proclus Diadochus was head of the Athenian Academy and one of the last and greatest lights of classical Neoplatonism. He believed himself to be a reincarnation of Nicomachus of Gerasa (see no. 3). The biography by Marinus depicts Proclus as an almost superhuman being, endowed with wealth, beauty, precocity, self-control, industry, and sanctity. In an age of naive syncretism, he was a deeply philosophical synthetist, expanding Platonism into a vast system capable of accommodating all religions without contradiction. The only opposition to his work came from the Christians, mainly because he refused to take seriously their expectation of the world's imminent end. Ironically enough, it would be his world that would end soon, as the Athenian Academy ended its nine-centuries' career shortly after his death.

Proclus was a prolific writer on Platonic theology, astronomy, hymns, and literature. He also had a gift of healing, and was devoted to Asclepios, patron of the therapeutic art. Marinus tells us that he had produced a commentary on the *Timaeus* by the age of twenty-three. What we have here is no doubt a more mature development of his early work.

In this extract from Proclus' commentary on the *Republic*, he clarifies the status of the Sirens, who in Plato's myth stand upon the spheres uttering each a tone. Proclus distinguishes the Sirens from the Muses, while introducing two further species of them. Readers of Proclus are invariably reminded that he was writing about eight hundred years later than the works on which he commented, and that whatever continuity the Academy may have had during that time, it certainly does not justify our taking Proclus' views as true to Plato's unwritten intentions. On the other hand, Proclus was familiar with the living tradition of the Orphic religion, whose origins are far older than Plato, as well as with the initiatic teachings of the various ancient Mysteries that were never written down: both matters about which Proclus and Plato knew more than anyone today. So we accord respect, at least, to Proclus' claims.

The commentary on the *Timaeus*, as it survives, covers only the first part of Plato's text (up to 44b), dealing with the creation of the World-Soul. Proclus has preceded our extract by situating the creative process described by Timaeus within a broader and truly universal context. He therefore reminds us that creative actions which might seem to be primordial and autonomous are in turn indebted to yet higher models, to which the Demiurge looked in dividing the soul (Taylor's edition, vol. 1, p. 74). Another of Proclus' concerns is to maintain the idea of the soul's unity, despite its division: "The essence therefore of the soul is at one and the same time a whole, and has parts, and is one and multitude" (ibid., p. 75).

In this extract Proclus attempts to explain the intervals into which the world-stuff is divided, with unprecedented and unparalleled thoroughness. It is an extremely dense and difficult text, to which the notes may offer some clarification. Besides his mathematics and philosophy, there are also mythological allusions to Apollo and Dionysus, Muses and Sirens, which were eagerly seized on by Renaissance theorists. Running through the whole work is Proclus' ambition to make a "Harmony of Orpheus, Pythagoras, and Plato" (to quote the title of one of his lost books) by commenting on Plato's writings in the light of Orphic and Chaldaean theology and Pythagorean arithmology.

Although himself a militant pagan, Proclus' writings were to become a recurrent influence on medieval philosophy. Given Christian clothing in the sixth century by Dionysius the Areopagite (see *MM&M*), Proclus' theology was a powerful influence on John Scotus Eriugena (see no.16). Extracts from his *Elements of Theology*, translated into Latin circa 1180, inspired Alain of Lille, one of the last inheritors of Chartres Platonism. No matter that the extracts, called *Liber de causis*, were attributed to Aristotle! After Proclus' works, including the *Timaeus* commentary, had been translated by William of Moerbecke in the later thirteenth century, they contributed to a fresh current of mystical Neoplatonism, represented by Albertus Magnus and his German and Dominican pupils and successors, Dietrich of Vrieberg, Berthold of Mosburg (who wrote an exposition of Proclus' *Elements of Theology*), and Meister Eckhart. Thence Proclus forms a link to Nicholas of Cusa and the early Renaissance.

Sources: Proclus, *Commentaire sur la République*, translated by A.-J. Festugière (Paris: Budé, 1970), pp. 192–196. Translated by the Editor. *Proclus on the Timaeus of Plato*, translated by Thomas Taylor (London: the Author, 1820), vol. 2, pp. 75–86.

On the Sirens as Souls of the Spheres

> And on each of its circles there was seated a Siren on the upper side, carried round, and uttering a single sound on one pitch. But the whole of them, being eight, composed a single harmony.
>
> (*Republic*, 617b 5-8)

Section 16. Here Plato introduces the divine and incorporeal Hypostases which preside over the movement of the Spindle. First among them he names the Sirens, mounted each upon her own ring which she turns round. Each one gives forth a single sound whose one tone, being in perfect agreement with musical ratios, signifies the unity of the sound and the total simplicity of the noise which they make. For 'single sound' means that this musical operation has an unchangeable character, not passing from one melody to another, because each Siren always emits the same note. And 'one pitch' shows that the sound of such and such a quality causes one and only one pitch to resonate (for it is by virtue of tension that a sound is called a pitch).

Finally, since the rings and the Sirens are eight in number, he says that there results from this ensemble a single harmony, like the interval of an octave, which can be perceived as consisting either of eight notes or of seven intervals. It is as if the Sirens' workings resembled the sounds from which results the interval of an octave, which is the most complete of all.[1] The intervals between the sounds correspond to the disposition of the Sirens, for they begin at the bottom with the *nete* and end with the highest one, the *hypate*. For it is proper that the higher bodies should move more quickly, although their returns to the same point seem to take a longer time because the ratios between their orbits are greater than the ratios between their motions. These are in fact things upon which all seem to agree.

However, since Plato has ordered matters as described above, we must now say what the Sirens are. Those who say that they are the

Muses, adding to the present eight an extra one to make them up to nine, are not being faithful to the text. Besides, the fact that the Sirens are involved in the Circle of the rings shows that they are inferior to the Muses.[2] [Lacuna of about two lines in text.] This is why, in truth, the Sirens have their movements harmonized, and impart to each of the rings on which they are mounted its properly rhythmic movement.

What is their essence, then, and how do they rank? Since they are anterior to bodies and appointed directly to the rings, they must evidently be souls. Moreover, the fact that they are involved in the Circle of the rings shows, I presume, that they have motion, passing from one place to another. True, if it were not a myth Plato would have said that the rings are involved in the Circle of the Sirens.[3] But, as is the custom of mythographers, he has reversed the order and said that the Sirens are involved in the Circle of the rings. Now if they also move with a circular movement—an incorporeal movement—I suppose that they must necessarily be souls endowed with intellective life: for circular movement is the image of the Intellect, as the Athenian Stranger showed [Laws X, 897c], and it is by Intellect that souls are moved in a circle.

If this reasoning is correct, the present text agrees with what is said in the philosophizing of the Timaeus [35a] on divine souls, namely that they are arranged in harmonic ratios. For if, as Plato has said, their movement is harmonized, they must contain as their essence harmonic ratios, which is certainly what Timaeus says; and if they are carried round in a circle they themselves must be in some way circular, as Timaeus also says [36c]. Moreover, if each gives forth a single sound and one tone, they must be essentially logoic[4] souls who perform simple and unadulterated operations, unlike the operations of our own souls which must resort to reasonings and various sorts of conjecture in order to know the Real. And again, if the Sirens themselves all make a chord together, they form as it were a choir around a single coryphaeus, the World-Soul.[5]

Plato has named these souls 'Sirens' in order to show that the chord which they impose on the rings is, nevertheless, of a corporeal nature. He has called them 'celestial Sirens' to distinguish them from the Sirens connected with generation, which elsewhere he advises one to avoid in one's voyage, after the example of the famous Ulysses of whom Homer speaks [Odyssey XII, 165ss].[6] But these Sirens down here begin with a dyad: the poet says 'the voice of both[7] the Sirens' as if there were two of them. The others by contrast, begin with a monad: for it is the Siren of the outermost circle who stands at the head of the other seven.

One must expect an appropriate number to be subordinate to this dyad. Now if the number subordinated to the celestial monad is 7, then the corresponding number below the dyad connected with generation must be twice times seven.[8] (Often the Theologians also double the celestial zones in the sublunary regions [*Chaldaean Oracles* 22,3].) Consequently there are also Sirens in Hades, touching whom Plato has said plainly in the *Cratylus* [403d] that it is not agreeable or even possible for them to leave Hades, where the ingenious Pluto holds them enchanted. Thus according to Plato there are three races of Sirens: the Celestial ones, dependent on Zeus; the Terrestrial, dependent on Poseidon; and the Subterranean, dependent on Pluto. It is a property of all these races to produce a harmony of the corporeal order,[9] whereas the gift of the Muses is above all an intellectual harmony: which is why they are said to prevail over the Sirens and to be crowned with Sirens' feathers.[10] And in fact they are the principle of ascent for the Sirens, and attach them to themselves, suborning to their intellective activity the powerful wings with which the Sirens are endowed.[11]

Commentary on the Harmonies of Plato's Timaeus

The mode, however, of unfolding it[12] should accord with the essence of the soul, being liberated from visible, but elevating itself to essential and immaterial harmony, and transferring from images to paradigms. For the symphony which flows into the ears, and which consists in sounds and pulsations, is very different from that which is vital and intellectual.[13] No one, therefore, should stop at the mathematical theory, but should excite himself to a mode of survey adapted to the essence of the soul; nor should he think that we ought to direct our attention to interval, or the differences of motions. For these are assumed remotely,[14] and are by no means adapted to the proposed subjects of investigation. But he should survey the assertions by themselves, and consider how they afford an indication of the psychical middle, and look to the demiurgic providence as their end. In the first place therefore, if you are willing thus to survey, since wholeness is triple, one being prior to parts, another consisting of parts, and another being in each of the parts, as we have frequently elsewhere demonstrated:—this being the case, Plato has already delivered the wholeness of the soul which is prior to parts. For he made it to be one whole prior to all division into parts, and which as we have said, remains what it is, without being consumed in the production of the parts. For to be willing to dissolve that which is well harmonized is the province of an evil nature.[15] But the dissolution is effected by consuming the whole into the parts. In what is now delivered, however, he constitutes it a whole from parts, consuming the whole mixture into the division of its essence, and through the harmony of the parts, rendering it a whole de novo, and causing it to be complete from all appropriate parts.[16] But he shortly after teaches us the wholeness which is in each of the parts, dividing the whole soul into certain circles, and in each of the circles inserting all the reasons,[17] which he had already made manifest to us in what he had before said. For he had said, that in each of the parts there are three [i.e. same, different, and essence] in the same manner as in the whole. Every part therefore, as well as the whole, is in a certain respect a triadic whole. Hence it is necessary that the soul should have three wholenesses,[18] because it animates the universe, which is a whole of wholes, each of which is a whole according to the wholeness which is in a part. So that the soul animating the universe in a twofold respect, both as it is a whole, and as consisting of total parts, it requires two wholenesses, and transcends the things that are animated, having something external to them, so as circularly to cover the universe, as Timaeus says, as with a veil. By the wholeness, therefore, which is prior to parts, the soul entirely runs above the universe, but by the remaining two connects the universe and the parts it contains, these also being wholes.

In the next place, it must be observed, that Plato, proceeding from the beginning to the end, preserves the monadic and at the same time dyadic nature of the soul. For he reduces the hyparxis[19] of it to essence, same, and different, and distributes the number of it according to a twofold division, beginning from one part, into duple and triple numbers. He also surveys the media or middles, in one of them comprehends the other two,[20] and according to each of these unfolds twofold sesquialter and sesquitertian ratios, and again cuts these into sesquioctaves and leimmas. In what follows likewise, he divides the one length into two, and the one figure of the soul into two periods. And, in short, he nowhere omits the monadic and at the same time dyadic, and this with the greatest propriety.[21] For the monadic alone pertains to intellect, on which account also intellect is impartible. But the dyadic pertains to body, whence in the generation of the corporeal-formed nature, Plato began from the duad, fire and earth, and arranged two other genera of elements between these.[22] The soul, however, being a medium between intellect and body, is a monad and at the same time a duad. But the cause of this is, that in a certain respect it equally participates of bound and infinity;[23] just as intellect indeed, is allied to bound, but body rather pertains to infinity, on account of its subject matter, and divisibility ad infinitum. And if after this manner, some refer the impartible and the partible to the monad and indefinite duad,[24] they speak conformably to things themselves; but if as making the soul to be number, in no respect different from monadic numbers, they are very far from asserting that which happens to the essence of the soul. The soul, therefore, is a monad and at the same time a duad, adumbrating by the monadic, intellectual [and] bound, but by the dyadic, infinity; or by the former, being the image, indeed, of the impartible, but by the latter being the paradigm of partible natures.

In addition to these things also, it is requisite to survey, how a two-fold work of the Demiurgus is here delivered. For he divides the soul into parts, harmonizes the divided parts, and renders them concordant with each other. But in effecting these things, he energizes at one and the same time Dionysiacally [i.e. Bacchically] and Apolloniacally. For to divide, and produce wholes into parts, and to preside over the distribution of forms, is Dionysiacal; but to perfect all things harmonically, is Apolloniacal.[25] As the Demiurgus, therefore, comprehends in himself the cause of both these Gods, he both divides and harmonizes the soul. For the hebdomad is a number common to both these divinities, since theologists also say that Bacchus was divided into seven parts:

Into seven parts the Titans cut the boy.

And they refer the heptad to Apollo, as containing all symphonies. For the duple diapason first subsists in the monad, duad, and tetrad, of which numbers the hebdomad consists.[26] Hence they call the God Hebdomagetes, or born on the seventh day, and assert that this day is sacred to him:

For on this day Latona bore the God
Who wears a golden sword.

Just as the sixth day is sacred to Diana. This number, indeed, in the same manner as the triad, is imparted to the soul from superior causes; the latter from intelligible, but the former from intellectual natures.[27] And it is also imparted from these very divinities [Apollo and Bacchus], in order that by a division into seven parts, the soul may have a signature of the Dionysiacal series, and of the fabulous laceration of Bacchus. For it is necessary that it should participate of the Dionysiacal intellect; and as Orpheus says, that bearing the God on its head, it should be divided conformably to him. But it possesses harmony in these parts, as a symbol of the Apolloniacal order. For in the lacerations of Bacchus, it is Apollo who collects and unites the distributed parts of Bacchus, according to the will of the father [Jupiter]. In these numbers also, the three middles are comprehended.[28] These therefore being three, adumbrate not only in the soul but every where, the three daughters of Themis.[29] And the geometric middle, indeed, is the image of Eunomia.[30] Hence Plato in the Laws says, that she adorns polities, and disposes them in an orderly manner, and he likewise celebrates her as the judgment of Jupiter, adorning the universe, and comprehending the true political science. But the harmonic middle is an image of Dice or Justice, distributing a greater ratio to greater, but a less to lesser terms. This however is the work of justice. And the arithmetical middle is an image of Peace. For it is this, as it is also said in the Laws, which imparts to all things the equal according to quantity, and makes people at peace with people. For the solid analogy [i.e., the triplicate proportion] prior to these, is sacred to their mother Themis, who comprehends the powers of all of them. And thus much universally concerning these three middles.

These three middles, however may be said in a way adapted to what has been before observed, to be the sources of union and connexion to the soul, or in other words, to be unions, analogies, and bonds. Hence also Timaeus denominates them bonds. For prior to this, he had said that the geometric middle is the most beautiful of bonds, and that the other middles are contained in this. But every bond is a certain union. If therefore the middles are bonds, and bonds are the unions of the things that are bound, that which follows is evident. Hence these pervade through all the essence of the soul, and render it one from many wholes, as they are allotted a power of binding together things of a various nature.[31]

As however they are three, the geometric middle binds every thing that is essential in souls. For essence is one reason, proceeding through all things, and connecting first, middle, and last natures, just as in the geometric middle, one and the same ratio, pervades perfectly through the three terms [of which the proportion consists]. But the harmonic middle connects all the divided sameness of souls, imparting to the extremes a communion of reasons, and a kindred conjunction. And sameness, indeed, is seen in a greater degree in more total, but in a less degree in more partial natures. And the arithmetical middle binds the all-various diversity of the progression of the soul, and is less inherent in things which are greater, but more in such as are less, according to order. For difference has dominion in more partial, just as sameness has in more total and more excellent natures. And these two middles have something by which they communicate with each other, in the same manner as sameness and difference. As essence also is the monad of the latter, so the geometrical middle is the monad of the former. The geometric middle, therefore, is the union of the essences in all the thirty-four terms;[32] the harmonic of the equally numerous samenesses; and the arithmetical, of the differences. Hence all these extend through all the terms, or how could a certain whole be produced from them, unless they were as much as possible united to each other? Essentially indeed, by the geometric middle, but in another and another way by the remaining two. On this account also the arithmetical and harmonic middles become the consummation of the geometric middle, in the same manner as sameness and difference contribute to the perfection of essence. For because the arithmetical and harmonic middles subsist oppositely with reference to each other, the geometric middle connects and, as it were, weaves together their dissention.[33] For the harmonic middle indeed, distributes as we have said, greater ratios to greater terms, and less ratios to less terms;[34] since it evinces that things which are essentially greater and more total, are also more comprehensive in power than such as one of an inferior nature. But vice versa, the arithmetical middle, distributes less ratios to greater terms, but greater ratios to less terms.[35] For difference prevails more in inferior natures, just as sameness on the contrary, has greater authority in superior than in inferior natures. And the geometric middle extends the same ratio to all the terms;[36] imparting by illumination union to first, middle, and last natures, through the presence of essence to all things.

The Demiurgus therefore imparts three connective unions to the soul, which Plato denominates middles, as binding together the middle order of wholes. And of these, the geometric middle collects the multitude of essences, and causes essential progressions to be one: for one ratio is the image of union. But the harmonic middle, binds total samenesses, and the hyparxes of them into one communion. And the arithmetical middle, conjoins first, middle, and last differences. For, in short, difference is the mother of numbers, as we learn in the Parmenides. These three, however, viz. essence, sameness, and difference, are in each part of the soul, and it is requisite to conjoin all of them to each other through a medium and colligative reasons.

In the next place, we say that the soul is a plenitude of reasons, she being more simple than sensibles, but more composite than intelligibles.[37] Hence Timaeus assumes seven ratios in the soul, viz. the ratios of equality, multiplicity, submultiplicity, the superparticular and superpartient ratios, and the opposites of these, the subsuperparticular, and subsuperpartient,[38] but not the ratios which are compounded from these. For these are adapted to corporeal reasons, since they are composite and partible. The reasons in the soul, however, proceed indeed into multitude and partibility, yet together with multitude, they exhibit simplicity, and the uniform in conjunction with a distribution into parts. Hence they are not allotted an hypostasis in the monad, and the impartible, in the same manner as intellect. For intellect is alone monadic and impartible. Nor does the multitude of them proceed into composite reasons. And multiple ratio indeed is in one way only partible,[39] viz. according to the *prologos* or greater term: for the *hypologos*, or less term, is without division, and is not prevented from being unity. But the superparticular, is divisible in a twofold respect, viz. according to the *prologos* and *hypologos;* but is impartible according to difference.[40] And the superpartient is partible, both according to the prologos and hypologos, and according to difference.[41] So that the first of these, is divisible in one way only, the second bifariously, and the third trifariously. But equality is impartible.[42] The soul therefore constitutes the universe by these ratios; the corporeal-formed nature indeed, by that which is trifariously partible; the nature of superficies by that which is bifariously partible; every linear nature, by that which is partible in one way only; and by the impartible the impartibility which comprehends all things.[43] For there is something impartible in partible natures. These things, therefore, are truly asserted.

It is necessary however, to survey these after another manner; premising, that numbers which are more simple, and nearer to the monad, ought to be conceived as more primary than those which are more composite. For Plato also, having arranged one part prior to all the rest, refers all of them to this, and ends in terms which are especially composite and solid. Having therefore premised this, I say that equality and the ratio of equality, has the relation of a monad to all ratios. And what the monad is in quantity per se, that the equal is in relative quantity. Hence conformably to this, the soul introduces a common measure to all things which subsist according to the same reasons; which measure likewise, brings with it one idea the image of sameness. But according to the submultiple, and multiple ratio, it governs all the whole series of things,[44] connectedly comprehending them, and exhibiting each total form of mundane natures frequently produced by itself in all mundane beings. Thus, for instance, it produces the solar, and also the lunar form, in divine souls, in daemoniacal and human souls, in irrational animals, in plants, and in stones themselves, and adorns the most universal genera by the more partial series.[45] And according to the superparticular and subsuperparticular ratios, it adorns such things as are wholes in their participants, and which are participated according to one certain thing contained in them.[46] But according to the superpartient and subsuperpartient ratios, it adorns such things as are wholly participated by secondary natures, in conjunction with a division into multitude. For of animal indeed, man participates, and the whole of this form is in him, yet not alone, but the whole is in him according to one thing, viz. the human form; so that it is present to its participant with the whole, and one certain thing which is a part of it. But what are called common genera, participate of one genus, yet not of this alone, but together with this of many other genera also, which are parts, and not a part of that one genus. Thus, for instance, a mule participates of the species from which it has a co-mingled generation. Each species, therefore, either participates of one genus according to one, and thus imitates the superparticular ratio which contains the whole, and one part of the whole; or it participates that which is common and many things besides, and thus imitates the superpartient ratio, which together with the whole possesses also many parts of the whole. And besides these there is no other participation of species or forms.[47]

Looking also to these things, we may be able to assign the specific causes of those natures which subsist according to one form, as, for instance, of the sun, the moon, and man; as likewise of those that subsist according to many forms, together with that which is common. For there are many things of this kind in the earth, and in the sea; such as animals with a human face and the extremities resembling those of a fish, and animals in the form of dragons, but with a leonine face; these having an essence mingled from many things. All these ratios therefore, are very properly antecedently comprehended in the soul, as they define all the participations of forms in the universe. Nor can there be any other ratios of communion besides these, since all things receive a specific distinction according to these.

Again, therefore, the hebdomad of ratios corresponds to the hebdomad of parts. And the soul is wholly through the whole of itself hebdomatic, in its parts, in its ratios, and in its circles.[48] For if the demiurgic intellect is a monad, but the soul primarily proceeds from intellect, it will have the ratio of the hebdomad to it. For the hebdomad is paternal and motherless.[49] And perhaps equality imparts to all the psychical ratios, a communion of the equal, in order that all may communicate with all. But the multiple ratio affords an indication of the manner in which the ratios that are more single, measure those that are multitudinous, the former wholly proceeding through the whole of the latter; those that are impartible measuring those that are more distributed into parts.

The superparticular however, and subsuperparticular ratios, indicate the difference according to which whole ratios do not communicate with whole, but have indeed a partial habitude, yet are conjoined according to one certain most principal part of themselves. And the superpartient and subsuperpartient ratios afford an indication of the last nature, according to which there is a certain partible and multiplied communion of the psychical ratios, on account of diminution and inferiority. For the more elevated of these reasons are united wholly to the whole of each other. But those of the middle rank are conjoined, not through the whole of themselves, but according to the highest part. And those of the third rank, partibly coalesce according to multitudes.

I say, for instance, essence communicates with all the ratios, measuring all their progressions: for nothing in them is unessential. But sameness, being itself a genus, especially collects the summits of them into one communion. And difference particularly measures their divisions and progressions. The communion, therefore, of the psychical ratios, is every where exhibited. For it is either all-perfect, or alone subsists according to the summits, or according to extensions into multitude.

Farther still, in the next place let us survey, how the seven parts are allotted their hypostasis.[50] The first part then, is most intellectual and the summit of the soul, conjoining it to *the one* itself, and to the hyparxis of the first essence. Hence also it is called *one*, as having the form of unity, and the number of it proceeding into multitude, is detained by union. It is likewise analogous to the cause and center of the soul. For the soul abides according to this, and does not depart from wholes. And the tetrad indeed, is in the first monads, on account of its stability, and rejoicing in equality and sameness.[51] But the ogdoad is in the monads of the second order, on account of diminution, and the providence of the soul which extends as far as to the last of things, and that which is most material. And the triad is in the monads of the third order, on account of the circumduction to the all-perfect of the multitude which it contains. And at the same time, it is evident from these things as from images, that the summit of the soul, though it has the form of unity, yet is not purely one, but this also is an united multitude.

Just as the monad [1],[52] is not indeed without multitude, yet at the same time is unity. But *the one* of the Gods is one alone. And *the one* of intellect is more one than multitude, though this also is multiplied. But *the one* of the soul is similarly one and multitude; just as the unity of the natures posterior to it, which are divisible about body, is more multitude than one. And *the one* of bodies, is not simply one, but the phantasm and image of unity. Hence the Elean guest, or stranger, says that every thing corporeal is broken in pieces, as having an adventitious unity, and never ceasing to be divided.

But the second part [2] multiplies the part prior to it, by generative progressions, which the duad indicates, and unfolds all the progressions of essence. Hence it is said to be double of the first, as imitating the indefinite duad, and the intelligible infinity.

And the third part [3] again converts all the soul to its principle; and it is the third of it which is convolved to the principles. This, therefore, is measured by the first part, as being filled with union from it; but is conjoined more partially to the second part. And on this account it is said to be triple of the first, but sesquialter of the second; being half contained indeed by the second, as not having an equal power with it, but perfectly by the first.

But again, the fourth part [4], and besides this the fifth [9], evince that the soul peculiarly presides over secondary natures. For these parts are the intellectual causes of the incorporeals which are divided about bodies, as they are planes and squares; the former being the square of the second, and the latter of the third part. And the fourth part, indeed, is the cause of progression and generation, but the fifth, of conversion and perfection. For both are planes; but one is from the second part, subsisting twice from it, and the other from the third part, thrice proceeding from it. And it seems, that the former of these planes, is imitative of the generative natures which are divisible about bodies; but that the latter is imitative of intellectual conversions. For all knowledge converts that which knows to the thing known; just as every nature wishes to generate, and to make a progression to that which is inferior.[53]

And the sixth [8] and seventh [27] parts, contain in themselves, the primordial causes of bodies and solid masses: for these numbers are solids. And the former of them, indeed, is from the second, but the latter from the third part. But Plato in what he says converting the last to the first parts, and the terminations of the soul to its summit, places one part as octuple, but the other as twenty seven times the first.

And thus the essence of the soul consists of seven parts as abiding, proceeding, and returning, and as the cause of the progression and regression of the essences divisible about bodies, and of bodies themselves. If, also, you are willing so to speak, because the soul is allotted an hypostasis between impartibles and partibles, she imitates the former through the triad of the terms, but antecedently assumes the latter from the tetrad.[54] But the whole of her consists of all the terms, because the whole of her is the center of wholes. It is possible also to divide these parts according to the duple order, if you assume the summit of the soul, and consider the permanency, progression, and regression of it, and also the conversion to it of things proximately posterior to it, and the last subjection of solids, or rather the diminution of the cause of them, according to the duple ratio. For you will find that the whole of this co-ordination pertains to the prolific duad. But again the regression of itself to itself, and of the natures proximately posterior to it, and of those that rank in the third degree from it, to the uniform and collective essence of wholes, subsist according to the triple order. The arithmetical therefore and harmonic middles, give completion to these intervals, which are essential, and surveyed according to existence itself; some of them as we have said, binding their samenesses, but others their differences.[55]

Farther still you may also say, in a way more proximate to the things themselves, that the soul according to one part is united to the natures prior to it, and this part is the summit of the soul; but according to the duple and triple part, it proceeds from intellect, and returns to it; and according to the double of the double, and the triple of the triple, it proceeds from itself, and again returns to itself, and through itself as a medium, to the principles of itself.[56] For through being filled from these principles it is prolific of secondary natures. And as indeed, the progression from itself, is suspended from the natures prior to itself, thus also the conversion or regression to itself is suspended from the regression of the beings that are prior to it. But the last parts, according to which it constitutes the natures posterior to itself, are referred to the first part; in order that a circle without a beginning may be unfolded to the view, the end being conjoined to the beginning, and that the universe may become animated, and at the same time endued with intellect, the solid numbers being co-arranged with the first part.

Moreover, he says that from these middles, sesquialter and sesquitertian ratios, and also sesquioctaves, become apparent. What else, therefore, does he intend to indicate by these things, than the more partial difference of the psychical ratios?[57] And the sesquialter ratios indeed present us with an image of partible communion, but according to the first of the parts. The sesquitertian ratios, of partible communion according to the middle terms. And the sesquioctaves of this communion, according to the last terms. Hence also, the middles or proportions are conjoined to each other, according to the sesquioctave ratio.[58] For as being surveyed according to opposite genera, they have the smallest communion; but they are appropriately conjoined to both extremes.

Timaeus also adds that all the sesquitertian ratios are filled by the interval of the sesquioctave, in conjunction with the leimma; indicating that the terminations of all these ratios, end in more partial hypostases, in consequence of the soul comprehending the causes of the last and perfectly partible essences in the world, and pre-establishing in itself, the principles of the order and harmony of them, according to the demiurgic will.[59] The soul therefore, possesses the principles of harmonious progression and regression, and of the division into things first, middle and last; and is one intellectual reason or ratio, receiving its completion from all ratios. And again, that all the harmony of the soul, consists of a quadruple diapason, diapente and tone, is consonant to these things.[60] For since there is harmony in the world, and also in intellect and in soul, on which account Timaeus says, that the soul participates of harmony, and is harmony,—hence the world participates of harmony decadically, but the soul tetradically, and harmony preexists in intellect monadically. And as the monad is the cause of the tetrad, but the tetrad of the decad, thus also, the intellectual harmony, is the supplier of the psychical, and the psychical of sensible harmony. Hence Timaeus conceived that the quadruple diapason is adapted to the harmony of the soul. For the soul is the proximate paradigm of the harmony in the sensible world. Since however, five figures and five centers in the universe, give completion to the whole, the harmony diapente also, imparts to the world the symphony which is in its parts. Since, likewise, the universe is divided into nine parts, the sesquioctave produces the communion of the soul with the world.[61]

And here you may see that the soul comprehends the world, and makes it to be a whole, according to cause, as one, as consisting of four, and also of five parts, and as divided into nine parts, harmonizing and causally comprehending the whole of it. For the monad, the tetrad, the pentad, and the ennead, procure for us the whole number, according to which all the parts of the world are divided. Hence the ancients assert that the Muses and Apollo Musagetes, preside over the universe; the latter supplying the one union of all its harmony; but the former connecting the divided progression of this harmony, and rendering their number concordant with the eight Syrens mentioned in the Republic.[62] Thus therefore, in the middle of the monad and the ennead, the universe is adorned tetradically and pentadically; tetradically indeed, according to the four ideas of animals,[63] which the paradigm comprehends; but pentadically, according to the five figures, through which the Demiurgus distributed all things, himself, as Timaeus says, introducing the fifth idea, and arranging this harmonically in the universe.[64]

13

◆

Boethius
c. 480–524 or 525

Boethius' treatise on music, arguably the most influential one ever written since it carried classical theory into the Middle Ages and Renaissance, is represented here by only a short extract. But it cannot go without mentioning that elsewhere in the treatise Boethius has given the definitive classification of music into three parts: *Musica mundana* or music of the spheres; *Musica humana* or music of the human body and soul; *Musica instrumentalis* or vocal and instrumental music. The present collection is devoted to Boethius' first category, whereas *Music, Mysticism and Magic* was largely concerned with the second, and virtually all of musicology with the third alone.

Boethius, like so many of our classical writers, was a compiler and a synthesizer of what was already ancient wisdom. Living a precarious existence at the very tail-end of Roman civilization, he worked in Ravenna at the court of Theodoric, king of the Ostrogoths. Before his brutal execution, Boethius had been trying to establish a canon for the Seven Liberal Arts, based on surviving works of Plato, Aristotle, and (for music) Nicomachus of Gerasa and Ptolemy (see nos. 3 and 5). His book *The Principles of Music* treats scale construction and interval calculation in great detail, but does not proceed to the discussion of the higher musics which was probably intended. In a single short chapter, Boethius summarizes two theories of planet-tone correspondence. One is from Nicomachus, the other a conjecture from the passage in Cicero's *Dream of Scipio* on which I have already included Macrobius' commentary (no. 11). The first scheme assigns to the moon the highest, the other the lowest tone of the scale. The lack of consensus as to which way the planetary scale should run is matched only by the confusion, from the Middle Ages onward, as to which way the Greeks meant the scale itself to go.

Source: Boethius, *De Institutione Musica*, Book I, chapter xxvii, translated by Calvin Bower in *"The Principles of Music*, an Introduction, Translation, and Commentary,"* Ph.D. diss., George Peabody College for Teachers, 1967, pp. 93-95. Used by kind permission of Professor Bower.

To What Stars the Various Strings Are Compared

Book I, Chapter 27. At this time it seems appropriate to add that those strings from the hypate meson to the nete in the above tetrachords[1] are a reflection of the order and differentiation in the heavenly spheres. For the hypate meson is attributed to Saturn, the parhypate to the orbit of Jupiter. The lichanos meson relates itself to Mars, and the mese to the sun. The trite synemmenon relates to Venus, whereas Mercury rules the paranete synemmenon. The nete reflects the orbit of the moon.[2]

But Marcus Tullius[3] gives a different order: for in the sixth book of his *De re publica* he says the following:

> And nature is so borne that low sound emanates from its outermost part, whereas high sounds emanates from its other part. Therefore the high celestial movement, that of the stars, whose revolution is faster, moves with a high and excited sound; whereas the movement of the moon and that of the very low celestial bodies move with a very low sound. Now the earth remains still as the ninth body, and it always holds to this first position.

Therefore Tullius regards the earth as silent, that is, immobile. After this comes that which is nearest to silence, that is, the moon; and it gives the lowest sound. Thus the moon may be presented as the proslambanomenos, Mercury as the hypate hypaton, Mars as the hypate meson, Jupiter as the parhypate meson, Saturn as the lichanos meson, and the highest heaven as the mese.[4]

Which of these tones are immobile, which are totally movable, and which are in part immobile and in part movable, these things will be explained in a more fitting place, that is, when I discuss the division of the monochord ruler.

II

MEDIEVAL

14

◆

Hunayn
803 or 808-873[1]

Hunayn, or Honein Ibn Ishak al-'Ibadi, was a Nestorian Christian who, after acquiring fluency in Greek, devoted himself to making translations from Greek into Arabic for his Muslim patrons and into Syriac for his Christian ones. During his later life he was chief physician to the court of Caliph al-Mutawwakil in Baghdad, and active in the famous "House of Learning" (*Bayt al-hikma*), an academy of translators commissioned to retrieve the knowledge of the ancient world. Hunayn's innumerable translations into Arabic include the Old Testament (taken from the Greek Septuagint) and the medical works of Galen.

The "Maxims of the Philosophers" (*Nawadir al-falasifa*) is a collection of sayings, anecdotes, letters, and so forth, culled from many Greek sources. The original is lost, and later Arabic extracts are unpublished; therefore the main current source is a medieval Hebrew translation by Judah al-Harizi (1170–1235).

Hunayn's aphorisms on music belong to a distinct genre, of which they may be the first. Al-Kindi (d. 874) and the tenth-century Ikhwan al-Safa' (the "Brethren of Purity"; see no. 18 of this sourcebook) preserve similar collections, but only a few maxims recur in all of them. Hunayn's classical sources, when they can be identified, are typically the later compilers such as Athenaeus, Plutarch, and Gellius. He himself saw no reason to credit them, but these selections bear out a physician's interest in the effects of music on the body and psyche and a complete acceptance of the doctrine of musical ethos as an image of states of the soul. As such, the *Maxims* were in accord with the program of the Brethren of Purity, and many were taken into their Encyclopedia.

As the earliest writings on music known from the Islamic domain, Hunayn's collections set the tone for later theorists. Besides their general acceptance of the Platonic world-harmony, naively described in Chapter 19 of this extract, the Muslim writers are always concerned for the moral effects of instrumental and vocal music. The attention of Western theorists, on the other hand, is typically on how the music works as a self-contained system, and perhaps this is connected with the fact that Western music, unlike that of Islam, has always been changing in style. But the interest in musical phenomenology among the Arabs and Persians led to a profound metaphysics of music, and to its use in practical mysticism by the Sufis—a phenomenon to which the West can offer no parallel (see *MM&M* for several extracts from Sufi works).

Source: Hunayn, *Kitab Adab al-Falasifa (Maxims of the Philosophers)*, sections on music, translated by Isaiah Sonne in Eric Werner and Isaiah Sonne, "The Philosophy and Theory of Music in Judaeo-Arabic Literature," in *Hebrew Union College Annual* 17 (1942–43), pp. 524–532. Used by kind permission of the publisher.

Maxims of the Philosophers concerning Music

Section I, Chapter 18, 1. Ammonius[2] reports that once, on the occasion of a banquet given by King Heraclius in his son's honor, a great meeting of philosophers took place, wherein the king asked the musician to be present at the conversation, in order to ascertain what the philosophers would utter on the topic of music.

One of them said: "Music is such a sublime subject that the dialectical faculty is inadequate to its presentation, leaving the philosophers powerless. But the soul perceives that subject through the effect of melody. As soon as this sublime form becomes manifest, the soul rejoices in it, and yearns for it. Pay therefore attention to the soul, hearken to it, and restrain yourselves from the contemplation of the affairs of the transient world."

2. The second said: "The excellence of music is evident by the fact

that it appertains to every profession, like a man of understanding who associates himself with everybody."

3. The third said: "Music coming from outside, moves the soul. Coming from inside, music moves the strings."

4. The fourth said: "The connection of tones in melodies is like the connection of air with air. But when one tone detaches itself, by being more or less than the other tones in pitch, there follows a disintegration, and it is like smoke and wind, moving right and left."

5. The musician was asked: "Why do the vibrations of the tones, the breaking of the notes, and their trilling in the throat render the recital sweet and pleasant, while something plainly told is not so pleasant?" He replied: "These breakings and vibrations make the recital acquire sweetness and grace, just as water that streams from the summits of mountains through the necks of rocks, tastes sweeter and is more refreshing than water that is gathered in the bowl of a lake or of a fountain."

6. One of the philosophers, whenever he was at a banquet, used to say to the musician: "Please move the soul toward its noble faculties, such as modesty, rectitude, kindness, courage, clemency, righteousness, and generosity."

7. Once a philosopher went out for a walk accompanied by his disciple. They heard the voice of a guitar.[3] The philosopher said to his disciple: "Let us approach the guitar; perhaps we can learn some sublime form" (Platonic idea). But as they came closer to the guitar, they perceived a bad tone and an inartistic song. The philosopher then said to his disciple: "The magicians and astrologers assert that the voice of an owl indicates death for man. Were this true, the voice of this man should indicate death for the owl."

8. The fifth said: "Living in solitude, the soul sings plaintive melodies, whereby it reminds itself of its own superior world. The soul will then compose sublime compositions and rhythmical melodies. As soon as Nature sees this, and becomes aware of it, she presents herself by all sorts of her images (sensuous beauties) introduced one by one, to the soul, until finally she succeeds in recapturing it. The soul will soon forsake that which constitutes its own essence, will be busy with the affairs of nature (worldly pleasures) and, unceasingly drawn away with Nature, will become with all of its faculties, entirely submerged in the ocean of nature."[4]

Chapter 19. At the time of Antophilus, an assembly of philosophers in great number met at the house of Favorinus the sage.[5] The king sent his secretary to listen to the wise sentences of the sages and to collect and deposit those sayings in his treasury of wisdom.

1. Favorinus, the sage, said: "He who is capable of making such an

accord between the motions of the soul and nature until they vibrate together like the accord of the motions of the four strings in a musical instrument—he will be the joy of the world, and its pleasures will be in harmony with his own pleasure. When he wishes to be joyful, his memory encompasses the pleasures of the world, pondering by what type he may obtain his desire."

2. The second said: "Melody possesses a sublime virtue; the dialectic faculty is inadequate to express it by means of dialectic terms. But the ear, as the natural organ of music, brings forth melody and, by its effect, brings the melody into the souls. These receive the melody from the ear by virtue of its specific nature. When the souls hear melody, they rejoice in it, so that as soon as the melody disappears, they remember it and yearn for it. They do not find rest until they have repeated the melody many times, by which repetition the soul finally finds rest, pleasantness, and relief."

3. The third said: "The rejoicing of the soul in a pleasant voice is of two sorts:

a) "Either the soul runs to and fro in search of the forms (Platonic ideas) out of its own essence, viz., the soul will submerge in its own ocean;

b) "or whatsoever of nature (the external world) may reach the soul, the latter will turn it into its own contents, will compare it with its own essence, and having achieved accord with it, the soul will rejoice in it, and will express it through a sublime form."

4. The fourth said: "The superiority of man over other animals consists in speaking and reasoning. A man, therefore, who remains silent and is without reason—becomes a beast."

5. Plato used to say: "Love between souls does not have to be restrained as does love between bodies."

6. He also used to say to the musician: "Show us the song of trees in their blossoms and the song of flower-beds in their various perfumes."

7. Plato used to say: "The swallow and all kinds of birds, as well as all kinds of horses and camels long and yearn for the voice of music."

8. King Alexander, the Great, did not drink much, and he seldom frequented banquets. Only when he was about to ponder a campaign against his enemies or an arrangement of his army for a battle, he would, as a rule, order that the strings of the harp be sounded. But as soon as his soul was submerged in the ocean of thought and meditation, he struck the shield between his hands with his scepter.

He also said: "I found music useful in horse-racing, in arranging the shields, and in commanding the march of troops. I never stood in a battle without knowing by the rhythm of my soul—viz., the harmony of its motions, on the one hand, and the rhythm of my opponent on the other hand—whether I should win the battle or lose it."

9. Aristotle said: "These are the effects of music: It awakes the remote counsel, brings closer the stray thought, and strengthens the tired mind. Music, therefore, causes the return (to the soul) of that which was lost; it makes us pay attention to that which was neglected, and that which was turbid becomes clear. He who has been exposed to this beneficial influence participates in every counsel and opinion, and finds the right one without error. He also will fulfill his promises without delay."

10. Solon said: "I saw rams, during a song, blowing of trumpets, and dances, laying their heads low upon the ground, until they were asleep because of the pleasure that they felt in their soul."

11. Alexander, when he was a young man, sat once with his father[6] and his courtiers in a tavern. A musician sang a song of love and coition, which led to the cohabitation of a courtier with a maid servant of the king. The king was angry, and said to the musician: "Do you not know that it is written: 'Bodies of free men shall not be coupled with the bodies of slaves, lest their offspring be bad ones'?   And it is also written: 'You shall not drink wine, lest it alter your character and corrupt your mind.'"   It was at that time that Aristotle said: "Were it not for his overwhelming magnificence of soul, we should say that the king were devoid of noble education accompanied by modesty. But we know that only the grandeur of his soul and the nobility of his spirit were the reason of his anger." The disciples then asked Aristotle: "Why was the king angry?"   The master replied: "Because of the cohabitation between a noble person and a vile one, as well as because of the drinking of wine by a man who does not know its benefit and its danger."

12. Aristotle once heard a guitar player performing a melody, through which one could distinguish between virtues and vices. Aristotle then said: "How could nature possess the capacity to understand this, were it not for the soul?"

13. He also said: "Reasoning leads knowledge to a known object; but music leads to spiritual knowledge."

14. Titus (Suetonius?) reports that it was a custom of the Romans that, whenever the illness of a patient became aggravated, they would let him hear a melody, whereupon he would feel better.

15. Plato once said to a guitar player who accompanied the music with his voice: "This voice is material, we do not need it." "Master," asked his disciples, "are you not material?" "Yes," replied Plato, "but my body is a servant of my intellect."

16. Plato said: "There are three sorts of pleasure. One of them consists in the sounding upon the harp which stirs up joy. The second

pleasure is life which emanates from nature. The third is something intermediate between the previous two and consists in a movement toward (?) a living body. Before it is set in motion it remains a spirit (?) which belongs to the species of rest; but when it is set in motion, memory awakens it."

17. He also said: "Nature has to be under the leadership of the soul, for virtues and noble deeds belong to the soul."

18. He further said: "The form (idea) of evil, as long as it is moving hither but has not yet appeared, awakens fear. As soon as it becomes manifest, it awakens sadness. Likewise, the form of good, as long as it is moving hither but has not yet appeared, awakens joy. As soon as it becomes manifest, it awakens pleasure."

19. Aristotle once said to a pupil of his who was a musician: "Stimulate the form (disposition) of courage, do you understand?" Later, Aristotle said to him: "I see in you signs of understanding." "How then?" asked the pupil. "The proof of understanding," replied Aristotle, "is joy, and I see that you are joyful."

20. Plato said: "The happiness of the intellect consists in assiduous occupation with wisdom. The beatitude of the good person (summum bonum) consists, likewise, in conceiving of a multitude of the higher substances (the ideas). Always hold therefore to the conceptions of the intelligibles, and they will render intellect constant for you."

21. He also said: "Indication of the understanding of something is enjoyment. When connected with some sensations, such enjoyment becomes manifest."

22. Alexander asked Aristotle: "What may be the reason for my seeing men move in such accord through music that I cannot perceive any difference or division between them, nor will the movement of one precede that of the other?" Aristotle replied: "This is love (eros), namely its intellectual expression. For the intellectual lover does not need material expression to render sweet his words for his beloved object; but he will render sweet his heart, and he will signal to his beloved, by means of a smile on his lips or a wink of his eyes as well as by other invisible movements like those with the eye and the pupil and with hints. All of these are speaking interpreters but they are spiritual. The animal alone, in order to manifest his faint love, uses as his interpreter a material expression."

23. Plato said: "A man who is sorrowful and sad should listen to the melodies of a willing soul. For as soon as the soul becomes melancholy and sorrowful, its light goes out, but when the soul rejoices, its light shines and its brightness becomes visible. It (the soul?) will manifest affection in proportion to the receptive capacity of the recipient (of the

melodies?), which capacity is according to his purity and to the extent that he is clean of adulation and resentment."

24. Finally he said: "The professions are of three sorts: (1) Professions in which there is more speech than action; (2) others in which there is more action than speech; (3) others again in which both speech and action concur in equal measure. A specimen of professions in which there is more speech than action, is the telling of stories and fables, which is accomplished by words and not by acts. The profession in which there is more action than speech is represented by the physician, whose deeds outweigh his speech. It is in the profession of music that action corresponds exactly to speech (sound). Music, therefore, is the best profession, since its speech (sounds) coincides entirely with its action, as in the case of a guitar player whose melody corresponds to his motions."

Chapter 20, 1. Archytas said: "Truly we have made the strings of the lute four, corresponding to the natures (temperaments) of which man is composed, which natures are four.[7] We have established that the *hissing* string (*Sorek*), termed "Zir," corresponds to courage, and courage corresponds to the yellow bile; that the *deutero* string (*Mishneh*), termed "Matnah," corresponds to justice, and justice corresponds to blood; that the *triple* string (*Mesulas*), termed "Matlat," corresponds to righteousness, and righteousness corresponds to the white humor (phlegm); that the *mute* string (*Illem*), termed "Bamm," corresponds to forgiveness and generosity (?), and forgiveness and generosity (?) correspond to the black bile (melancholy). From the second string follow gladness and joy; from the third string follow fear and cowardice; from the fourth string, corresponding to the black bile, follow sorrow. Sorrow and joy issue out of that which stirs up the respective temperament in the compound of the above mentioned natures. We have compared the *hissing* string and the third string to the yellow bile and the white humor, which correspond to the seasons of heat and of autumn, and are like courage and cowardice. We have further compared the second string and the fourth string to the blood and the black bile respectively, which are like the seasons of winter and spring, and they are like joy and sadness. For joy and sadness, like courage and cowardice, represent changing affections, as the fingers and the hands change their position on each of the strings of the guitar. To courage belong: Sovereignty, generosity, and kindness. To cowardice belong: Abasement, avarice, and vileness. To joy belong: Pleasure, love, and graciousness. To sadness belongs: Suspension of desire."

2. One of the sages said: "The apple of the eye is the mirror of the soul."

3. Another said: "A singer has to show by his song the essence of the soul; and a lute has to be attuned to a melody appropriate for it."

4. Euclid said: "Music is an art which connects every thing pertaining to the same species. Subduing the natures (temperaments), music stirs up that which is at rest, and brings to rest that which is restless."

5. Ephorus (?) mentioned a general principle derived from the experience of war, and said: "A warrior has to drink a strong drink when he reaches the battle line. If he has done so, he will be fit; if he does not do so, the fire in him will be extinguished, motion will stop, and the body will become cold to such an extent that he will appear like a man trembling and shivering. The channels (i.e., the blood vessels) will be destroyed, and the warmth barred. But if he drinks and kindles ardor, that ardor will move by rhythm and burn. When rhythm has inflamed his ardor, the form (i.e., the Platonic "eidos") of courage appears. For the movement of war is conducted by the rhythm of music. This is a statement well known to the heroes of war, although not every brave warrior is able to explain it, unless he is acute and intelligent in the secrets of war. Many of the courageous men, therefore, used to drink a little wine when they came into the ranks, in order to stir up movement, and to get rid of fear and sorrow which they may have to meet. In this way, right at the beginning, as soon as they reach the battle line, they will be provided, by virtue of the heat resulting from wine, with the heat of the elements. Wine, therefore, is only an occasional cause, making it possible that the thought of the musician (i.e., the musical rhythm) should move him (the warrior) toward courage. But it is courage that moves the courageous man. For when the mind has reached its final perfection, that sublime form becomes manifest."

6. He also said: "A small quantity of wine stimulates the mind (i.e., its rhythmical function), and causes pleasantness of speech as an effect of its rhythmical perfection; for every thing which is balanced is pleasant. But it may also be that the pleasantness of speech derives from the fineness of the assembled spirits."

◆

15

Aurelian of Réôme
fl. 840–850

Aurelian of Réôme (present-day Moutiers-Saint-Jean) dedicated his treatise for the use of church singers to the Benedictine abbot Bernard (locum tenens 840–849), a grandson of Charlemagne. The *Musica Disciplina* is the oldest surviving medieval music treatise and the first to describe the eight church modes. It is based on Boethius, Cassiodorus, Isidore of Seville, and a music treatise ascribed to Alcuin. Chapter 8 contains a digression in which Aurelian says that the modes seem to imitate the eightfold cosmic motions, but declines to judge whether the reverse is true, that is, whether the celestial music follows modal rules. Only Kepler (see no. 34) would later propose that it does.

It is interesting to see the cosmic and classical associations of music present at the very cradle of medieval learning. It seems quite natural for Aurelian to use this part of his treatise to show off his knowledge of Greek, and his astronomical lore. In *Music, Mysticism and Magic* a later extract is reproduced from the same book, on instances of angelic music becoming audible on earth. But here he is concerned with the Muses, rather than the angels, and with the eighth heaven of the Zodiac rather than the abode of the Christian elect. What his mysterious

Greek names for the modes are, no one has been able to explain. Since they consist mainly of vowels, they may have something to do with the vowel-incantations so common in the Greek magical papyri, which Ruelle (see Bibliography) interpreted as musical notation.

Source: Aurelian of Réôme, *Musica Disciplina*, translated by Joseph Ponte (Colorado Springs: Colorado College Music Press, 1968), pp. 20–24. Used by kind permission of the publisher.

Concerning the Eight Modes

Chapter VIII. We have said that in music there are eight modes; and through these every melody seems to hold together, as though with a kind of glue. A tone, although a rule (mode), is the smallest part of music, just as the smallest part of grammar is the letter, and the smallest part of arithmetic is the unit. Just as speech arises and is governed by letters, and the multiplied accumulation of numbers by units, so every melody is governed by the boundary line both of its sounds and of its modes.

A tone is defined thus: it is the difference and quantity of the whole musical system, which consists of an intonation of the voice, or a tenor. Their names, as used here, took their beginnings from authority and from their order. Four of them are called authentic, a term that refers to their excellent sound, because, as it were, a certain guidance and instruction is furnished by them to the other four. The higher are also called the first; the lower, the second. In the Greek language we call an author or a master or a model *authentic*; we also call authentic very old and worthwhile books, since they, in accordance with their value, can furnish instruction and authority to others.

The first of them is called *protus*, a term that in our language means first: hence we call the first martyrs, Abel in the old law and Stephen in the new law, the protomartyrs. The second is *deuterus*, that is, the second; for in the same Greek language a repetition or a summary is called *deuterosis*; whence also Deuteronomy, second law or legislation, receives its name. The third is called *tritus*, which similarly, because it is third in order, is called by the name of three. The fourth is called *tetrardus*, which took its name in the same way in which the others did, from its order, because it is clear that its domain occupies the fourth place. For four is *tetra* to the Greeks; whence also the name of God is called *tetragrammaton*, because it is said to be written with four letters; whence, too, *tetrarchia*, that is, the fourth part of a kingdom. All four of those that are joined to them are called plagal, a name that is said to mean side or part, or their inferiors; for obviously they are, as it were, a certain side or certain parts of them, since they are not separated from them completely; and they are inferior because their sound is perceived to be lower than that of the higher ones.

Since there are eight of them, they seem to imitate celestial motions. Philosophers say that there is a higher circle of the sky which is called the Zodiac, that is, the standard bearer, and which by another name is called *aplanes*, that is, without wandering, because it moves in a straight course, to the right, from East to West, as the eyes prove. This whole course, they say, is finished in twenty-four hours, which are considered twice twelve, with individual signs: their names are the Ram, the Bull, the Twins, the Crab, the Lion, the Virgin, the Scales, the Scorpion, the Archer, the Goat, the Water-bearer, the Fish. Under this circle are seven stars that are called planets, or the wanderers, since they move with a motion contrary to the above, to the left, from West to East, as is sufficiently clear in the waxing or waning of the moon. The names of the planets are Saturn, Jupiter, Mars, the Sun, Venus, Mercury, the Moon; these, just as they differ in the amplitude of their orbits, so do they in the amount of time. For Saturn completes the Zodiac in thirty years; Jupiter, in twelve; Mars, in two; the Sun, in one; Mercury, in three hundred and twenty-nine days; Venus, in three hundred and forty-eight days. It must be noted that of the twelve signs never more than six can be discerned at the same time, since, as one of them is rising, another presently sets; and thus it happens that nine of these signs are seen succeeding each other throughout a whole night, the remaining three not appearing: the one where the Sun is, the one that follows the Sun, and the one that precedes the Sun, as it remains in each sign thirty days and two and one-half hours.

These motions, then, of the stars are eight, seven of the planets and one of that which is called the Zodiac, which all say make the sweetest harmony of song, that is consonance. Even the Lord, in the reply that he made out of the whirlwind to Holy Job, called this the harmony of heaven.[1] Whether such music holds to the aforementioned rules, however, is not mine to say.

There are other things that writers on this art have discovered. They say that the whole theory of the art of music consists of numbers. They chose to divide philosophy into three parts, physics, ethics, and logic, which in Latin are called the natural, moral and rational parts.[2] The natural discipline is given over to four sciences, namely, arithmetic, geometry, music, and astronomy. In these, numbers, the measurements of the earth,[3] sounds, or the positions of the stars are examined; but their essence and their whole origin is in mathematics. Arithmetic is concerned with numbers that are stable and abstract. Geometry has to do with numbers that are stable and pertain to forms. Music has to do with numbers that are equally abstract, yet mobile and in proportion. Astronomy is concerned with calculation that is mobile and always pertaining to forms.

Scientific writers affirm that all the consonances of the art of music (are derived) from either multiplied numbers, or from one and one-half, one and one-third, or one and one-eighth. They are multiplied when a lesser number is contained by a greater either two or three or four times, and so on. From these is born the consonance that is called the diapason, or the one that is called the disdiapason, that is, the double diapason, depending upon whether one sound surpasses the other either twice or four times. They are one and one-half when the greater number has all the smaller number plus its half, as do two or three. From these is derived the consonance that is called the diapente. They are one and one-third whenever the greater number has all a smaller and its third part, as do three and four. From these is born the consonance that is called the diatessaron. The combination is one and one-eighth when the lesser number is contained by the greater with its eighth part, as, for example, eight and nine. A studious reading of the learned Boethius, whose words and diagrams we have set down above, will suffice to instruct whoever wishes to know the theory of such matters.

There were certain other writers who thought that the number of these eight modes had been taken from the nine Muses, whom, as we have already said, the poets represent to be the daughters of Jupiter. Since, clearly, eight would correspond to these eight modes, the ninth would be for discerning the differentiae[4] of the melodies. This Muse is said not to be represented among the number of the modes, but is reckoned by the name of these devices. Just as other parts of speech overlap in the adverb, so in this Muse do the differentiae, which are left over, and which have manifold varieties.

There used to be some singers who complained that there were certain antiphons that could not be fitted to any of the rules. Hence your pious and august forefather (grandfather?) Charlemagne, father of the whole world, ordered that four modes be added, whose names are inserted here: *Ananno, noeane, nonanneane, noeane.*[5] It was because the Greeks were boasting that they had perfected the eight modes through their genius, that he chose to round out the number to twelve.[6] Then, the Greeks, so that they could be our equals and could have a common philosophical rank with the Westerners, and lest by any chance they should be found inferior in station, likewise added four modes, whose lettering I have thought it worth while to write down here: *Neno, teneano, noeano, annoannes.*

Even though such modes have been invented in modern times by both the Latins and the Greeks, and notwithstanding the different lettering they have, melody always reverts to the first eight of them. For just as no one is able to increase the eight parts of the discipline of grammar and add more parts, so no one is able to produce a greater number of modes. Hence it is not necessary to disregard the limits of our forefathers and to assail this proverb: "Pass not beyond the ancient bounds that your fathers have set for you" (Proverbs 22:28), especially when, up to the present time, in which these (ostensible modes) have been found, every *ordo* (liturgical scheme) both of the Greek and of the Roman Church has passed through these former modes in the antiphons, responses, offertories, and communions.

The Names That Are Assigned to These Modes

Chapter IX. The mind is wont to be concerned about the meaning of the names that are assigned to these modes, as *Nonannoeane* is to the first mode, *Noeane* to the second, and each one of the others. I asked a certain Greek how they would be translated into Latin. He answered that they were untranslatable, but that among the Greeks they were exclamations of one rejoicing. The more extended the melody, the more syllables are assigned, as in the first authentic, which is the beginning, six syllables are joined, namely, *Nonannoeane*. In the second authentic and in the third authentic, since they deserve less, only five syllables are assigned, namely, *Noioeane*. In their plagals, however, the lettering is uniform, namely, *Noeane*, or according to some, *noeagis*. When he had remembered, the Greek added that they seem to have a parallel in our language to the cries that those who plow or drive wagons are accustomed to utter, except that *Noeane* is merely a cry of one rejoicing and of expressing nothing else; and it contains a modal melody.

16

John Scotus Eriugena
c. 815–c. 877

Eriugena ("Erin-born") represents the high culture of Ireland at a time when learning was eclipsed in most of Europe. This remarkable man knew enough Greek to translate Dionysius the Areopagite; he imbibed Platonism from the sources of late Antiquity, and reconciled it with a Johannine Christianity, essentially mystical and monistic, which envisioned the whole of creation as emanating from the One, by the agency of the Holy Spirit, and all creatures as eventually attaining salvation or reunion with their source. "Therefore," he wrote, "it does not disturb me to hear that the fairest harmony will arise from the punishment of evil will and the rewarding of the good, for punishments are good when they are just, and so are rewards when they are more in the nature of gifts than payments for what is earned; just as I perceive low, high, and intermediate sounds making a certain symphony between them through their proportions and proportionalities."[1] To Eriugena, the cosmos is a theophany and in every respect a symbol of higher reality; the physical world itself has been created out of compassion, in order to lead back mankind to the angelic state from which it originally fell. The study of the world of phenomena, so irrelevant in the opinion of the Latin Church Fathers, is here sanctified as part of this process of return.

Eriugena's cosmology seems to have developed independently, and is quite idiosyncratic. The Moon, Sun, Saturn, and the Fixed Stars turn around the Earth, while the other planets Mercury, Venus, Mars, and Jupiter circle the Sun. Therefore the only fixed distances are those between Earth, Sun, Saturn, and the sphere of the stars, and any theory of planetary music must include the possibility of variable pitch. It is a tribute to the flexibility and originality of Eriugena's mind that he could accept this, and draw the necessary consequences. He is the only author prior to Anselmi and Kepler (see nos. 24 and 34) to have imagined a mobile harmony of the spheres.

Source: John Scotus Eriugena, *Commentary on Martianus Capella*, ninth-century manuscript in the Bodleian Library, Oxford (Ms. Auct. T.II.19), fols. 11'–15'. Translated by the Editor.

On the Harmony of the Celestial Motions and the Sounds of the Stars[2]

[11'] The sounds of the planets are eight, namely seven plus the one of the [starry] Sphere. The lowest of them is Saturn's, the highest that of the Sphere. The sound of the Sphere hence concords with the sound of Saturn in quadruple ratio, and they produce a double octave which, as in the organ or on strings, is the principle of principles and the highest of excellencies.[3] The sound of the Sun is between Saturn and the Sphere, like the *mese* between the aforesaid two strings. It is thus twice as high as Saturn and sounds the octave with it, while the Sphere is twice as high as it and together they make another octave. Here one must admire the wonderful virtue of Nature; for what anyone can accomplish on four-stringed lyres [12] is achieved in the eight celestial sounds. But the method by which it is done must be sought out with diligent investigation.

First understand, therefore, that the three planets which are situated above the Sun have the lower sounds:[4] not unsuitably, since they also move in the wider spaces of the universe and with less speed than the Sphere, to which they pursue a contrary direction, being prevented from having such velocity. But those beneath the Sun are more distantly removed from the Sphere's speed; they traverse the lesser spaces of the universe and give forth higher sounds. And hence it is not their positions but the ratio of proportion of their sounds[5] that causes the celestial harmony, especially since there can be no ratio in the universe localized upward or downward. For in the quality of sounds the depth and height and the accompanying middle varieties cause the different symphonies.

Therefore the diatonic genus, for example the Sun-Saturn octave, is made by duple proportion. Similarly the octave from the sphere to the Sun produces another. And hence the Sphere, being in quadruple proportion with Saturn, sounds the double octave. Note [12'] that every octave consists of eight sounds, seven spaces, and six whole-tones. Therefore first among the lower ones is Saturn, to whom Jupiter is next joined by a tone. Jupiter is similarly a tone from Mars, Mars a semitone from the Sun, and from the Sun to Saturn there is a fourth, in sesquitertia proportion. In the same sounds the Sun also makes a sesquialtera consonance with Saturn, thus: from Saturn to Jupiter a tone, from Jupiter to Mars a tone, from Mars to the Sun a tone—and you have simultaneously a fifth[6] and a fourth between the Sun and Saturn. And do not be surprised that the Sun fits with the other planets in multiple proportion. We therefore say that it connects with Saturn in three ways, namely by the coupling of duple, sesquitertia, and sesquialtera [proportions].

As you see, the sounds do not always relate by the same intervals, but according to the altitude of their orbits. No wonder, then, that the Sun sounds an octave with Saturn when it is running at the greatest distance from it; but when it begins to approach it, it will sound a fifth and when it gets closest, a fourth. [13] Considered in this manner, I think it will not disturb you when we say that Mars is distant from the Sun sometimes by a tone, sometimes by a semitone. For what prevails in strings according to their length and shortness, extension and remission, is also the case with organ pipes, in which it is the longitudinal measure that causes the distance of the pitches. It is the same with the planets, according to the altitude of their orbits and their distance or nearness to the Sun; and what we say of the Sun should also be understood of all the other planets relative to one another. For they are not always separated, nor do they approach each other, by the same intervals, on account of the condition of their orbits. And hence it is to be believed that all the musical consonances can be made by the eight celestial sounds. I do not mean just in the three genera—diatonic, chromatic, and enharmonic—but even in others beyond the conception of all mortals.

Thus an octave is to be found, as it seems to the senses of mortals, in the points above the Sun and in the places beneath it. The Sun is joined to the Sphere by an octave. And first it has a fourth with the Moon: The Sun sounds a tone [13'] with Venus, Venus another tone with Mercury, the Moon a semitone with Mercury. But a fifth sounds in the same spaces when Venus yields a tone with the Sun, Mercury a tone with Venus, the Moon a semitone with Mercury, and the sphere of the Moon another tone. Note that these tones, since they are computed from Earth to a sphere (such as the tone from Earth to the Moon) are not in the proportions of notes [*vocum*] but in the intervals of the places of the tones, of which there are many species. Therefore by the tones [one understands] the intervals of the stars, i.e., how far any one is distant from another, and how far the Moon is from the Earth.[7] These tones are varied according to the diversity of orbits and circles. Martianus defines this species of tones, saying: A tone is a space with a legitimate quantity, which species is called in music an interval [*diastema*].

There are tones of times, in their constituent length and brevity. There are tones of spirits, in the density and thinness of voices. There are the harmonic tones now under consideration, in the depth and height of sounds from which every proportionality of harmonies is constituted. Therefore just as in an organ [14] one does not consider what place a pipe is in, but according to what sort of voice it has, and to how many and to which other ones it is joined, and what proportions any one pipe makes when coupled with different ones, it creates different harmonies; even so it is not the position of the stars but their sound that composes the celestial harmony.[8]

It is truly no wonder that the pitches of the stars are changed according to the distances of their orbits, since even their colors change for the same reason; and we see that a string placed in a shorter or longer space, or stretched and loosened, does not give the same pitch even though it is the same string. For a 'tone' receives its name from extension: it is Greek, and derived from the verb *teino*, i.e., stretch. Because it is properly reckoned in eighths[9] it is used by musicians, being the common measure of all proportions. And do not wonder that we have said that it is a tone from the Moon to the Sun, for we do not consider here the tunes of places, that is to say spaces, but the consonances of voices. For we commenced rationally from Saturn, the lowest of all sounds, and by proportional ascent rose to the midpoint of the Sun's sound, then, extending our reason higher, we reached the Moon's, highest [14'] of all the planets' notes (not without cause, since it occupies the narrowest course of all the circles.) Not being able to ascend higher, reason led us to the highest motion of all sounds, that of the celestial sphere. Thus we coupled the highest of all the planets' sounds, the highest and most distant of the whole, to a motion in tonic proportion. Therefore the cause of error to many is ignorance of the tones, believing that the tone by which the Moon is distant from the Earth pertains to the proportions of celestial voices, and not first observing that a musical tone must be between two sounds. For Earth, being

stationary, makes no sound. Therefore there is no musical tone between Earth and the Moon. Hence, musical intervals are never measured in numbers of Stadia, but only by the rational ascension of distances according to the rules of numbers. For it is one thing to measure
126,000 stadia from the Earth to the Moon, another to call the 24 [15]
unities between the number 192 and 216,[10] where the tone is, 126[000]
stadia. This eighth part of the lesser number is the tone, 24. We use
this particular example so that it will become clearer what we are trying
to assert.

In a chorus where many are singing at once, what is considered is
not the position in which any one is placed, but the proportion of his
voice. For wherever he is placed, if he sings with a very low voice, he
will necessarily hold the lowest proportion of all the voices. For the
same reason, the person with the highest voice will necessarily maintain
the highest of all sounds wherever [she] is in the chorus. The same is
understood of the rest: not their local position but their proportional
singing. In the universe of melody, therefore, it is vain to measure the
pure celestial music by the ratios of local intervals. Herein there is
nothing to be observed but the ascent and descent of depth and height,
for as the depth ascends on high by decreasing until it is stopped in the
heights, so the height descends similarly by decreasing until it establishes its terminus in the depths.

17

Regino of Prüm
d. 915

Like Aurelian of Réôme, his near-contemporary from across the Rhine, and like innumerable theorists who have written on *musica mundana* and *musica humana*, borrowing in the first instance from the late Latin writers and then from one another, Regino begins his book by paying his respects to the hoary legends of music's virtues and powers. The custom of beginning a treatise with a chapter in praise of music and its celestial genealogy (known as a *laus musicae*) carried seeds of the ancient wisdom far into the era of skepticism and general disinterest.

Regino was primarily a chronicler, who undertook to catalog and correct the plainsong melodies in use at his native town, Trier. His *Epistola de harmonica institutione* ("Letter on Harmony") was written in about 901 as an introduction to his edition, the *Tonarius*. Regino's summary serves to illustrate the early medieval state of learning and attitude to the planetary music, devoid as it was of any precise astronomical knowledge. Later we will find a far more sophisticated but no less derivative example in Jacques de Liège. In the Renaissance the *laus musicae* either dwindled to a couple of paragraphs, or became the pretext for a virtuoso display of classical learning.

Source: Regino Prumensis, *De harmonica institutione*, adapted by the Editor from the translation by Sister Mary Protase LeRoux in "The 'De harmonica Institutione' and 'Tonarius' of Regino of Prüm," Ph.D. dissertation, Catholic University of America, 1965, pp. 32–35.

Ancient Beliefs about the Harmony of the Spheres

5. The Pythagoreans[1] argue the presence of music in the heavenly motions thus: how, they say, could the heavenly apparatus, so rapid in its course, move in silence? Even though it does not reach our ears, it is still quite impossible that such headlong speed should lack sound, especially since the courses of the stars are arranged in so convenient and well-adapted a way that nothing so enmeshed and conjoined can be imagined. Some are higher, others lower, yet all are turned with an equal impulse so that their unequal and disparate orbits fall into a determined order. From this it is argued that there is a harmonious arrangement in the heavenly motion.

Without sound there can be none of that consonance which reigns in every measure of music; and sound never occurs without some attack or beat. Again, there is no attack without previous motion. Some motions are faster, some slower. A slow and sporadic motion will cause low sounds; a rapid and firm one will necessarily give forth high ones, and in an immobile object there is no sound at all. So the musicians define sound thus: A sound is the continuous beating of the air to the point of audibility. Both high and low pitches therefore arise from manifold motions.

Based on these conjectures, both astrologers and musicians maintain that the spaces between the outermost sphere and the circles of the seven planets are filled by all the musical consonances. They say that the highest sound of all occurs from Saturn up to the celestial sphere, and the deepest one from the Earth to the circle of the Moon. But others think differently, namely that the lower sound is between Saturn and the sphere, and the higher one between the Earth and the Moon, because what is narrower and shorter must necessarily sound higher, and conversely what is longer, lower. So between the circle of Saturn and the Moon, the whole domain of the planets is filled with a variety of different tones and all the consonances of music.

Martianus, in his book *The Marriage of Philology with Mercury*, represents all these as taking place in the grove of Apollo,[2] the Sun himself being the moderator of the celestial music. For, he says, the more eminent summits (i.e., the higher branches), being more distended (i.e., greatly stretched), result in (i.e., resonate with) a high pitch (i.e., a subtle and graceful one). But whatsoever is adjacent or near to Earth, namely the more drooping and nether branches closer to the soil, shudder (i.e., strike and re-echo) with a a low and hoarse din. And the ones in between (i.e. the middle parts of the same wood) join with each other in a double *succentus*. A *concentus* is like the friendship of voices in chorus: but a *succentus* is a blending of divers sounds, different yet in wonderful agreement, as we see in organum. He says a double *succentus*, and sesquialtera, even sesquitertia, touching on the three consonances: octave, fourth, and fifth. Octaves are consonant without distinction, i.e., joined without an added interval. Here he deals with the whole tone, although limmata or semitones have intervened; here he deals with the two semitones. And from these things the whole of music is made. But all these matters will be demonstrated more clearly in their place.

Let us not omit that even the strings or cords are likened to the heavenly music.[3] The *hypate meson* is attributed by musicians to Saturn. The *parypate* is exactly like the circle of Jupiter. The *lichanos meson* they assign to Mars. The Sun possesses the *meson*. Venus has the *trite synemmenon*. Mercury rules the *paranete synemmenon*. The *nete* is a model of the lunar circle. Thus Boethius. Now Cicero gives the contrary order: in *The Dream of Scipio* he asserts thus, and Nature so operates, that from the highest to the lowest part sounds from high to low. Consequently the outermost sphere, the star-bearer, with its swifter motion gives forth a high-pitched tone, whereas the lunar sphere, the lowest, has the deepest tone. Of course the earth, the ninth and stationary sphere, always clings to the same position. Cicero therefore makes the Earth as it were silent, being immobile. After this, being nearest to silence, the Moon gives forth the lowest sound, as if the Moon were the *proslambanomenos*, the lowest string of all, Mercury the *hypate hypaton*, Venus the *parypate hypaton*, the Sun the *lichanos hypaton*, Mars the *hypate meson*, Jupiter the *parypate meson*, Saturn the *lichanos meson*, and the celestial sphere the *mese*. The Moon's note compared to that of Mercury sounds a tone; in relation to Mars' a fifth; with the Sun, a fourth; with the celestial sphere, an octave. There you have a summary of the whole of music in the motion of the heavens. So let this brief account of celestial music suffice. If you wish to know more about it, read the second book of that excellent philosopher Macrobius on *The Dream of Scipio*. We would just add that not only the heathen philosophers but also vigorous commenders of the Christian faith give their assent to this heavenly harmony.

18

✦

The Ikhwan al-Safa'
(Brethren of Purity)
tenth century

The Brethren of Purity were a community in Basra (southeastern Iraq) of whom little is known beyond their great monument: an encyclopedia (the *Rasa'il*) in fifty-one or fifty-two volumes, encompassing the whole of human knowledge. Like all the early Muslim scholars, the Brethren of Purity were anxious to rescue all that they could of the scientific and philosophic learning of the Graeco-Roman civilization, and in this they were rather more successful than their Christian counterparts, being closer to the sources and free from certain prejudices. The resultant encyclopedia became one of the foundations of Islamic learning, and is still being studied today in the many languages of the Muslim world. In the West, it is available in Dieterici's German.

What makes these Brethren particularly interesting is their grasp of the Perennial Philosophy as a continuity of revelation in all ages and races, appearing in Hermes Trismegistus, Pythagoras, Plato, Abraham, and Jesus as well as in Muhammad and the Imams who succeeded him. They were also "Hermetic" in the sense that they embraced a cosmology of many levels of being, linked by correspondences and permeated with one divinity, in which the human being is a microcosm whose destiny is to rejoin its inner divinity which is also the One God. In another sense, they shared the practical and scientific outlook of Hermetism—Hermes was, after all, the god of technique and craft—in their curiosity about the world and their program for understanding it in all its aspects. In their treatment of music there is none of the contempt for the craft that we find in most of the Greek theorists, from Plato to Boethius, and their encyclopedia itself is a practical as well as a theoretical work.

Having given in *Music, Mysticism and Magic* two chapters on the powers and effects of music, I add here a more "cosmic" chapter that takes as its starting point the four strings of the *oud* or Middle-Eastern lute, whose musical symbolism has been expressed briefly in the Maxims of Hunayn (no. 14). There is nothing in Western theory before the Renaissance to compare with this vision of the world's many domains linked with precise correspondences, all founded upon numbers both harmonic and inharmonic.

Source: The Epistle on Music of the Ikhwan al-Safa', translated by Amnon Shiloah (Jerusalem: Tel-Aviv University, 1978), pp. 43–49. Used by kind permission of Professor Shiloah.

The Four Strings of the Lute and Their Parallels

Chapter 10. We return to the subject we are treating and we say: the musician philosophers limited themselves to four strings for the oud, no more, no less, so that their productions would be comparable to the natural things that are under the lunary sphere and in the image of the science of the Creator, may He be exalted.[1] This we have expounded in the treatise on arithmetic. In effect, the first string is comparable to the element of fire, and its sonority corresponds to the heat and its intensity. The second string is comparable to the element of air and its sonority corresponds to the softness of air and its gentleness. The third string is comparable to the element of water and its freshness. The fourth string is comparable to the element of earth and its sonority corresponds to the heaviness of earth and its density. These diverse qualities are given to bodies in terms of their regional relations and the effects that their notes exercise on the mixtures of the temperaments of those who listen to them. In effect, the sonority of the first string reinforces the humor of yellow bile, augments its vigor and its effect; it possesses a nature opposed to that of the humor of phlegm, and softens it. The sonority of the second string reinforces the humor of the blood, augments its vigor and its effect; it possesses a nature opposed to that of the humor of black bile, it refines it and makes it more tender. The sonority of the third string reinforces the humor of phlegm, augments its force and its effect; it possesses a nature opposed to that of yellow bile, it is suited to break its irascibility. The sonority of the fourth string reinforces the humor of black bile, augments its vigor and its effect, is opposed to the humor of blood and attenuates its boiling.

If the notes produced by these strings merge into one harmony in the melodies that correspond to them and if these melodies are used during the various parts of the day and the night[2] whose nature is opposed to that of the illnesses and ailments in force, these melodies would appease the illnesses and ailments in question, would break their violence and alleviate the pains of the sick, for things similar in their natures, once multiplied and united, gain in strength, exercise an evident effect and overcome what is in opposition to them.[3] Similar phenomena are familiar to people in the case of wars and disputes.

We have clearly established, according to what we have said, the characteristic features of the science employed by the musician philosophers in the hospitals, and the application they make of it at moments opposed to the nature of the illnesses, the affections and the ailments [treated]. It is because of this that they restricted the number of strings of the oud to four, no more, no less.

The motivation conducting the philosophers to establish the thickness of each string in the proportion of 4:3 with that of the string above it, is explained by the fact that they wished to imitate the science of the Creator—great are His praises—and reproduce the Signs of His art in the natural productions.[4] In effect, physicists advance the theory that the diameter of each of the spheres of the four elements, that is fire, air, water, earth, is, from the point of view of its quality, in the proportion of 4:3 with the diameter of the sphere below it. By quality I mean the subtlety and thickness relative to the spheres in question. They say then: the diameter of the sphere of ether, I mean the sphere of fire, which is beneath that of the moon, is in the proportion of 4:3 with the diameter of the sphere of frigidity (zamharir); the diameter of the latter, in the proportion of 4:3 with the sphere of air; the diameter of the latter is in the proportion of 4:3 with the diameter of the sphere of water; the diameter of the latter is in the proportion of 4:3 with the diameter of the sphere of earth. This proportion means that the substance of fire, from the point of view of subtlety is in the proportion of 4:3 with that of air; that the substance of air, from the point of view of subtlety, is in the proportion of 4:3 with that of water and that the substance of water, from the point of view of subtlety is in the proportion of 4:3 with that of earth.

As for the reasons explaining why they [the philosophers] stretched the *zir* (high string) which is comparable to the element of fire and whose sonority is comparable to the heat of fire and its sharpness, below all the strings, and the *bamm* (low string) which is comparable to the element of earth, above all the strings, then the *mathna* (second string) after the *zir* and the *mathlath* (third string) after the *bamm*, there are also two reasons for this: (a) the note of the *zir* is high and light, it moves upwards, while that of the *bamm* is low and heavy, it moves downwards. Now, this permits them to blend better and to unite. It is the same with the two other, intermediary, strings, the *mathna* and the *mathlath*. (b) The relation between the thickness of the *zir* and that of the *mathna*, [that] between [the thickness] of the *mathna* and the *mathlath*, [that] between [the thickness] of the *mathlath* and the *bamm* is the same as the proportion between the diameter of the sphere of earth and that of the sphere of air, [between] the diameter of the sphere of air and that of the sphere of frigidity (*zamharir*), [between] the diameter of this last and that of ether. This then is the reason that explains why they stretched the strings in the order indicated.

If they applied the relation of the eighth to the notes of the strings and preferred it to the exclusion of the fifth, the sixth and the seventh, this is because it derives from the number eight, which is the first cube. Besides, given that the six is the first perfect number,[5] and that the most eminent of forms is that which is composed of six faces, that is, the cube, the cubic form is thus the highest, because of the property of equality which is manifested in it, as we have demonstrated in the Epistle on geometry. In effect, the three dimensions of this form are equal. The cube has six square faces, all equal, eight plane angles, all equal, twelve parallel and equal intersections and twenty-four equal right angles which represent the sum of 3 x 8.[6] We have said that the more the created thing possesses the property of equality, the greater its eminence. Now, after the circular form, there is none endowed with more equality than that of the cube. It is for this reason that it was said in Euclid's last treatise that the form of the earth is probably cubic and that of the celestial sphere probably a dodecahedron defined by twelve [regular] pentagons.[7] We have shown the preeminence of the spherical form and of the number twelve in the Epistle on astronomy.

On the subject of the virtue of the number eight, the mathematician philosophers have advanced the theory that a harmonious proportion exists between the diameters of the celestial spheres and those of earth and air. The proof of this is that if we express the diameter of the earth by 8, that of the sphere of air by 9, then the diameter of the sphere of the moon will be 12, that of the sphere of Mercury 13, that of the sphere of Venus 16, that of the sphere of the sun 18, that of the sphere of Mars 21.5, that of the sphere of Jupiter 24, that of the sphere of

Saturn 27⁴/₇ and that of the sphere of the fixed stars 32.⁸

Based on these figures, the diameter of the earth and that of the moon will be in the proportion of 3:2; the diameter of the moon and that of the air will be in the proportion of 4:3; the diameter of Venus compared with that of the earth will be in the double proportion of 16:8 (2:1) and with that of the moon in the proportion of 4:3; the diameter of the sun will be in the double proportion of 18:9 with that of the air and with that of the earth in the proportion of 21:4, and with that of the moon in the proportion of 3:2; the diameter of Jupiter will be in the double proportion of 24:12 with that of the moon, with that of the earth in the triple proportion of 3:1 (24:8), and with that of Venus in the proportion of 3:2 (24:16); the diameter of the sphere of fixed stars will be in the proportion of 4:3 (32:24) with that of Jupiter, with that of Venus in the double proportion of 2:1 (32:16), with that of the sun in the proportion of 4:3⁹ (32:18), with that of the moon in the proportion of 2³/₄ (8:3) and with that of the earth in the quadruple proportion of 4:1 (32:8). As for Mercury, Mars and Saturn, they do not have a [harmonic] proportion. That is why these heavenly bodies are called maleficent.

These same philosophers [the mathematicians] also maintain that various proportions exist between the dimensions of the bodies of these heavenly bodies, proportions which are either arithmetical or geometrical or harmonic. These same proportions govern the celestial bodies and the body of the earth. At all events, some of them are noble and superior, while others are inferior. It would take too long to give a detailed commentary on this.

From what we have said, it appears that (a) the ensemble of the bodies of the Universe, that is to say, the spheres, the bodies of the stars, the four elements and their concentric constitutions, all these bodies have been established, composed, conceived and created in the respective proportions that we have expounded;¹⁰ (b) that everything that comprises the body of the Universe is like the body of one sole animal, of one sole person and of one sole state; and, finally that their director, their Creator, their constructor, their artisan and their Maker *ex nihilo* is One without associates, [Master without his like, Unique without equal]. Thus we have attained one of the aims that we have set ourselves in this Epistle.

These are other facts concerning the eminence of the number 8. If, my brother—may God assist you and assist us through the spirit [emanating] from Himself—you examine created things and the elementary nature of beings submitted to generation and corruption, you will find a multitude of octads, such as the qualities of the elements: the hot-wet, the cold-dry, the cold-wet and the hot-dry, which add up to eight, and they are the principle of natural things and the basis of begettable and corruptible things. In the same way, you will confirm the superiority of the number 8 by referring to the respective position of the planets which gives rise to eight combinations, to the exclusion of all the others: conjunction (*mukarana*), opposition (*istikbal*), trines (*tathlithat*), quadratures (*tarbi'at*), and sextiles (*tasdisat*). These positions are also one of the causes that preside at the generation and corruption of the beings placed under the moon's sphere.

In the same way, if you reflect, you will see that the 28 letters of the Arabic alphabet, which are comparable to the 28 mansions of the moon, can be reduced to 8 letters which are articulated in the following way: *alif, lam, fa', ya', mim, nun, dal, waw* (a, l, f, y, m, n, d, w).[11] And the paradigms (*mafa'il*) of Arabic poetry are also of the same number, namely eight; these elements form the metrical feet. The kinds of rhythms of their music are also of the number of eight, as we shall explain in another chapter. In the same way it has been said that paradise is composed of eight degrees, that the bearers of the throne are of the number of eight, and that the hells have seven sections (*abwab*); we have shown the reality of this in the Epistle on "the last judgment and the resurrection."

In following the same analogy (*al-kiyas*), if, my brother, you examine these things, and if you reflect on the conditions of beings, you will find numerous dyads, triads, tetrads, pentads, hexads, hebdomads, octads, enneads and decads, and so on to infinity. If we insist on the octad, it is to awaken you from the sleep of heedlessness and the sloth of ignorance, and so that you will know that the partisans of the number seven (*al-musabbi'a*) in their avid study of the hebdomads and in assigning them the supremacy over all the other groupings, advance only a partial theory and a non-universal doctrine. It is the same with the dualism of the dualists, the trinity of the Christians, the tetractys of the physicists, the pentad of the *Khurramiya*, the hexads of the Hindus, and the enneads of the *Kayyaliya*.[12] This is not the doctrine of our noble brothers— may God assist them and assist us through the spirit [emanating] from Himself—whatever the country where they happen to be, their theory is universal, their speculation general, their science total and their knowledge extends to everything.

Let us now return to the subject that occupies us. We have thus clearly established in the light of what we have said, the characteristic features of the oud, the number of its strings, the proportions of their respective thicknesses, the number of the frets, how they are tuned, their respective proportions, the number of notes produced by the strings whether free or shortened and stopped, and, finally, the relations of these notes between themselves. In consequence, the most perfect of productions, the most coordinated of constructions and the most beautiful of compositions is that whose disposition and structure are governed by the most eminent proportion. It is because of this that most people find enjoyment in the audition of music and that the majority [of persons gifted with reason] approve its qualities, and make use of them in the gatherings of kings and chiefs.

19

$\blacklozenge$

Al-Hasan al-Katib
fl. c. 1000

Al-Hasan Ibn Ahmad Ibn 'Ali al-Katib lived in Syria and was a Shi-ite *katib*, meaning "secretary," or by extension a man of wide culture. His treatise can be dated from the fact that he cites the musical works of Al-Kindi, Al-Sarahsi, and especially Al-Farabi, pre-eminent theorist of the Islamic world, but does not mention Ibn Sina (Avicenna), who died in 1037.

Much of Al-Hasan's work is concerned with musica humana. He cites the hoary classical examples of the power of music—a practice as common among Muslim writers as it would become for Westerners in the Middle Ages and the Renaissance. We have already seen in Hunayn's maxims the earliest Arab example of this. Besides acknowledging his debt to Nicomachus of Gerasa (see no. 3), Al-Hasan must have received some ideas, very indirectly, from Ptolemy. These are contained in this extract, which is a succinct review of the three types of music as defined by Boethius: *musica mundana, humana*, and *instrumentalis*.

Al-Hasan was writing at approximately the same time as the Brethren of Purity were compiling their Encyclopedia (see no. 18 and *MM&M*). While his scope and his grasp of symbolism and psychology scarcely compare to theirs, he is nevertheless part of the same Pythagorean-Platonic tradition as it was being continued in the world of Islam. In music, this meant especially the Pythagorean insight into number, and the proportions between numbers, as being the key to the universe; also the moral and medical applications of *musica instrumentalis*.

Professor Shiloah states that the original is full of obscurities, especially the last paragraph of this chapter. His edition contains copious notes, giving the original Arabic of many terms. I have kept closely to his French version, but have divided some of the very long sentences.

Source: Al-Hasan Al-Katib, *Kitab Kamal Adal Al-Gina'*, translated into French by Amnon Shiloah as *La perfection des connaissances musicales* (Paris: Geuthner, 1972), pp. 69–74. Translated by the Editor. Used by kind permission of Professor Shiloah.

The Resemblance of the Soul, of Music, and of the Celestial Sphere

Chapter 9. We will expound only a few characteristic points of this doctrine, because it would be far too lengthy to make a complete study of it on account of the large number of persons who have made pronouncements on the subject, and the multiplicity of their statements.

Here is the opinion of al-Kindi and others on the similitude of the soul's activities and the apportionment of the consonant connections of notes. The commentary on the latter will be included in an appropriate chapter, on the systems.

Al-Kindi said: "The simple consonances are three in number—simple in this case being those whose extremes are alike and have a noble relationship between them;[1] they are the group of four notes (the fourth), the group of five notes (the fifth), and the group of the whole (the octave). Even so, the divisions of the soul are three in number: the rational, the sensible or sensitive, and the natural [vegetative].[2] The similitude of these three [groups and principles] is expressed thus: just as one finds in the place of the rational faculty also the sensitive and instinctive faculties—by the rational faculty I mean discrimination—, the interval of the octave embraces by its very existence the intervals of a fourth and a fifth. And again, wherever there is a group of five, a group of four is there too. The group that corresponds to the sensible principle is that of five notes, which is the nearest to that of the rational; the group which corresponds to the natural principle is that of four.

As for the elements constituting the rational soul, they are seven in number: a number equivalent to the elements in the group of the whole (the octave): comprehension, intelligence, memory, reflection or deliberation, estimation, syllogism, and knowledge.[3]

Comprehension is a matter of what the senses bring to the soul.

Intelligence is the formation of concepts in the soul consequent on images received from the senses.

Memory is the retention in the soul of images received by the soul.

Deliberation is the state in which the soul compares things by their exterior forms.

Syllogism is the state in which the soul establishes a comparison by advancing a correct argument: I mean by deducing conclusions from true premises.

Knowledge is the grasping of truth by the soul.

The sensitive soul is subdivided into four elements, a number equivalent to that of the elements of fifths: hearing, sight, taste, and smell.

The elements of fourths correspond to those of the vegetative soul, namely the beginning of growth, the end of growth, and dissolution or decline. This is how each species responds to that which is equivalent to it.

Given that harmony involves excellence in notes (that which is called *malas*), and disharmony their faults (which is called *amalas*); given that the virtues of the soul correspond to them and that its vices are compared to the imperfections of notes; given that harmony in music is the assemblage of sounds having consonant relationships with each other, and that disharmony is the contrary situation; given that the virtues of the soul are expressed in the tempering and harmonious composition of its parts: it would clearly seem that music causes the soul to pass through different states, by means of which it indicates that it resembles and harmonizes with it, if it is well ordered, and that it contradicts it if it is incoherent. Thus it causes it sometimes to experience desire and a joyous equanimity, sometimes abstinence and introversion, sometimes silence and ascesis—I mean sleep—and sometimes anger and passion, sometimes calm and immobility, sometimes movement and leaping, sometimes joy or sorrow, and sometimes security or fear.

Dionysos[4] made mention of certain modes[5] which he believed equivalent to the virtues of the soul. He assigned justice to the mode of the index finger on the second string (d_2), good understanding to the mode of the open third string (f_2), purity to the middle finger on the second string (e^b_2), intelligence to the index finger on the third string (g^3), solicitude to the open fourth string (b^b_3), passion to the middle finger on the third string (a^b_3), patience to the index finger on the fourth string (c_3), and virility to the middle finger on the fourth string (d^b_3). The savants of Antiquity had lengthy theories on this subject, of which we have no need.

As for the resemblance between music and the celestial spheres, in their position and their movements, it concerns the proper nature of these spheres and their position. Concerning the nature [of each sphere], there is the fact that its movement, namely the movement of the celestial sphere, has no beginning in its position. Concerning its position, there is the fact that it is possible to situate the beginning in a different place for each instance, i.e., that any position can serve as the point of departure, since the line of the meridian divides the sphere of the Zodiac into two halves at any point, with two opposite signs of which each one can serve as the extremity of the diameter of the Zodiacal sphere.

The Complete System which is qualified by a double octave (*bis diapason*) includes the greatest musical interval twice, for the greatest interval is that of the whole (the octave). The extremes of these two intervals are the *mafruda (proslambanomenos)*, the *wusta (mese)*, and the last of the high-pitched notes. By the *mafruda* is understood the note of the first string open (G), by the *wusta* the note of the index finger on the third string (g), and by the other the note of the ring finger on the fifth string (g'). All these notes are qualitatively the same because two of them heard together will form a homogeneous mixture for the ear. It follows that the Complete System is potentially a circle, since its extremity curves round and rejoins its origin.[6] The sphere of the Zodiac is divided into twelve parts which represent the houses of the Zodiac. We believe that this division was established (or defined) thus because the number 12 is divisible into halves, thirds, and quarters. These are the elements which are found in the division of the complete system, because the last note of the octave is half the first (2:1), the note of the fifth is in the ratio of one and a half to it (3:2), the note of the fourth is in the ratio of one and a third to it (4:3).[7] The excess of the first term above the second in the fourth and with the octave exists for the resemblance of the ratios of the sphere of the Zodiac and that of the Complete System; the study of these matters will be included in the appropriate chapter.[8]

20

$\blacklozenge$

Anonymous of the
Twelfth Century

This is a poem written in a late eleventh- or early twelfth-century hand in a manuscript of Boethius' *De Institutione musica*, to which the unnamed scribe has added an explanatory chart. The musicologist Jacques Handschin ascribes it to the French sphere of influence. The poem is set to an unaccompanied line of music, covering the unusually wide range of A to g': only a tone short of the two octaves which the text ascribes to the universe itself. The highest note is reached just as the text mentions the heavens, the summit of the visible cosmos and presumably of the audible one as well.

The author has taken the scale of planet-tones that Boethius (no. 13) ascribes to Cicero, with the Moon sounding the lowest note, and added a second octave to accommodate the angelic hierarchies. The poet was certainly familiar with Macrobius' Commentary on the Dream of Scipio (see no. 11), as were all the early medieval writers; with Macrobius, he emphasizes the number seven as the bond of all things. While the enclosure of the nesting spheres of the planets and fixed stars in a further series of angelic spheres was a commonplace of medieval Christian and Muslim cosmology, the logical expression of these spheres by a two-octave scale is not to be found again until the late sixteenth century, in Guy Lefèvre de La Boderie (see *MM&M*).

Source: Paris, Bibliothèque Nationale, Ms. lat. 7203, fols. 2'–3, transcribed and reproduced in Jacques Handschin, "Ein mittelalterlicher Beitrag zur Lehre von der Sphärenharmonie," in *Zeitschrift für Musikwissenschaft* 9 (1927), pp. 193-208. Translated by the Editor.

The Natural Concord of Notes with Planets

There is a concord of planets similar to that of notes.
From Earth to Heaven a divine order ascends.
Tully[1] thus enumerates them, rising from the bottom:
Moon, Hermes, Venus and Sun, Mars, Jupiter and Saturn.
In similar order you should sing your notes:
Give the Moon's first, which is proximate to Earth,
Then observe how much higher Mercury is;
This interval in the musical system amounts to one tone.
Venus, following, certainly marks off an interval worth a
    leimma;[2]
Then a tone to the Sun fills out the fourth.
And bellicose Mars with another tone completes a fifth.
Jupiter of the white locks sings his brief leimma,
And lofty Saturn, for his part, joins a tone to these.
The seventh tone reaches Heaven, after the manner of the
    [seven] days.
With these eight notes the order of the octave is completed.
Just as the heavy rules at the mese, *so does the light at the*
    heights.[3]
As far as the heavens is a duple [proportion], a quadruple
    above.
The duple consists of voices,[4] the quadruple of virtues.
There are seven different species of octave,
Three of the fourth and four of the fifth.[5]
By means of these, songs differ and their savours change.
Seven planets, "seven distinctions of notes,"[6]
Seven—[7]or gifts of the bountiful breeze,
And by sevens of days the solar year revolves.

Six for labor, and the seventh rest; it becomes life through
  eight.[8]
I believe there is life in the eighth, after seven thousand.[9]
This heptadic number is the bond of well-nigh all things.[10]

God

Seraphim	*Nete hyperbolaion*	a'
Cherubim	*Paranete hyperbolaion*	g'
Thrones	*Trite hyperbolaion*	f'
Dominations	*Nete diezeugmenon*	e'
Principalities	*Paranete diezeugmenon*	d'
Powers	*Trite diezeugmenon*	c'
Virtues	*Paramese*	b
Heaven	*Mese*	a
Saturn	*Lichanos meson*	g
Jupiter	*Parhypate meson*	f
Mars	*Hypate meson*	e
Sun	*Lichanos hypaton*	d
Venus	*Parhypate hypaton*	c
Mercury	*Hypate hypaton*	B
Moon	*Proslambanomenos*	A
Earth	Silence	

21

Isaac ben Abraham ibn Latif
c. 1220–c. 1290

Ibn Latif was a Jewish philosopher of Kabbalistic tendencies, living in
Spain during the golden age of religious tolerance under the Muslim
rule of that country. The Jewish educational curriculum was similar to
the Arabic, hence ultimately influenced by the formulations of late
Antiquity, especially the concept of the Seven Liberal Arts: Grammar,
Logic, Rhetoric, Arithmetic, Geometry, Music, and Astronomy. Often
these were expanded by adding Optics and Medicine.

In Jewish, as in Islamic and Christian philosophy, the thirteenth
century was a time of confrontation between two forces: on the one
hand, an alliance of dogma with Aristotelian rationalism; on the other,
a movement of mystic illuminism with Neoplatonic roots. Maimonides,
most eminent and influential of Jewish philosophers, had given the seal
of his disapproval to the idea of the Music of the Spheres, and as a
result, allusions to it are scarce in Jewish literature. Ibn Latif is reserved
about the details of his own beliefs, but his arithmology evidently places
him within the more mystical stream.

Source: Isaac ben Abraham ibn Latif, *The King's Treasures*, translated by
Isaiah Sonne in Eric Werner and Isaiah Sonne, "The Philosophy and

Theory of Music in Judaeo-Arabic Literature," in *Hebrew Union College Annual* 17 (1942–43), pp. 551–553. Used by kind permission of the publisher.

Music among the Liberal Arts

After the science of Geometry, follows the science of Music which is a propaedeutic one, leading to the improvement of the psychical dispositions as well as to the understanding of some of the higher intellectual doctrines. This was manifest in the case of Elisha when he said: "But now bring me a minstrel" (II Kings 3.15).[1] Moreover, this science is also propaedeutic to the science of Astronomy, the explanation of which is as follows: The science of Music envisages eight modes of melodies which differ from one another because of the expansion and the contraction, the height and the depth and other differences in musical instruments.[2] The eighth mode functions as a genus which comprehends the other seven modes, and this is the meaning of "For the Leader, on the Sheminith" (Psalm 12:1). The Psalmist has alluded to this cryptically by means of the number seven in the repetition of the word *Kol*[3] characterizing the Psalm, "Give unto the Lord, O ye sons of might" (Psalm 29:1), while the phrase, "All say, 'Glory'" (Psalm 29:9) alludes to the eighth tone which comprehends all of the others. I cannot explain any further. Returning to my task, I say that this science, Music, relates to the various movements of the spheres, that is, to the nine [seven?] planets and to the movement of the all-comprehending eighth sphere. The relation arises from the analogy between the various tones in the various spherical movements, involving as these do, direction, speed, retardation, withdrawal, deflection sideward, approach to the center and removal from the center and involving also the various activities of their respective stars.[4] In all of this, there is a subtle and profound analogy linking the two sciences. Only by those who are well acquainted with both of these sciences, can this analogy be grasped. Those who affirm the existence of heavenly tones corresponding to the musical tones partly follow our suggestion.

22

✦

Jacques de Liège
c. 1260–after 1330

The work of Jacques de Liège (Jacobus Leodiensis) is a reworking and an enormous expansion of Boethius' treatise, standing in relation to the Middle Ages much as *De Institutione Musica* did to late Antiquity. Jacques throws his net wide, to encircle every type of music and to present it in lapidary form. Like Boethius, he is especially concerned with the mathematical side—tunings and intervals—, but he is able to go much further, and offer a theoretical manual of polyphonic and mensural music as practiced in his own day.

Speculum Musicae is the longest of all medieval musical treatises, and remains untranslated. It commands little interest among scholars because of Jacques' reactionary attitude to practical music: he lost credit in the eyes of historians by vigorously opposing the change from the Ars Antiqua of the twelfth and thirteenth centuries to the Ars Nova of the fourteenth. In no way does this affect our reading of his chapters on the higher forms of music, which he unfolds with the expertise of a logical and broadly learned mind.

As every medieval scholar knew, Boethius had outlined three types of music: *mundana* (of the worlds), *humana* (of the human being), and *instrumentalis* (of instruments, including the voice). On the first two,

Boethius' book as it stands has little to say, and Jacques applies himself to make good the omission. *Music, Mysticism and Magic* included a translation of his chapter on human music; here are his chapters on the higher types. Jacques makes it plain that *musica mundana* as Boethius understood it is not just the "Harmony of the Spheres" or of space, to which it is conventionally equated, but also the harmonies of matter (the four elements) and of time (the seasons of the year); in short, of all things in the greater cosmos.

Jacques finds, moreover, that he must go further than Boethius, because the doctrines of the Christian religion imply that the starry spheres are not the terminus of the universe. Above the heavenly bodies, always in circular movement, are the unchangeable realms inhabited by the angels and the souls of the elect in perpetual contemplation of God. So just as the Christian cosmology added angelic spheres beyond the stars, Jacques crowns the Boethian scheme with a fourth music, *musica coelestis*. What he says about this highest of all musics is a little disappointing; but what could he have said without seeming to usurp the role of the Fathers and Doctors of the Church? He does at least hint that the essential distinction is between a static harmony (of the unchanging) and a mobile one (of the spheres), which suggests a parallel with the unalterable numbers of music, and their working out in time. But prudence, and perhaps humility, forebade further speculation into just what sort of musical archetypes might reflect the mysterious relationships of the Trinity and the nine orders of angels.

Source: Jacobi Leodiensis, Speculum musicae, edited by Roger Bragard, vol. I (Rome: American Institute of Musicology, 1955), pp. 36–49. Translated by the Editor.

The First Division of Music

Book I, Chapter 10. I. Music is divided, according to Boethius, into *mundana, humana,* and *instrumentalis* or sonorous, and all these treat of natural things.

But *mundana* concerns simple bodies, both incorrupt and corrupt, such as the heavenly bodies, the elements, and those things connected with them, and therefore treats of the macrocosm [*maiore mundo*].

Humana, on the other hand, concerns the microcosm [*minore mundo*] i.e., man, who among compounded things is most noble and perfect and wonderful in the union of his form with matter, inasmuch as the mixture is of something incorruptible with something corruptible. For the human soul has an incorruptible existence, that is to say an inexterminable one, because God made the soul of man. And elsewhere it says: "Do not fear those who kill the body, for they cannot kill the soul" [Matthew 10:28]. This soul of man, therefore, existing incorruptibly because it is not produced by the power of matter and is created after the image of God, is the most perfect of the forms which are united with matter and a corruptible body. Human music therefore consists of this wonderful union in which there is a wonderful connection and a wonderful concord, and of the parts thus united, i.e., of soul and body.

Instrumental music, however, concerns the sounds which manifest matter in sensible musical consonances and which are expressed by artificial and natural instruments.

But it seems that a further species of music may be added to this: one which may be called celestial or divine.[1] This regards things separate from motion and from sensible matter, both according to existence and according to intellect: that is to say, transcendent things pertaining to metaphysical or divine knowledge, of which something has been said in the Commendation of Music [Ch. 1]. And now we must speak more fully of this first one, then of the other types of music.

Celestial or Divine Music Should Reasonably Be Included Among the Types of Music

Chapter 11. It is not unreasonable, I think, to reckon celestial or divine music among the types of music: first, by reason of the description of music, taken broadly [*generaliter sumpta*]; second, by reason of its name; thirdly, by reason of divine praise.

It has been said above that music, taken broadly, extends to the knowledge of the harmonic modulation of any things whatsoever that are connected with one another by any modulation.

But harmonic modulation, taken broadly and speaking rationally, cannot be understood as merely that sensible modulation which exists in sensible sounds, causing a perceptible consonance to the ear; for we know that such harmonic modulation pertains only to instrumental music. Otherwise the other species of music would be lost; and it is clear from Boethius that the term "harmonic modulation" does not mean only the modulation of sounds, for he says: "Now if a certain harmony does not join together the diversities and contrary qualities of the four elements, how is it possible for them to unite in one body and machine?" [*De Inst. Mus.* I, 2] But this harmony of blending is not one of sounds, but arises out of the relative positions and proportions of their qualities. And Isidore said that this "world may be said to be put together according to a certain harmony." Likewise, according to Isidore, music extends to all things, since "nothing [exists] without it."[2] Therefore it expands even to transcendent and divine things, for if God has indeed made all things in number, weight and measure [Wisdom 11:20], then all things which have proceeded from the primal origin of things are formed by reason of numbers.

It now seems, regarding music in general, that harmonic modulation should not be restricted merely to the natural and corporeal things which concern *musica mundana* and *humana* and the sounds and notes which concern instrumental music, but should even [include] the First Mover, because *musica mundana* treats of the celestial spheres which are moved by separate movers, i.e. by intelligences whom the Philosophers call the movers of the spheres.[3] Now between the mover and the moved or movable a certain proportion is required, at least if the mover be of a finite virtue and the thing be moved naturally (which I say because the First Mover is altogether immovable, since being infinite in virtue he moves by his own free will). Besides, if the proportion or harmony [*coaptatio*] of the human soul with its body, which it moves as moving something joined to it, is called a harmonic modulation, why should not the harmony or proportion between those separate movers and the spheres which they move be called a harmonic modulation?

In any species of music, therefore, it will be appropriate to examine the proportion, order and concord of the separate movers with each other, with the first mover of all, and with the things which they move. And perhaps if Boethius had pursued his *musica mundana* in detail, he would have extended it to the mover of the celestial spheres.[4] For it was Aristotle's method to come to the knowledge of separate substances through movement. What is wrong, therefore, if we extend harmonic modulation as generally accepted not only to corporeal, natural and substantially numbered things, but also to metaphysical things, amongst which when numbered and compared a certain appearance of connection, order, concord or proportion is perceived? Thus it falls not only under two parts of theoretical philosophy, but under three, as stated above [Ch. 8].

Second, the same is evidenced by some of the names of music. It was stated above that music is named after the Muses, from whom wisdom was sought. But the things which pertain to the state of wisdom are metaphysical ones, chief among which are divine things and knowledge. Therefore music extends to these. Music is also said to come from "*muso, sas,*" which means "investigate something by thought." But what are these things of which it is so necessary to gain knowledge by thinking?

Things such as the separate substances, and most of all the first one, in the face of which our intellect, lowest in the class of intelligibles, abandoned by its natural [faculties], finds itself like the blinking of the eye in the light of day. To know even a little of these things, according to the Philosophers, is far better than to know a great deal about lower things; and according to Saint Augustine nothing more advantageous is to be found.

Third, the same is confirmed by examining the divine praises which are offered up. A certain type of music has been provided for wayfarers in this transitory world by which God may be praised in this Church Militant, both in himself and in his Saints; and in order to further this, the ministers of the Church receive temporal gifts. Shall the Church Triumphant of that other, incorruptible world lack a type of music suitable for the citizens of that Church, who receive such priceless gifts? Certainly not! In that celestial Church, therefore, there will be a place for music by which God may be praised incessantly by those citizens, as has been touched on above. And this type of music is as much more excellent and perfect than the others as it is more perfect in its object, and as the citizens singing it are in a higher state—a beatific one. And beatitude is the perfect state in which all good things come together, according to Boethius [*De Cons. Phil.* III, Prosa 2].

And now we will say something about these three types of music.

What the Celestial or Divine Music Is, and Why It Is So Called

Chapter 12. As claimed above, celestial or divine music considers the category of transcendent things, just as instrumental music considers that of sounds. Now transcendent things are metaphysical things.

Metaphysical derives from *meta*, which means "beyond," and *physis*, "nature," because it deals with things transcending natural, i.e., movable things. "Nature is a cause of movement or rest in that to which it belongs primarily by virtue of itself, and not secondarily through an accident," as is clear according to the *Physics* [of Aristotle, Book II, Ch. 1]. And since metaphysics considers being as such, whoever wishes to extend this type of music generally to metaphysical things may thereby compare substance to accident and the nine categories of accidents to one another, besides the general passions of a being within itself and to being itself; or compare the One to being and to the many, the same to the different, potency to act, like to like and unlike, equal to equal and unequal—indeed, wherever equality and inequality and proportion are found, and so with other things pertaining to metaphysics.

But since the nobler and more perfect metaphysical things are the separate substances, to knowledge of which metaphysics is principally directed (for the reason of being is best preserved in them, since they have more of act and are simpler⁵—God most of all, who is pure act and utterly simple), this type of music is named after those things and called divine, insofar as it concerns God. For the science of metaphysics is called divine so as to be named after its most perfect object, since a name should be derived from something worthier, and since it is proper for everything to be called after its final object. But it is called celestial because it considers separate substances other than the first: not those of the material heavens, which *musica mundana* treats, but those of the spiritual heavens, to which belong the good angels and holy men, of which the Prophet says: "The heavens tell of God's glory" [Psalm 18:1].

For as there are nine material and movable heavens (if the empyrean heaven is excluded),⁶ there are also nine spiritual heavens, just as there are nine orders of angels, with which good men will be united. Moreover, this celestial type of music is named after those citizens of heaven not so much because it is objective toward them, but because it is subjective with them. For they have this music in perfection, they who no longer contemplate God in a glass darkly through any exterior representation, but behold him directly, face to face [I Corinthians 13:12]. And since they find in God whatsoever they desire, and know no evil at all nor covet any earthly thing in the least, how could they cease from the praises of God in whom is their consolation, whole and entire? Hence Augustine says of these citizens in his book *De Musica* [VI, 16] that they could not be purged of the love of temporal things unless taken by storm by some sweetness in eternal things, as is found in the Psalm when it is said: "The children of men will put their trust in the protection of thy wings; they shall be inebriated with the richness of thy house, and thou shalt give them to drink of the torrent of thy delight" [Psalm 36:7–9]. And David with his lyre-player [*citharoeda*], who much desired to enter into their fellowship though still on earth, frequently exhorted those citizens to singing. "Sing," he says, "a new song; praise him in the congregation of his saints" [Psalm 149:1], in which are none but the holy ones, for "no iniquity shall enter in there" [Apocalypse 21:27]—there, where there is perennial festival, continual nuptials, a perpetual banquet. Surely music has its place in these. For "music is pleasant at a banquet" [Ecclesiasticus 49:1], according to the Scripture.

But just as the possessors of that celestial home may be said to be inebriated after drinking of the richness of the chief delight that is there, so they may also be said to be seized [*rapti*]. Hence Saint Paul, taken up for a short time to the fellowship of those citizens, says of himself as if of another: "I know a man in Christ who fourteen years ago, whether in the body or out of it I know not—God knows—was seized up in this way to the third heaven," and because of the magnitude of that vision and joy he adds, repeating that rapture: "And I know that this man, whether in the body or out of it I know not—God knows—was seized up into Paradise, and heard secret words which are not lawful for a man to utter" [II Corinthians 12:2–4]. And no wonder, if they are seized up and hear what is not lawful for a man to utter, that they see what cannot be fully spoken of, namely the concord and inseparable fellowship of the Divine Persons and their perfect union in one utterly simple essence, replete with every perfection, whereby the Son is in the Father and from the Father throughout all ages; the Father in the Son, not from the Son nor from any other; the Holy Spirit in both and from both through one enspiriting power. They learn of the inward divine emanations, the personal properties, ideas and attributed perfections; how these are distinguished by comparison either internally or externally of the relation to essence; and what the distinction of the Divine Persons is according to the nature of the thing or to reason alone. They see their order without [any Person being] prior or posterior, their equality and similitude. They see the ideal Forms and exemplars, and contemplate in that voluntary and eternal mirror, in that Book of Life, things which cannot be told us.

Neither do they fail to see in it the order of the Angels, the distinction of the three Hierarchies[7] and the natural properties with which angels are favoured; the connection and unshakable concord among themselves and with God, since their love is of the highest kind; which of them are the superior ones, which the middle, which the lowest; which assisting, which ministering; which receive more immediately the divine illuminations and revelations, and, if God should will it, tell us of them through intermediaries. Nor do they fail to see there the specific nature of everything else: their order, connection, and concord among themselves and with God. In all of these, compared in due fashion as they are able to compare them (mostly through the "meridional" knowledge which they have of things in the Divine Word, though that does not exclude the "vespertine" or natural knowledge which they have of things after their own kind), in these, I say, duly compared with one another, just as they find the most excellent harmonic modulation, so they also find the most perfect music. Therefore the best musicians are those who in their contemplation [*intuitive*] observe that eternal book. For in it there lies open and shines forth every proportion, every concord, every consonance, every melody; and whatever things are needed for music are written down there. But that type of music of which we are now speaking, which those celestial citizens possess to perfection, we wayfarers can only possess imperfectly, since it is only permitted us to know about divine things and other separate substances through the Holy Scriptures, through the Catholic Faith, through right doctrine, and through philosophy.

Let it not displease the reader if I have touched on other general matters pertaining to this type of music, because it is no shame to this science that it extends to separate substances, which are the loftiest objects one could address; nor, in turning from this, is one less capable of addressing other matters. For this music, taken broadly, may be applied to all those things in which it discerns the cause of its object, taken broadly. But its object, taken broadly, is the numerical comparison of all things whatsoever, not absolutely but with something else, i.e., with one another; hence the formal cause of its object, taken broadly, is that a harmonic modulation is to be found between any objects numbered and compared. As touched on earlier, I understand by this "harmonic modulation" a condition, a relation of some kind: of connection, order, proportion, concord, similitude or dissimilitude, equality or inequality, interdependence [*coexigentiae*] or imitability, or any other sort of condition. For since, according to Isidore, "no discipline can be perfect without music, for nothing [exists] without it" [*Sententiae de Musica*, Ch. 3], and since the formal cause of its object is harmonic modulation, it is necessary that such harmonic modulation be taken very broadly indeed so as to extend to all beings compared to one another; and this is to treat them as some condition of things compared to one another.

Accordingly, as metaphysics extends to every being insofar as the general cause of being is found therein, so music is considered to extend to every numbered being, by the said cause. And since, according to the Philosopher, there are three forms of knowledge working around the whole being: dialectical, sophistical, and metaphysical (which he calls philosophy), music, too, may be joined with these—but differently, because dialectical knowledge applies itself by argument to every probable matter, sophistical to every apparently probable matter, metaphysical to every true thing in which it finds the general cause of a being; whereas music, as we have said, applies to everything numbered. If, however, anyone wishes to learn more of the things that celestial or divine music has to compare, he should frequent the schools of sacred Theology, become a humble and diligent student there, acquire that knowledge by which he may be prepared to obtain celestial life, and thus perfectly pursue this type of music of which we have been speaking.

This type of music may not unsuitably be included under *musica mundana*, since every creature enclosed by the bounds of the world as a whole, existing as finite and limited, finds itself encompassed and defined here. But God, who is infinite and simply unlimited, even when he is in the world is not there in himself; because he was there, in truth, before it was produced by him and would still be there if there were no world, since the whole world cannot hold him, not even if there were a thousand of them. He is said, however, to be in the world in respect of those things which he has done in the world, principally as cause to its effect, which for the maintenance of its being requires his cause to be an actual presence and influence; and as far as this is concerned it is said that "He was in the world," and that forthwith a cause was supplied, because "the world was made by him" [John 1:10]. Hence in the creation of the world it is said that "the spirit of God was carried above the waters" [Genesis 1:2], for the intention of his creation is formed before his work. And the Prophet, speaking as from God's mouth, says: "I will fill Heaven and Earth" [Jeremiah 23:24], and David the Psalmist: "If I go up to Heaven, thou are there; if I descend to Hell, thou art there also," etc. [Psalm 138:8]. And as God coexists with all time, he is said to be always timeless. For as Plato says in the *Timaeus* [37c–d], he is not in time but is the reason time exists; thus he coexists in every place, and is not, in himself, in any place. Yet it is said that God is in the world and in every creature according to the three general modes of being: potency, presence, and essence. More particularly, however, he is said to be in holy men through the gift of Grace, and most especially in the human nature taken on by the Person of the Word, and that by an inexplicable union.

But I have distinguished this type of music from the *mundana* because Boethius applies *musica mundana* only to natural, mobile and sensible things (as will be explained forthwith), whereas the things which I have said pertain to this type of music are metaphysical things, transcendent things, separate from motion and sensible matter, even according to their being.

What Musica Mundana Treats of

Chapter 13. *Musica mundana*, according to Boethius [*De Inst. Mus.* I, 1], consists of a certain proportion of the orbs and motions of the celestial bodies in their order, connection, position, rising and setting, and in their natural relationship [*coaptatione*]. But hitherto [it has consisted] in a certain harmony proceeding from the rapid motion of their bodies. "For who," as Boethius says, "could suppose that such a swift heavenly mechanism could move dumbly and silently in its course?" [I, 2]. For certain Pythagoreans and Plato, following whose opinion Boethius speaks here (as does Macrobius on the *Dream of Scipio*), have maintained that since sound is caused by the motion of solid bodies touching one another (and the celestial spheres are utterly firm and compact, touch each other and move most rapidly), a sound will be caused here which, they assert, appertains to *musica mundana*. Their contention appears to be confirmed by what is found in Job 38, where it says: "Who shall explain the system of the heavens, and who shall make the heavenly music sleep?"[8] But the reason we do not hear its sound is said to derive either from its excessive distance from us, or because it has been with us from infancy, inborn and implanted, so that we have never heard silence. For as the Philosopher says in Book II of *Heaven and Earth* [Aristotle, *De Coelo* II, 9], hearing is [always] of something, so that when we perceive silence it is the privation of sound: hence one brought up in a mill does not hear its noise, nor do hammerers notice the sound of the air.

These opinions do not satisfy Aristotle, so in Book II of *Heaven and Earth* he says: "But it is obvious from these things that to say that there is a harmony of the broad heavens like the sounds made by sounding objects is indeed to speak lightly and contrary to the teachings. The truth, however, is not like this," etc. In refuting the said opinion, the Philosopher touches on two [points]: first, that if the broad heavens made a sound we would hear it even louder than the sound of thunder, since that sound is far greater and more rapid than the motion of thunder because of the magnitude of the things moving and the velocity of their motions. And that sound would not only inflict pain and injury on the senses of the listeners, the fineness of whose senses would be ruined through such disproportion, but it would also destroy inanimate things, as thunder does stones through its great force, even destroying harder bodies.

Second, the Philosopher proves that the orbs make no sounding music by their own motion, as the Pythagoreans say they do; in the first place, because the fixed parts in any whole do not produce any sound, however fast that whole may move (as demonstrated in a ship borne on the water, and in things fastened by nails when they rotate); for the things that make a sound with each other are distinct. The second proof is that three things are required to utter a sound: a striker, a thing struck, and a medium which is air, the material of sound or of the voice; but there is no air there, as is plain from the Philosopher. The Philosopher therefore holds that no sounding harmony is caused by the movement of the celestial bodies.

What then are we to say as an excuse for the Ancients? It seems that Boethius, who speaks according to the opinion of the Pythagoreans, does not understand by that harmony an audible and sensible music, for the latter does not appertain to *musica mundana*, nor even to *humana*, but only to *instrumentalis*. For if a true sound were made there, that sound would be made by a motion; but as the motion of the orbs, particularly the primum mobile, is the most rapid, the sound made therein would move and shake the other orbs as it diffused itself through them, and in the same way the sounds of the other orbs would unsettle the orbs beneath, from top to bottom. So they would shake move and unsettle exceedingly the whole [spheres] of fire, air, and perhaps water. But since air or water is the medium for transmitting sound, and there is no air between the orbs (whether or not there may be water), what [would transmit] that sound between the orbs which are not receptive to a foreign or extraneous impression? (The Doctors have various explanations of the waters which Holy Scripture says are above and beneath the firmament [Genesis 1:7]; but we cannot treat everything in detail.)

Therefore perhaps Boethius and the Pythagoreans understand by that music proceeding from the motions of the celestial bodies the connection, order, proportion, concord, or any other suitable relationship which the orbs have with one another in motion, position, luminosity, virtues, inequality or equality of movement. For there is an inequality of their movements as to speed or slowness, because the superior sphere moves faster than the inferior ones, and by the motion of the firmament the higher orbs move faster than the lower ones, such as Saturn compared to Jupiter. The planets, too, are said to move unequally in their own motions according to speed or slowness, such that one planet traverses the circle of the Zodiac faster than another, as the Moon compared to the Sun. But their motions are said to be equal in respect of time in that there is a carrying movement by which all the planets traverse their circle of the firmament in a natural day. Hence Boethius says: "For some are borne higher, other slower, and they are all turned with an equal energy so that a fixed order of their courses may be reckoned through their diverse inequalities. Thus this celestial revolution cannot lack some fixed order of modulation" [*De Inst. Mus.* I, 2].

Boethius does not say whether this modulation is vocal or instrumental [*sonora*]. For descending straightway to the elements, he says: "If a certain harmony does not join together the diversities and contrarieties of the elements, how is it possible for them to unite in one body and mechanism?" [loc. cit.]. And it is plain that this harmony of the elements is not a sound. And later, in the two disjunct tetrachords (the disjunction called *diazeuxis*), he ascribes the strings from *hypate meson* to *nete diezeugmenon* to the planets as if the order of these strings were modeled on the order of the planets. But since he says, "*hypate meson* is attributed to Saturn, *parhypate* to the circle of Jupiter" [I.27], it is plain that this is a metaphorical expression, and so it may be when he says that sounds and modulation arise from the celestial motions.

But whether the truth of the matter is that there really are sounds and sounding music there, God knows, to whom nothing is hidden and who can do these things and greater ones still, though they are unknown to us. And therefore His word to Job may be understood as it should be, literally or otherwise.

But according to Boethius it is not only the motion, order, position, etc., of the stars that pertain to *musica mundana*, but also the links of the elements and the agreeable variety of the seasons; for if the elements were not in mutual proportion as to their position and order, their active and passive qualities, they would not be naturally stable, and as much of the world-mechanism as is in them would be impaired. Therefore the Philosopher maintains[9] that one handful of earth contains ten handfuls of water; one of water, ten of air, one of air, ten of elemental fire; so they are proportioned in such a way that one does not altogether vanquish another nor convert it into itself. And this diversity of elements and variety of seasons is useful and brings about a variety of fruits: it makes one body of the year. Hence if any of these, which furnish things with such variety, is withdrawn, natural things will perish, and none will remain consonant. Therefore, according to Boethius, just as low strings are tuned so that none of them, however low, reaches complete noiselessness or silence, while none of the high ones, however tightly stretched, is so drawn out as to snap, so it is in the music of the world: no created thing is so feeble that its whole species can be altogether destroyed by the other creatures. For just as a creature is unable to create, so it also cannot annihilate, however much it may corrupt, any of those beings which are composed of matter and form; since those things which are separate from matter, like the separate substances, are said to be incorruptible and perpetual.

Hence the Commentator[10] said: "The removal of matter is the cause of perpetuity and incorruptibility." Such things when separated from matter cannot be annihilated by the power of any creature, but only by that of God who created them; for if he should withdraw his influence by which a thing maintains and conserves its being, things will return to nothing, just as the Psalmist says: "But when thou turnest away thy face they are troubled; thou takest away their breath and they die, and return to their dust" [Psalm 103:29]. Hence Augustine says: "All things would revert to nothingness if it were not for God's maintenance." Therefore the elements and other beings are friendly to one another, one giving assistance to another even as regards the change of seasons. "For what winter confines, spring releases, summer heats, autumn ripens, and the seasons in turn bring forth their fruits and help the others to bring forth theirs" [Boethius, op. cit., I, 2]; thus a condition of love and concord is innate in all creatures, as far as concerns the natural part received from their Creator, to help each other to preserve themselves and remain in existence.

Musica mundana therefore consists principally of three things: first, the comparison of the celestial bodies as to their motion, nature, and position; second, the various comparisons of the elements as to their qualities and position; third, the differences of time: in days, in the changes of night and day; in months, in the waxing of the moon; in years, their divisions which succeed one another as winter, spring, summer and autumn. And as a consequence, this type of music is included in many sciences, for example in natural [science]. For the part of it that treats the celestial bodies is contained in [Aristotle's] book *On Heaven and Earth*, and the one which treats the elements is contained in many books on nature: in Books 3 and 4 of *On Heaven and Earth*, in the books *On Generation and Corruption*, and in the book *On the Properties of Elements*.

This type of music is also contained in Astronomy insofar as it considers the motions, position and nature of celestial bodies, comparing them to one another; and the part of it that concerns the course of the sun and moon and the divisions of time and the year is contained elsewhere, in the appropriate science. However, these matters are considered on different grounds in this type of music and in the sciences referred to—as appears, in any case, from what has been said above.

And now that the principal points have been made, let this suffice on *musica mundana*.

23

✦

Ugolino of Orvieto
c. 1380–1457

Ugolino was a prominent Italian churchman, resident first in Forli, then obliged for political reasons to move to Ferrara. Here he wrote his *Declaratio Musicae Disciplinae* ("Explanation of the Discipline of Music"), probably in the early 1430s. Ugolino's is one of the last treatises of the medieval type. Although what he has to say is fully in accord with most of the authors in this collection, his approach is not Platonic but Aristotelian, and his vocabulary is the specialized one of Scholastic philosophy.

Ugolino says at the outset of his work that the highest form of music is *musica coelestis*, because that includes all three categories as defined by Boethius. By "celestial" music he apparently means the actual songs of the Heavenly Host, ordered in perfect beauty and proportion. But Ugolino's view of how one might ascend thither is not through mysticism, but through the use of intellect or reason, man's noblest faculty. He thinks that we must work upward from the known to the unknown, meaning in the present case a progression from plainchant and polyphony to the speculative world of pure, sense-free number-forms. Therefore he does not discuss *musica humana* and *musica mundana* at the beginning of his treatise, as the Renaissance writers would do before proceeding to practical music theory; instead, Ugolino reserves them for the end, after his readers have been prepared with a course in plainsong, measured music, and interval-ratios, each considered a stage more speculative than the last. Presumably he would not have approved in the least of our turning straight to the end of his book.

With characteristic Scholastic pedantry, Ugolino determines what *musica humana* and *mundana* are not, before saying what they are. His final definition of them is as broad as it can possibly be, for "music" must eventually include the whole of creation. The only authorities he draws on are Aristotle and Boethius, whose definitions supply the data for his logical developments. These extracts show how the mystical idea of a higher realm of music could survive even the ascendance of the neo-Aristotelian philosophy.

Source: Ugolino Urbevetani, *Declaratio Musicae Disciplinae*, vol. III, edited by Albert Seay (Rome: American Institute of Musicology, 1962), pp. 94–8. Translated by the Editor, with the kind assistance of Professor Seay (now deceased).

Concerning the Subject of Musica Mundana

Chapter III. Now on the subject of *musica mundana* it should first be noted that *musica mundana* is threefold. First, there is that of heaven and the planets, of the seasons and elements, in which the double apparatus of the world is contained, i.e., the major and minor, on which division the author of *De Sphera*[1] has touched in his treatise on the sphere. It is also contained in these because the world is understood in three ways: in one way as the sphere of the elements, according to which those living on the earth are said to be in that world; second, as the aggregate of all the superior spheres from the shell of the lunar orb as far as to the last heaven; and third, as the aggregate of both. The world as proposed here is not understood in the first and second ways, but in the third way.

Also, it is contained in these because the order in which the seven upper spheres are situated is this: the first sphere is the sphere of the Moon, the second of Mercury, the third of Venus, the fourth of the Sun, the fifth of Mars, the sixth of Jupiter, the seventh of Saturn, the eighth is the sphere of the Primum Mobile, at whose motion all the lower ones are moved, though in contrary motion, on account of the proper motion attributed to each sphere, and also on account of the motions conferred on the lower spheres by the deferent motions. This is shown by the theory of the planets and by the treatise according to Thebit[2] on the movement of the eighth sphere in astrology.

Also many things are contained in these which do not belong to the present speculation, namely the different motions of the celestial spheres, which of them move faster and which slower; but these refer to another faculty, therefore we will leave them to their own faculty.

But we must not omit the following: namely, that it is not possible for the higher bodies of the spheres to move in silence and without sound, for how could anyone believe that the great mass of such a huge machine should move silently in its course, notwithstanding that its sound never reaches our ears?[3] Everyone may wonder for themselves about why this is so, for there are many causes, e.g., the disproportionate distance, the remission of the sound, the lesser hardness of the bodies of the spheres, and many others which we omit in the cause of brevity. For we have said enough in the first book on this consonance of the heavens, of which Boethius in his *De Musica*, Book I, Chapter 2, and Macrobius in his *Saturnalia*[4] have said much. Pliny also, in the first book of his *Natural History*,[5] maintains this, for he says that either the universe is measureless, and therefore the continuous whirling of such a rotating mass exceeds the sense of hearing—and I might well say that the ringing of the revolving planets and their orbs is even greater than that of the Primum Mobile surrounding them—, or else a certain sweet music of incredible beauty would never escape us who live within it, either by day or by night; in which case it may be understood that the celestial music is made from subtler things, without the presence of sound.

If on the other hand it is poured out secretly from higher things to lower ones, as far as to our hearing, although we do not sense it because of familiarity, like those who live around the Catabathmon,[6] i.e., the falls of the Nile; whilst if anyone were born in another world, if it were possible (as Saint Augustine affirms) and then should come into this world without any [lacuna] it would please him exceedingly, whereas earthly music will please us very much because it is made from the more corporeal elements: this is scarcely to be found free of incongruity.

And here it is to be noted that just as many a living being [*anima*] acts by itself unconsciously, like hair and nails as they grow, thus also many things happen by themselves, unheard and hidden by their nature, such as the sounds of the planets. For these are the actual words of Pliny. And the Peripatetics confirm and believe this, even though Aristotle in his book on the properties of the elements[7] asserts the opposite, saying thus: The celestial bodies are not born able to make a sound.

Having explained these things, we say that the consonances of heaven and the planets are not an adequate subject in or of *musica mundana*. The conclusion is proved since many other things are considered in *musica mundana* besides the consonance and the consideration of heaven and the planets, therefore the conclusion is true, the consequence is noted and the antecedent proved, since this considers the consonance of heaven, the planets, and also the elements. Also the consonance of the elements is not an adequate subject of *musica mundana;* this is shown just as in the second statement immediately preceding.

Also the consonance of heaven and the planets, of time and the elements together, is not an adequate subject of *musica mundana;* this is shown by the three preceding statements. Also the consonance of heaven and the planets, of time and the elements is similarly not a subject of demonstration in the whole *musica mundana*, nor any one of them. This is proved since there are many subjects of demonstration in all of music, therefore the conclusion is true, the consequence holds, and the assumption is proved, since there are as many subjects of demonstration in the same as there are demonstrations; but as the demonstrations are many, so the subjects are many, therefore, etc.

Now heaven and the planets, time and the elements, are the subject of operation in the whole *musica mundana*. This is proved since *musica mundana* operates about heaven and the planets, time and the elements, by adapting their consonances and agreeable relationships, according to Boethius, *Musica*, Book I, Chapter 2, therefore, etc.

Also the proper coordination and mutual assemblage of heaven and the planets, time and the elements, is the subject of attribution in the whole *musica mundana*. This is proved because all things considered in *musica mundana* are attributed to the proper coordination of heaven and the planets, time and the elements. Therefore the conclusion is true, the consequence holds, and it is assumed from Boethius, *De Musica*, Book I, prefatory chapter.

Also the soul is the subject of information of *musica mundana*. This conclusion is proved, since the soul is the subject of all behaviors and knowledges. Therefore the conclusion is true, the consequence holds, and the antecedent is proved, since knowledge is the intellectual behavior of an acquired conclusion, demonstrated in demonstration through the necessary predispositions; and so it is about universals that are in the soul, therefore, etc.

24

◆

Giorgio Anselmi
before 1386—between 1440 and 1443

Anselmi was a citizen of Parma, a physician and writer on astrology and astronomy as well as on music. His musical treatise, dated 1434, takes the form of three days of conversations on the three divisions of harmony, which for him were *celestis, instrumentalis*, and *cantabilis:* celestial, playable, and singable. Like other late medieval theorists, he did not feel obliged to respect Boethius' threefold division into *mundana, humana,* and *instrumentalis.* Anselmi's second and third days deal with the usual matters of mathematical interval calculation and modal theory, including suggestions for improving notation which would later interest Gafori (see no. 29), the owner and annotator of one of the manuscripts of the work.

Whereas the planetary music of a Jacques de Liège (no. 22), for instance, was still largely a metaphor for the harmony of their motions, and as such reducible to mathematics, Anselmi seems rather to anticipate the polyphonic planet-songs of Kepler (no. 34). Each of Anselmi's planets has its own variation of tones, and the whole ensemble produces a harmony in accordance with the laws controlling the structure of the World-Soul. With Anselmi, moreover, we meet the first of our theorists for whom the establishment and elaboration of correspondences is a consuming interest; it will become the very essence of Renaissance Hermetism. This impulse is part of a new attempt to expand the conception of the cosmos by moving from a straightforward chain of being—the hierarchy of the Middle Ages, stretching from the lowliest stone to the highest of the angels—to a more subtle construction in which, as it were, the chain is looped and folded. One might compare Anselmi's scheme of correspondences to the simple ladder of the Twelfth-Century Anonymous (no. 20).

Probably neither Anselmi nor Dante (who is assumed to be his inspiration), and certainly not the Kabbalists and Sufis, actually believed the angelic hierarchy to be physically situated in the planetary spaces to which they are here assigned. That would be a simplistic view. The planets, rather, reflect on their own level of being the orders which the angels manifest on a superior level. And here is where music comes in: nowhere else in nature does one find a better aid for grasping this conception of corresponding orders on different levels of being than in the musical scale, which replicates itself every octave in a similarity that is yet not an identity.

Source: Giorgio Anselmi Parmensis, *De Musica*, edited by Giuseppe Massera (Florence: Olschki, 1961), pp. 97–106. Translated by the Editor.

On the Heavenly Harmony

Part I. [131] Now it is in truth that heavenly harmony[1] (of which I wish I deserved to speak worthily, or were able sufficiently to praise) that the soul should reproduce as far as possible: such I believe to be the teaching.   [132] The tireless soul of the whole world,[2] indeed, sings with the same [harmony] its ceaseless praises to the eternal, most high and all-beneficent Governor by means of the celestial motions, with which the holy throngs of blessed spirits, sweetly echoing, contend in song and in the ineffable beauty of their rivaling hymns.

[133] But perhaps you prefer to call these spirits separate intelligences. It was surely most fitting that all things should rejoice in him, the best, most glorious, and excellent of all, full of ineffable grace, and to give continual thanks in musical songs, since he wished that every creature should participate in him, at least as far as it was able.   [134] And the surest indication of this is that the World-Soul, by which the heavens are turned and all living beings animated, is confined by definite laws.

This is attested by the numbers which mark the connections within it, since, as we have said, they are actually the same as those in which our audible harmony sounds.

[135] When God Almighty devised the construction of this corporeal world out of formless matter, it was necessary for him to add a ruling and moving soul—for no bodies can move by themselves without animation—which though incorporeal could still fashion the world and everything contained in it. [136] But for the created world to be such, is it not necessary for its soul to have the same proportion as itself? For every body is composed in three dimensions, namely longitude, latitude, and depth: the kind of composition which in arithmetic denotes a cube. [137] Each such body contains six sides, eight corners, and twelve edges.[3] A cube is produced if any square is multiplied by its own root, as for example the square 4 multiplied by its root 2 gives the number 8.

[138] Thus God, in making the sensible world-body, first took one part away from the substance [1], then surrounded it with its double [2]; on these he superimposed a third, triple of the first but sesquialtera of the second [3]; then he added a fourth to the third, double the second and sesquitertia of the third [4]. And next he added a fifth to the fourth, thrice the third [9]. There was a sixth above the fifth, twice the fourth [8]. He then brought forth a seventh above the sixth, three times the fifth [27]. Between the limits of the duple and ternary quantity, there must fall relationships of sesquialtera, sesquitertia, and sesquioctave, as well as limmata and dieses. Surely, therefore, it is plain that the solid body of the celestial machine, both visible and audible, assembled in these proportioned quantities in which all harmony consists, and solemnly moved, must forthwith utter a sound of wondrous beauty; the more so since the soul which moves all things with a perfect intelligence must move the universal body of the world, not by chance or accident but in proper order.

[139] Is not our human soul,[4] which is never at rest, distressed when its own body hears dissonant motions either within itself or without, and pleased by consonant ones? How it shrinks to hear the thunder, when it sees the lightning flash! When a lyre-string is stretched more tightly than it should be, is it not for a while altogether dissonant, which the ear strongly rejects; and similarly, is not the whole tibia unmusical if an aperture is open or closed more than is proper? [140] What pleasure the soul, both human and animal, takes in the tibia, the lyre, or any other kind of music! The occasions abound wherein a skillful piper, giving forth his melodies, has assisted laborers; or when [music] has been the remedy for the bites of the constricting viper; [141] or has chosen to transmute those present from peace into fury, or contrarily

has mollified raging men by changing its tune.   [142] Are not warriors incensed to battle by the sound of trumpets?   [143] Nurses know how to quieten squalling infants with a song.   [144] But it also calms the distress of those who weep; so that no intelligent man can doubt that this soul of ours has a close kinship with harmony, that its whole structure is musical, and that its communion with the body cannot possibly exist without concord.   [145] Therefore if any of our humors exceeds its share,[5] or breaks out in a fever, the whole [organism] suffers, all the energies are affected by its aberration, and all health is undermined. Then, driven by the many excesses of the wanton humors, health takes leave of its beloved body, and, once dismissed, returns injured.

In this matter it is well to keep to Plato, that greatest and best of philosophers, and Aristoxenus following him, the disciple of Aristotle, and the other philosophers who compounded the soul from those numbers through which all harmony is formed.   [146] For if our own soul governs this transitory, mortal human body with such wisdom and care, and when one sees the eternal motion governed so regularly, how could you ever suppose that the sublime soul of the universe, animating the eternal heaven and every bodily thing, could move such vast bodies aimlessly, without order, without measure, without mind, and accomplish this without any knowledge? But it never errs, and always endures, else the monstrous noise of such bodies in collision would be heard and felt.

[147] After all, how would those blessed spirits,[6] whose nature is to rejoice and be glad, fare with such an unseemly noise in their bodies? And would not our souls be plucked clear away from their bodies, with which they usually enjoy harmonious song?   [148] And how, I conclude by asking, could the Greatest and Best, the kindly Governor, the first and unfailing mover of all things, permit this excellent machine to move toward a dreadful collision; and what pleasure would he take in useless and absurd silence? Far better that this beauteous world should resonate, as one may suppose, harmonically, and give forth many a song, so that nothing should be wanting to complete this creation, the most wonderful and orderly of all.

So now as the heavens move, the happy throngs of spirits and our own souls, too, on reaching that place must sing wonderfully harmonious songs.   [149] And the very soul of this great heavenly motion, varying its ideas, now more forcibly, now more subdued, will conform in its harmonious sound to those divine spirits, so that all the different consonant notes may combine with one another; but far from being content with a single sound, it continually varies the consonances, sounding now a fourth, now a fifth, now an octave, and other [intervals].[7]

[150] A single sphere does not always produce the same harmony, but manifold *phthongoi, limmata, dieses* and *commata;*[8] so that the blessed spirits must be imagined not only with the sound of their own sphere, but also with those situated nearby: now leading in song, now following, now pursuing, now accompanying, and playing in wonderful harmony in an ever more graceful game.

[151] Since any sphere gives forth songs not only consonant with its neighbors but also with those spirits presiding over it, it would be fitting, seeing that the heavenly harmony thrives on motion, that within each one the many movements of its parts should also make harmony, each to each.   [152] Certainly the deepest sound is the one made by the sphere which is moved by its own motion; and the highest consonance that which it causes by the motion we call diurnal.[9] The chief ones in between these are in the seven planets, when they traverse the spaces of heaven which are called the stations of aspects. Those little circles which the Greeks call epicycles are the semitones, and the dieses and commata the bodies of the stars when they traverse them in an epicycle. As their natural motion is slower, so they give forth lower sounds, and by diurnal motion they produce higher ones.[10]

[153] The harmony which the agitations of heaven produce by their natural motions[11] resembles the diatonic [genus], and the note which the sphere of Saturn makes compared to that of the sphere of Jupiter is like a twelfth: for the former completes its orbit in the natural course in about 30 years, Jupiter in about 12. The note of Jupiter's sphere compared to Mars' is as it were a double octave. The sphere of Mars compared to that of the Sun, Venus, and Mercury is like a twelfth; and from the sphere of the Sun and its followers to the Moon there sounds the great consonance of the triple octave.

[154] The modulation originating in the diurnal motion[12] appertains rather to the chromatic [genus]: for all the stars, both wandering and otherwise, turn their faces from hour to hour now to the East, now to the South, now to West or North; and revolving in the regions of the whole celestial wheel they proceed like the chromatic [genus].

[155] Most similar to the enharmonic [genus] is the sound which the spheres make together with one another,[13] or the association of the seven planets in related spaces, on reaching the harmonic limits known as stations of aspect. These are the opposition, sextile, square, and trine. In the same limits, a space diametrically related to another one sounds the octave, and likewise a sextile sounds an octave with a trine; then if the sextile is compared to the square, a fifth will sound forth; if the square to the trine, a fourth.

[156] Yet it is not a single mode that all the heavens sing, with the blessed spirits who inhabit them, but, diverse as they are, one which is

as varied as it is harmonious. Indeed, their songs sound grander and more beautiful by their very diversity. These spirits preside according to their ranks, and the whole force of harmony flows forth from them, owing to their correspondence with the spheres.

[157] These are in truth the spirits which Socrates in the *Republic* of Plato called Sirens, each sitting, as he said, upon one ring.[14] 'Siren' in fact means a singing god, but he actually meant to signify spirits, as ceaseless in their song as the spheres are in their motion.   [158] Our theologians more correctly call these spirits angels, and distinguish nine orders of them, giving to each order a name. Those who are called simply by the name of Angels, i.e., the messengers of the divine will upon earth, are said to inhabit the circle of the elements as guardians of mankind and witnesses of all we say and do.   [159] The second order is the one they call Archangels, presiding over the Moon's orb: their duty is to be the special messengers of the divine will outside of nature's bounds.   [160] The third order is named Virtues: through them God displays to the world his great and portentous miracles, which are the universal signs of the most important things to come: their seat is the heaven of Mercury.   [161] The fourth order is called Powers: their seat is the heaven of Venus, and they restrain the evil and unclean spirits to prevent them from doing harm in the world of generation, particularly to human nature, for which they are always eager with their snares.   [162] The fifth is the order known as Principalities: their seat is the sphere of the Sun; they preside over the hosts of angels whose duty is either to withdraw from or to attend on Almighty God, and also to assist the rulers of the world when their monarchies prevail in good government.   [163] The sixth is the order called Dominations: these inhabit the sphere of Mars, and are the captains of the hosts of angel warriors—hosts which obey their captains as soldiers their generals—restraining those who fight unjustly on earth and helping just men.   [164] The seventh is the order known as Thrones— the seats of the Thrones are situated in the orb of Jupiter—and through them God proclaims his laws, decrees, and judgments.   [165] The eighth order is called Cherubim, the interpretation of this name [requiring] vast knowledge: they dwell in the sphere of Saturn, and their hosts are replete with all possible wisdom on account of their proximity to the highest wisdom.   [166] The ninth order is named Seraphim, i.e., the burning ones: they exceed all the other orders of angels in wisdom and power and bliss, and also in joy; they are perfect in every virtue above all created things, and are called 'burning' because, having a fuller participation in the divine light, they are vouchsafed a more intense flame of love and joy and perfection; they are also said to veil the Divine Face (for the spirits and souls beneath them could otherwise

not bear it), as if defending it. Their seat is the eighth sphere.   [167] 'Angel' is the name of their function, not of their nature, for they are speedy and altogether tireless in their journeying, having no need to perceive time, and hence by pictorial license they are imagined winged.

No spirit is permitted to dwell in this sphere. It is the realm of eternal heaven, inaccessible to any creature of a lower one. This circle truly includes the melody of all those beneath it, utterly excelling all harmony. Certain of the poets[15] therefore placed there the muse Urania, the one who embraced and surpassed every charm of the nine Muses, said to be the daughters of Jupiter and Memory.   [168] If indeed the theologians assert that the tenth heaven is the seat of God Most High, they say that the ninth has no occupant: for there is not a thing beneath God Most High which either deserves to or can stand before his divine and ineffable majesty, nor can such overwhelming energy be either suffered, nor the idea of it conceived, since everything in comparison to it is like the blinking of the eye in the light.

[169] Here, dear George,[16] you have the celestial harmony in brief, and would that I were equal to it in eloquence or in fact: certainly it is too great for tongue or mind to attain, yet I believe there is nothing greater, more serious or more worthy of the understanding. But the human ear shrinks from hearing the divine voices. If, however, the mind should turn toward the heavens and toward God their director, it will deem nothing more pleasing to the intellect and certainly nothing more desirable, and the same mind will long to be loosed from these bonds and to join with those most holy spirits from whom it originally issued forth, and to whom it will certainly not be inferior if its merit should suffice.

25

◆

Isaac ben Haim
c. 1467—after 1518

Born in the flourishing Jewish community of Jativa, Valencia, Isaac ben Haim ben Abraham ha-Cohen was expelled from Spain in the persecution of 1492. In his book *Etz Haim*, ("The Tree of Life") he tells of his journey to Constantinople and of constant wanderings around Italy, working as a rabbi in Bologna, Rome, and other centers. No trace of him appears after 1518.

The book contains a treatise on the art of poetry, the last witness to a long period of creativity among the Jews of Spain. In the course of his essay, Isaac ben Haim touches several times on music, and especially on the role of instruments. Although mentioned many times in the Bible, these had been excluded from the synagogue after the destruction of the Second Temple in 72 C.E. What was their purpose, and what is the purpose of music? The following extract attempts to answer these questions.

Primary to Isaac's thought is the assumption of three musics: those, in ascending order, of instruments, of voices, and of the celestial spheres. In his discussion of the lowest type, Isaac ben Haim leans on a theory well known in the Arab world (see nos. 14, 18 and 19): that of the four humors as corresponding to the four strings of the lute. Each string allegedly excites one of the humors and its corresponding emotion. But Isaac attaches no intrinsic value to such music, or even to vocal music except insofar as the latter turns the singer or listener toward God. More important is the fact that music can induce the state called in this translation "reflection": a condition of the intellect, which, far from being identical to the brain and its thoughts, is the organ through which man can know the divine. We might paraphrase his often convoluted arguments by saying that music can lead one to a state of contemplation, free from all sense-impressions.

From the point of view of God, Isaac tells us, earthly music is not a source of pleasure. God enjoys the simple purity of the planetary music, but he allows us our vocal and instrumental musics in order to lead us in devotion to him. All of this is explained in a style common to all the religions of the book: phrases from Hebrew Scriptures, Bible, or Quran are used as support for the argument, sometimes wildly out of context, at other times adorned with fantastic etymologies. The Judaic literature, however, is notably poor in discussions of the harmony of the spheres, and this example, initially translated into French by Amnon Shiloah from the sole manuscript (Bodleian Ms. Heb. f. 16, Neubauer no. 2770), is a precious witness to the currency of such ideas among the cultivated Jews of Spain.

Source: Amnon Shiloah, "La Musique entre le divin et le terrestre," in *Anuario Musical,* vol. XXXVIII (1983), pp. 6–12. Translated by the Editor. Used by kind permission of Professor Shiloah.

Music between Heaven and Earth

1. Know that it was in view of the spiritual preparation of the Levites for speculation by means of canticles and hymns that they were enjoined to sing and play on musical instruments well-regulated melodies; so that these things should be manifested before the Creator, great be He, with comprehension and reflection; for it is under the aspect of comprehension and reflection that man honors the Lord, and not by matter which is contrary to His essence, great be He.

2. Since the instruments of music are tools of matter, not of spirit, their effects are not enjoyed by the Creator, great be He.   3. However, He decreed that they should be used to prepare man for intellectual things, and to help the poet-musician[1] the better to render the

elements of his activities according to order and reflection, and the better to comprehend the course of music when it moves upward or downward, or when it develops on a high or low note.   4. When the musician reflects on these matters, the words sung will not be uttered without intention or haphazardly.

5. The instruments of music also provoke contrary effects in the listener: sometimes joy and sometimes anguish, sometimes courage and sometimes feebleness and fear.   6. The reason is that the sounds which engender these different effects are connected to strings that correspond to the four elements.   7. To each element corresponds a sound which gives birth in the listener's heart to an appropriate effect.   8. Thus the sound which corresponds to the element of earth refers to black bile and engenders anguish; the sound corresponding to water relates to phlegm and engenders fear; the sound corresponding to air [relates to blood and][2] engenders [joy] in the listener's heart; and the sound corresponding [to fire relates to] yellow bile [and engenders] courage.   9. For these contrary sounds are produced by the instruments and the music, causing the expression appropriate to a single element to arise.

Moses said to Joshua, "It is not the noise of a song of victory, nor the noise of a song of defeat, but the noise of singing redoubled that I hear" [Exodus 32:18].   10. This remark of Moses to Joshua signifies that Moses has seen in his spirit, through the medium of the sense of hearing, the golden calf, the dances, and "the noise of the people in tumult" [Exodus 32:17]: he meant by the word be-reoh (tumult) his abomination.[3]   11. In point of fact, idolatry is called the sacrifice of the dead, which is an abomination without place, that is to say, a spiritual one, according to the sayings of the wise: the spirit of impurity, and not concrete and material impurity.   12. And Joshua said to Moses, "There is a noise of war in the camp" [Exodus 32:17], and Moses replied to him: it is not a noise which indicates arousal to battle, nor a noise attesting to a cry for pity; it is a noise of singing which resounds.   13. By "a noise of singing redoubled," as I read and understand it, he means a voice bearing witness to joy and disquietude simultaneously. 14. In fact, the multitude which Moses led out of Egypt, without the authorization of the Creator, great be He, was joyful because it had returned to the way of evil, while those faithful to God— Aaron, his sons, and the sons of Israel—were disquieted and afflicted by Moses' delay, by the profanation of the name of God, and by the way in which the calf had been made so quickly.   15. Because there must have been two opposing groups, the musicians who saw them could not direct their melodies to undiluted joy, for fear of changing their humor and of corrupting by thought the melody of joy, transforming it to anguish.[4]

16. Thus, because melody and music produce contrary effects, it is not possible to say of God, great be He, who is simple and immutable, that He enjoys music, or that it causes in Him joy, courage, pain or fear.   17. If nonetheless He has willed these, it is in order that the song attached to them should stimulate thought and reflection, to which one attains by the medium of the intellect and its action; it is this that signifies taking pleasure.   18. In fact, the intellect resembles Him, great be He; it is in His image and likeness, and placed in man.

19. It is not a valid objection to point out that because the sounds of melodies and of music produce contrary effects, namely the four opposite actions, the Creator, great be He, has no need of them. If we admit that joy and courage stimulate the intellect and prepare the heart of man for understanding (as it is said: "Now, as soon as the minstrel played music, the hand of Iahve was upon Elisha" [II Kings 3:15]), what interest would one have in pain and fear, which are contrary to understanding?   20. The response to this objection is that all the actions of music are approved by the Creator, great be He. Joy and courage are intended for glorifying Him in jubilation and well-being; pain and fear are intended for the prayer that should be addressed to Him, great be He, when humility and fear awaken in our hearts because of our sins. 21. For in that case it is the intellect that awakens disquietude in the interests of the soul, and it is not fear that simply stimulates the intellect.   22. For that is not necessary for him who is befouled by the stain of his sins; he does not speak with wisdom and his words are not reasonable; "a broken and contrite heart, O Elohim, you do not despise" [Psalm 57:19].   23. This is where the fear engendered by music is good, in that it prepares the heart for humility; the intellect loves it, and God also, great be He, appreciates it.

24. However, it is said that joy and courage stimulate reason and bring reflection within reach.   25. Thus in the case of anguish, we find that the Creator, great be He, has recourse to its true sense, as it has been said: "My heart will weep in secret because of (your) pride, and my exile will weep, it will run with tears because the flock of Iahve is led away captive" [Jeremiah 13:17].   26. The words "your pride" signify that the object of Israel's pride has been taken from them and given to the nations.[5]

27. The text just cited says the same as we find in *Berakhot*: "My son, what voice do you hear? I hear, said he, a voice that coos like a dove, saying: 'Woe to me for having destroyed my house, burned my Temple, and scattered my sons among the nations; woe to the Father who has

exiled his children, and woe to the children who are banished from their father's table.'   28. He said to him: Elijah! By your life and by your head, it is not only now that the voice sounds thus, but also every time that the people of Israel assembles in the synagogues and the religious academies, and responds: 'Amen, glorious be His Name.'   29. Then the Blessed-be-He shakes his head and says: 'Happy is the king who is thus glorified in his house; woe to the father who has exiled his sons, and woe to the children who are banished from their father's table.'"6   30. The exaltation of the Creator, great be He, which is brought about by the proclamation of "Amen, glorious be His Name," and which surpasses that of all other praises together, indicates that this response of Israel strongly arouses His compassion, for in so doing they neglect their own interest and honor.

31. In fact, the celebrant intones "Let His Name be exalted and sanctified," etc., "Let His salvation come quickly and the advent of His Messiah be nigh," etc. Now, Israel does not think of imploring what might suit her own aspirations and honor, namely the speedy coming of the Messiah, but thinks rather of His Glory, great be He, which is profaned by the Nations, and of His Name which cannot be complete until the remembrance of Amalek be wiped out, as it is said: "Because one hand is against the throne of Iah, war between Iahve and Amalek from generation to generation" [Exodus 17:16].   31a. Then, when the enemy Amalek is utterly wiped out and his cities destroyed for ever, after the divine decree, the Eternal in His complete Name will reign for ever, and His complete throne will be set up for judgment.

32. But because the children of Israel say: "Amen, let His Name be exalted," thus seeking to honor Him, the Lord, great be He, shakes His head and is filled with compassion because of this great praise and adoration that we pay Him without seeking to be rewarded.   33. It is thus manifest that the Creator, great be He, uses disquietude to express His mercy and to lament over our fate; for our part, we use it to adore Him with humility and to express our anguish over the pro- fanation of His honor, although we praise and glorify Him with this sentiment of anguish.

34. It is in this sense that the Levites sometimes have need of this sentiment; by the intermediary of their intellect, they awaken disqui- etude through their musical presentation in His honor, great be He, saying: "By the waters of Babylon" [Psalm 137:1], and "O God! the heathen are come into Thy inheritance" [Psalm 79.1]. Now the music associated with these psalms engenders disquietude and fear in the heart. 35. When they prepare their intellect for worship by means of specu- lation, they give birth to joy after the model of the superior world, where power and jubilation reign.

36. The Creator, great be He, has no need for musical instruments that are material and changeable, since He has at His disposal the simple and pleasant sounds that are produced by simple and immutable spheres. 37. It is in reference to these that it is said: "The heavens are telling the glory of God" [Psalm 19:1], (that is,) by means of the pleasant expression of their movement. "And the firmament shows forth His handiwork," (that is,) by the alternation of sounds. 38. In fact, since the number of spheres is seven, as the treatise *Hagigah* proposes, and since one is greater than the other and turns around it, it follows that the sound produced by the greater will be different from the sound produced by the smaller. 39. Since also the spheres are simple, the sounds that they engender are simple and distinct; they are all unique of their kind and exemplars of their type. 40. Their case is similar to that of "Wisdom who has set up her seven pillars" [Proverbs 9:1], symbolizing the seven sciences[7] whose objects of speculation differ one from another. Nevertheless, they all depend on speculation and rest on the action of the simple intellect.

41. Thus the melodious sounds engendered by the revolutions of the seven simple spheres awaken speculation and intellect, and not a material excitement, from which God preserve us. 42. Consequently one cannot attribute to Him, great be He, the taking of pleasure in material musical instruments. However, recourse to them has the purpose of exhorting the intellect of man, and preparing it for the adoration of its Creator, for by their mediation the singers perform their hymns and canticles in tune and with intelligence, and not haphazardly and blindly. 43. And King David, peace be upon him, revealed in his song the meaning of the music of the spheres, saying: "One day speaks a word about it to another" [Psalm 19:3]. 44. He meant by this that the particular sound of the diurnal movement is not the same as that of the nocturnal movement (but they remain identical in themselves from day to day).[8] 45. If nevertheless our hymns change each day, it is because they (the spheres) do not change, whereas our hymns change each time because we are always changing. 46. Thus each time the worshiper changes, the praises change accordingly. 47. Given that God renews creation each and every day, we proceed according to what is said: "One day to another" following speaks a different "word"; but it is not a matter of different knowledge and reflection. 48. It is the same when "One night gives" a different "knowledge to another." It is a matter of concept and not of word, for the word (the abstraction) and the actualization depend on the day, while the expression of the heart and the concept are necessarily expressed at night during the hours of sleep, when one is free from all preoccupations.[9]

49. Truly all things in this lower world are to be found in the

celestial world.   50. Nevertheless, all that is above is simple, while all that is here below is dependent on the composition of humors and elements, as we find in the commentary of the treatise *Hagigah*[10] on the verse: "I am the holy one in your midst and I will not come into a city" [Hosea 11:9].   51. The commentary on it is thus: because I am Holy in your midst in the diaspora, I will not come into the city of Jerusalem on high until I have restored the Jerusalem here below.   52. Moreover, all the speculations and hymns of praise here below are found above, but they are simple.   53. It is not the case with the reflection aroused by the playing of instruments, which changes the mode of these expressions and activations from one day to another and from one night to another; the consequences of its actualization are equal by day and by night.

54. King David, salvation be upon him, indicates to this effect that in celestial music are also found the four opposing actions which awaken the material humors that we have mentioned in connection with material music, that is, music played by corporeal instruments.   55. However, the effects engendered by spiritual music are four simple ideas, to which he alludes in saying "One day speaks a word about it to another."   56. He means by this that the illumination of the intellect which corresponds to "day" is the cause of joy and courage; "speaks about it" corresponds to joy and happiness, with reference to the Aramaic translation of "Exult and cry out with joy;" "a word" corresponds to courage, making *omer* (word) derive from *vayitmarmar* (it is irritated with).[11]   57. "And night to night" indicates the darkening of the intellect which corresponds to the obscurity of night, because *yehaveh* (gives) corresponds to black bile and to pain.   58. In fact, the Aramaic word *haveh* is the translation of *nahash* (serpent), and the serpent represents the earth and black bile which is at the origin of anguish.   59. And "knowledge" corresponds to pity, because the seat of knowledge is the brain, where the element of water is found in greater quantity than in the other members of the body; while water is the origin of phlegm and of nobility.

60. In addition, he (David) makes the division between day and night correspond to the divine attributes of mercy and justice.   61. The contrary effects of melodies correspond in their simple aspect to these attributes.   62. When one glorifies Him with intellectual praises, exempt from matter, joy and courage are drawn up on the side of the attribute of mercy, for it is said: "Strength and joy are in His place" [I Chronicles 16:27].   63. As for disquietude and feebleness, they will be on the side of the attribute of justice, as it is said: "See, the people of Ariel have cried out in the street, the messengers of peace

weep bitterly" [Isaiah 33:7], because of the verdict concerning the destruction of the Temple.

64. He also says that celestial music is distinguished by an even greater difference, as it is said that "There is neither speech nor language" [Psalm 19:4]. That means that in this celestial music there is nothing else, apart from what we have indicated in connection with the four contrary actions; and (as the verse continues) "their voice is not heard" signifies that their action never suffers any transformation, that is, it is not corrupted, contrarily to terrestrial music which is subject to corruption.   65. In fact, it may happen that a string breaks or becomes soft or hard, stretches or is too tight. In consequence, the listener loses the thread of his listening, and finds himself obliged to prepare at every moment for it to fail.   66. This is why (David) says that on high there are "no speech, no words, and their voice is not heard"; everything is always equal, without any alteration in the expression of the truth, and no vain action takes place.

67. In the circumstances, given that there is such a difference between celestial and terrestrial music, far be it from us to attribute to Him, great be He, any joy in listening to the material music here below. God permits it in order to stimulate the preparation of the intellect of the poet-musician, so that he may better depict the things with which he wishes to praise his Creator by means of his intellect. 68. For the intellect is in His image and after His likeness; for there is no difference between the intellect of man when it acts, and the intellect of the angel.

III

RENAISSANCE

26

$\blacklozenge$

Marsilio Ficino
1433–1499

It is arguable that the crucial date for the beginning of the Renaissance was 1439, when George Gemistos, called "Plethon," visited Florence as an envoy from the Eastern Orthodox Church, and there met Cosimo de' Medici. Plethon was actually a pagan, with the secret ambition of restoring a Neoplatonic government and the cult of the Olympians. But during this embassy it was sufficient for him to implant in the Florentine intelligence the vision of a perennial wisdom, faithfully transmitted from distant ages by the *prisci theologi* (ancient theologians) Hermes Trismegistus, Zoroaster, Orpheus, Pythagoras, and Plato, as well as by Moses and Jesus Christ.

Thus the rich banker Cosimo was inspired, first, to collect all the manuscripts of Greek philosophy that his wide network of business contacts could bring him; second, to commission their translation from Greek into Latin, beginning with the *Corpus Hermeticum* attributed to Hermes Trismegistus himself. The translator was Marsilio Ficino, who completed his work on the *Hermetica* shortly before Cosimo's death in 1464, then proceeded to translate the complete works of Plato. Ficino had been appointed at a young age as head of a revived Platonic Academy, and his own paganism, though ultimately controlled by Christian

faith and practice, extended to magical invocations of the planets, singing the Orphic Hymns (see no. 7) to his own accompaniment on the
lira da braccio.

Music, Mysticism and Magic presented the chapters from Ficino's *De
vita coelitus comparanda* on the magical properties of music which were
the philosophical basis for his rites. Here is a lesser-known letter that
touches instead on the themes of the present collection, especially the
metaphysical reasons for the power and quality of the musical intervals,
and the relation of music to astrology.[1]

Source: Marsilio Ficino, Epistola ad Domenico Benivieni, in *Supplementum
Ficinianum*, edited by Paul Otto Kristeller (Florence: Olschki, 1937).
Translated by Arthur Farndell. Used by kind permission of Mr. Farndell.

A Letter concerning Music

Marsilio Ficino to Domenico Benivieni,[2] illustrious philosopher and
master musician: greetings.

Plato thinks that music is nothing other than harmony of mind:[3]
natural, in so far as its powers are consonant with the powers of mind,
and acquired, in so far as its motions are consonant with the motions
of mind. He thinks that its reflection is that music which modulates
notes and sounds to charm our ears. He believes that the Muse Urania
presides over the former and Polyhymnia over the latter.[4] Hermes
Trismegistus says that both have been assigned to us by God, so that
through the former we may continually imitate God himself in our
reflections and dispositions and, through the latter, we may regularly
honor the name of God in hymns and sounds.[5] Pythagoras was accustomed to call him a master musician who had attained both, and he and
his followers have acknowledged this in both word and deed.

So greetings, Domenico, you master musician.[6] As for your long-
standing questions to us on some of the principles of music, you really
know the answers yourself. Nevertheless, since you so wish, accept a
brief statement of them again in our letter.

RATIOS

As you are aware, musicians consider the principal ratio to be that of
2:1. This produces the diapason, the perfect consonance of the octave,

the consonance which poets entitle Calliope. The second ratio is considered to be that of 1½:1. This produces the diapente, the almost perfect harmony of the fifth, the number to which the lyric poet ascribes the nectar of Venus. The third ratio is that of 1¼:1. From this is born the gentle harmony of the third, recalling Cupid and Adonis. The fourth ratio is that of 1⅓:1, by means of which the fourth now reverberates, as if midway between a consonant and a dissonant, blending something of Mars with something of Venus. In particular, the third, the fifth and the eighth, which are more pleasing than the rest, remind us of the three Graces. The ratios which unfold at will beyond the double can be reduced to the likeness of those we have mentioned. Here I would just mention that the ratio of 1⅛ :1 produces a tone, while a smaller ratio produces a semitone.

Advancing step by step on this principle, the notes proceed from the low one, which Orpheus calls hypate, up to the high one, which he calls neate, by way of the intermediate ones, which he calls dorians. To begin with, the low note, because of the very slowness of the motion in which it is engaged, seems to stand still. The second note, however, quite falls away from the first and is thus dissonant deep within. But the third note, regaining a measure of life, seems to rise and recover consonance. The fourth note falls away from the third and for that reason is now somewhat dissonant; yet it is not so dissonant as the second, for it is tempered by the charming approach of the subsequent fifth and simultaneously softened by the gentleness of the preceding third. Then after the fall of the fourth the fifth now arises; it rises, mark you, in greater perfection than the third, for it is the culmination of the rising movement; while the notes that follow the fifth are held by the followers of Pythagoras not so much to rise as to return toward the earlier ones. Thus the sixth, being composed of the doubled third, seems to return to it and accords very well with its yielding gentleness. Next, the seventh note unhappily returns, or rather slips back, to the second and follows its dissonance. Finally, the eighth is happily restored to the first, and by this restoration it completes the octave together with the repetition of the first and it also completes the chorus of the nine Muses, pleasingly ordered in four stages, as it were: the still state, the fall, the arising, and the return.

The followers of Pythagoras think that a chorus of this kind is round, yet ovate rather than spherical.[7] Within it, as if uniting the breadth of the first note to itself through the more pointed end, the eighth now produces a single note from itself and the first. And just as the eye perceives ovate roundness as a single shape though it be broader at one end than at the other, so the hearing takes in as one the note which resounds from the low one and the eighth, and which rises sweetly and gradually like a pyramid from a generous bass to a high treble.

We believe this is why Nature has bestowed this sort of shape upon the instrument of hearing and a similar shape upon the instrument of speech, and why it likewise has done all it could to bestow a similar shape upon the instruments of music. There is no doubt that the closer they are in shape to an oval or pyramid the more harmonious they are.

THE GENERAL CAUSES OF HARMONY

Next, we must ask why all musicians make especial use of those ratios that we have described above. They acknowledge them in different ways, on different occasions—in the size of pipes, in the mass or weight of other instruments, in the tension or length of strings, and finally in the vehemence of action and the speed of motion, as well as in their opposites. The followers of Pythagoras and Plato consider the One itself the most perfect and the most pleasing of all. Next in importance they place steadfastness in the One; then, thirdly, actual restoration to the One; and finally, an easy return to the One. At the other extreme, they consider disconnected multiplicity the least perfect and most distressing; second to this is movement towards multiplicity, I mean a multiplicity which finds it difficult to return to the One.

Now that we have laid these foundations, let us build what I may call the house of music. If you stretch two equal strings on a lyre absolutely equally, you will say that they are at one, and hence you will hear unison. But if one of the strings be stretched more than the other, there will now be a departure from the one. For example, if you add a tenth part more, this kind of departure from the one occurs by means of that part which can restore the wholeness of the one only with considerable difficulty, because it needs the addition of nine parts for full restoration. Consequently, the ears are brutally offended in that sound because of its excessive distance from the one. And if you add a ninth part rather than a tenth, this is also very far distant, for it needs eight parts to effect the return. The principle will be much the same if you add an eighth part instead, or a seventh or a sixth or a fifth, since fractions of that kind still have difficult access to the whole itself.

However, if you proceed to stretch one of the two strings above the other by a fourth part, this is the point where the ear is in some way delighted, since easy access to the one appears here—the addition of three parts being sufficient for this fourth to complete the whole. Now three parts are easily added to one to achieve unity, for the number three is considered by many to be indivisible, all-embracing, and the most perfect of all, in which particulars it corresponds to unity. Indeed, the ratio of $1\frac{1}{4}$ to 1 produces the melody of the third note. Moreover, if you proceed from the outset to increase the tension by a third part, the harmony of the fourth will delight you, for a third part easily recreates unity itself as a whole since it completes it by the addition of two parts. Now two are easily added to one, and easily come to rest in one, for duality is the first departure from one. However, this superior harmony produced by the third part will charm you the more fully because duality is reduced to unity purely on the basis of three. Similarly, if from the outset you correctly stretch one of the two strings by a half more than the other, this ratio of 1½ to 1 certainly produces the harmony of the fifth and gives greater delight because from there the return to one is very short and rapid. For when one part has been added to it, it becomes whole, since a whole is made from a half and a half.

Now one is easily added to one, and through them both there is a blending into one. But if after stretching one of the strings, you now increase the tension on the other by exactly the same amount, you certainly do not take your stand further out in the movement away from one, as in the previous examples, but you instantly recreate that full unity which had to some degree been dissolved. At this point, therefore, the ratio of 2 to 1 now fills the ears with wonderful pleasure by means of the octave, the most perfect harmony of all. One must remember that in general the hearing is in all places soothed by unity and always offended by duality, as if by division. And so, whenever it most clearly perceives two notes as two, then is it most offended. But when it perceives this less, then less offense arises; and when least, least. Hearing indeed longs for unity, since it is one itself and also arises from one; but it desires a unity perfectly blended from the many and held together in the same relation as that by which it naturally effects a certain unity from the many. Finally, since hearing itself consists of a multitude of natural parts which blend fully together into one form, it readily welcomes a number of notes when they are brought perfectly into one note and into harmony. This occurs particularly when one of the two notes absorbs the other into itself or makes it continuous with itself, and they are able to achieve this solely by virtue of those ratios which we have been discussing.

THE PHYSICAL CAUSES OF HARMONY

Almost all philosophers consider that pleasure arises from a correspondence of object with sense. For the moment, I make but passing mention of the fact that the followers of Plato, in their scheme of the senses, match sight with fire, hearing with air, smell with a vapor blended from air and water, taste with water, and touch with earth; and they think that wondrous pleasure appears when the proportions of something, perceptible through its qualities and degrees, match up and harmonize at every point with the proportions which constitute the nature of sense and spirit. The nature of pleasure itself is a question which we have dealt with at great length in our book on 'Pleasure.' And so—not to digress further from our purpose—the followers of Plato locate in the constitution of hearing one degree of earth; also, one of water, but with a third more; one and a half degrees of fire; and lastly, two of air. Hence they consider that the power to arise most strongly is that of the ratios of $1\frac{1}{3}$ to 1, $1\frac{1}{2}$ to 1 and 2 to 1.

THE ASTRONOMICAL CAUSES OF HARMONY

There are those who trace such things back to a loftier plane and, in the manner of the Pythagoreans who affirm a celestial harmony, derive the principle of harmony from some celestial power or some celestial correspondence. And while alluding to their view that the extent, or depth, of the celestial spheres as well as their intervals and the rapidity and slowness of their movements are determined by those ratios which we have described, I certainly cannot pass over in silence the fact that if you start out from the very head of the twelve celestial signs and then wish to move through those that follow, you will find that the second sign falls away from the first in some way.[8] And just as with notes we find the second dissonant from the first, so here we find that the second sign is in some way dissonant from the first. But then the third sign, as though it were the model for the third note, looks upon the first constellation with that friendly aspect which astronomers call sextile. The fourth sign, although dissonant, is but moderately so, as they say, and in the view of musicians this is the nature of the fourth note. Then the fifth constellation looks benevolently upon the first with a very friendly and agreeable aspect, thereby providing a model for the fifth note in music. Astronomers give the name trine to an aspect of this kind and consider it most beneficial.

But what shall we say of the sixth constellation by which is indicated the soft and, so to speak, frail consonance of the sixth note? Although astrologers in judging a natal star consider this frailty to be undoubtedly bad, the ancient theologians think it useful, since man himself is in truth mind itself, while the body is the prison both of mind and of man, and the frailty of a prison will be useful to anyone shut in by the prison. After this, the seventh constellation, which they call angular, being very vigorous in its discord, which is set against the first constellation, and in its open hostility seems to prefigure the seventh note of music which with its vigorous, even violent, tone is now most clearly discordant from the first note. There follows the eighth constellation which, though appearing unfavorable to the common mind because it is allotted to death by astrologers, is nonetheless most fortunate for the celestial mind in the view of the ancient theologians, since it finally unbars for it the earthly prison, freeing it from elemental dissonance and restoring it to celestial consonance. It is not without good reason, therefore, that it denotes the absolute consonance of the eighth note, the consonance that returns to the beginning. If someone now asks about the ninth sign, let him understand that it turns out to be related to the first as the fifth is, and that it indeed looks at the first with a trine and kindly aspect; in the view of astronomers wisdom and the god Pallas are expressed by it, and in the view of musicians, the nectarial Venus of the fifth note. Now what of the tenth constellation? It displays ambition, which astrologers see as the foundation of human discord, musicians as the discord—human and middling, as it were—of the fourth note. Then the eleventh, the sign of human friendship, demonstrates the friendly melody of the third note. Lastly the twelfth, allotted to hidden enemies and to prison, expresses the dissonant falling away of the second note from the first.

27

◆

Ramis de Pareja
c. 1440–1491 or later

This Spanish theorist who passed his later life in Bologna and Rome was the first to dissent from the authorities, hitherto unquestioned, of Boethius and Guido d'Arezzo. Ramis (or Ramos) was an aggressively original and practically-minded thinker whose work could not be ignored by the learned world, however much the latter was irked by it. While in matters of practical music he was completely emancipated from dependence on previous theorists (whose work he knew thoroughly, if only to refute it), Ramis did not cut himself off from the more arcane speculative tradition, as one can see here from his uncritical transmission of it.

The correspondences that Ramis draws up between Muses, modes, tones, and planets are derived quite logically by combining Martianus Capella's Muse-sphere correspondences with Boethius' planet-tone scheme. The combination was to become a popular one; both Gafori and Cornelius Agrippa (see *MM&M*) were to adopt it. Ramis' effort is on the same lines as Anselmi's, only the divine beings here are the nine Muses of revived paganism, not their christianized equivalents, the nine orders of angels.

Source: Ramis de Pareja, *Musica Practica* (Bologna, 1482, facsimile reprint Bologna: Forni, 1969); edition by Johannes Wolf (Leipzig: Breitkopf & Härtel, 1901), pp. 56–60. Translated by the Editor.

The Conformity in Mode of Heavenly, Human, and Instrumental Music

Chapter 3. Instrumental music is closely conformable and similar to the human and heavenly [*mundana*] sorts. In the former case, the four modes affect the four humors of man. Thus modes I and II are ruled by the phlegmatic humor, III and IV by the choleric, V and VI by the sanguine, and VII and VIII by the melancholic.[1] This first mode moves the phlegm when one awakens from sleep, and thus its form is depicted with a crystalline color; for the crystalline heaven is held to be made from the waters, the element which creates phlegm. We call it crystalline, and not the color of other waters, because not all phlegmatic persons benefit through contemplating pleasant sounds. But gifted people with minds clear as crystal can be burdened with phlegm through a little food, drink, or other extrinsic accidents; and the sleepiness, idleness, or sadness thus induced may be alleviated by playing in the first mode. Its plagal form, however, behaves in a contrary fashion. For the first tone, as Lodovicus Sancii says, is mobile and adaptable to every emotion, hence most desirable for songs.

The second, however, is serious and doleful, most appropriate for sorrowful and grievous songs such as the *Threni* or Lamentations of Jeremiah. We have no doubt that sorrow leads to sleepiness through the motion of the phlegm, hence 'their eyes were heavy from sorrow.' The Pythagoreans had the custom of employing the Hypodorian when about to resolve the cares of the day in sleep, that a calm and quiet slumber might restore them; but on rising they would banish the stupor and confusion of sleep with the Dorian, knowing for a certainty (as Boethius says) that the whole compound of our soul and body is knit by a musical link, and that the pulse of the heart moves according to the emotions felt in the body.

Now the third mode moves the choler by exciting and provoking to anger. Therefore it is depicted with an igneous color, for it is harsh and vehement, having in its course stronger leaps, as we will explain in the proper place. This mode is most suitable for arrogant, irascible, and haughty men, harsh and cruel; and they rejoice in it. Of this Boethius says that rougher peoples enjoy the harsher modes of the Thracians, but gentler peoples the more moderate modes. He tells of the Taorminian youth incensed by this mode who was about to break down the doors of a prostitute's house, but was calmed by the Hypodorian.[2]

The plagal or fourth mode is said to be attractive, chattering, and most suitable for flatterers who charm men with sweet words when they are present, yet stab them in their absence. So this mode is luxuriant yet without beauty, even when used in combination.

The third authentic mode [Mode V] rules the blood. Therefore Saint Augustine calls it delectable, modest, and joyful, gladdening those who are sad or anxious, recalling those who have fallen or despaired. It is depicted with a sanguine color. Boethius says of it that the Lydians, the most happy and joyous of men, love this mode, especially their women-folk. The Russians are said to descend from them, great lovers of choruses and dancing.

The plagal version [Mode VI] is not so good: it is a weeping mode, fit for those who are easily provoked to tears, because it has the most coadunate notes, as stated in the same place.

The fourth authentic mode [Mode VII] is partly lascivious and partly joyful, having some energetic leaps and representing the behavior of youth. Therefore it rules melancholy, sometimes resisting it, sometimes increasing it, according to the mixture that is made with other modes, as we will explain shortly. On this account it is shown with a dirty, semicrystalline color. Its plagal is sweet and slow, or fretful, according to the mode chosen, as Ambrose says.

The seventh and eighth modes thus activate the melancholy humor by their melody, restoring sad and lethargic people to normal: the authentic by arousing them, the plagal by gladdening them.

From this can be seen the correspondence between instrumental and human music. A familiarity with Cicero[3] shows, moreover, that the heavenly music also conforms most closely to the instrumental. For the order of the planets is disposed from the *proslambanomenos* to the *mese*, such that the moon is *proslambanomenos*, Mercury *hypate hypaton*, Venus *parhypate hypaton*, the sun *lichanos hypaton*, Mars *hypate meson*, Jupiter *parhypate meson*, Saturn *lichanos meson*, and the starry heaven *mese*.

If the moon is indeed *proslambanomenos* and the sun *lichanos hypaton*, then evidently these two planets a fourth apart will have their melodies arranged with the moon in the Hypodorian, the sun in the Dorian mode. Obviously the moon increases the phlegmatic and humid qualities in man, while the sun dries up the same. So these two planets, the chief and brightest of all, rule the first and second modes, i.e., the first authentic and plagal forms. The Dorian, the first authentic mode, is justly compared to the sun, because it holds first place among all the modes, as the sun among planets. All exhalations from the earth, all vapors of the sea are raised by the solar rays, by means of which meteorological phenomena are created. Thus there is a clear collusion between sun and moon: one illuminates the night, the other drives the night away; the Hypodorian induces sleep, the Dorian dispels it. Together they make, by their position and form, the consonance of a fourth.

Mercury rules the Hypophrygian: the mode of flatterers, who praise equally the evildoers and those who are wise and upright; they are easily swayed to either side, to lamentation or rejoicing, to excitement or sedation. Such is the nature of Mercury, who with good people is good, and with bad people is worse.

Mars holds to the Phrygian, an altogether choleric and irascible mode; he connives to destroy everything good in the world with his anger. Mercury, conjunct or in a certain aspect with him, becomes as bad as Mars: but the one wounds with the sword, the other with the tongue.

The Hypolydian is attributed to Venus who is beneficent, yet also feminine, because she sometimes arouses holy tears.

The Lydian is properly compared to Jupiter, the greater benefic, who creates sanguine and friendly men, mellow and cheerful, since it always denotes joy. In relation to Venus it sounds a fourth, and these two agree in their benevolence, differing only in their notes: for the lower is not so sweet or agreeable as the upper one.

The Mixolydian is attributed to Saturn because it tends to melancholy.

The Hypermixolydian is reckoned the Castalian source of all, because it is given to the starry heaven or firmament. More than any other, this mode has an innate beauty and loveliness; it is free from all qualities and suitable for every use. Guide and Odo say that it represents glory, and do not stoop much when they explain that for seven different ages we toil, longing for the octave to bring repose from all our labors.

All of this shows the correspondence of human and heavenly music with the instrumental kind. But to what has been said superficially thus far, much will be added in the second and third books. What we have written will serve to satisfy the reader for a while and dispel his ignorance of the subject. It shows by comparison and authority that every mode has a distinct quality of its own.

In order to demonstrate this more assuredly, let us set out those to which music owes its origin, namely (as Hesiod[4] has it) the nine Muses, daughters of Jupiter and Mnemosyne. She who sings of war we commit to Mars, and thus to the Phrygian mode; she who commemorates tragedies or victims to Saturn, hence to the Mixolydian; she who sings of Joy, to Venus. And thus to each of the Muses we assign the proper place, following the authority of Martianus and Macrobius.[5] To each other we also give a verse, to show her appropriateness for music.

Thalea remains silent, like the Earth. Clio we will attribute to the moon; Calliope we will say is Mercury's; Terpsichore we affix to Venus. The sun will blanch Melpomene. Erato will rouse up Mars; Jupiter will make Euterpe kind and glad; Saturn will sadden Polyhymnia. Lastly, the starry heaven will give comeliness and peace to Eurania.

Let us now make a circle from the first, i.e., from silence, to the last, then passing all the others and curving back to the second, and so create the Hypodorian. In just the same way, let us proceed with the rest, not ceasing to make a spiral until we have reached the last Muse.[6] To go further would be superfluous, being a repetition of the first. We follow Roger Caperon in calling the note above *nete hyperboleon* '*crisis*,' and the one below *proslambanomenos* '*coruph*.' In pagan days I believe this comprised the whole of the art, and when it reached *coruph* that was the end. For it was accepted that this Muse kept silence, while the last one had the highest note. We should beware of altering anything in an author on account of antiquity. Our first note will therefore be *proslambanomenos*, our last *nete hyperboleon*, and we show the whole system in the accompanying figure.

The disposition of this figure shows why they are named *tropi*, because each one is procreated from another. They proceed in the following way: one raises by a tone the first octave-species (the notes enclosed between *proslambanomenos* and *mese)*, diminishing by a tone *hypate hypaton* and making all the others a tone higher. A whole order will result, a tone higher than the previous one. The whole higher arrangement thus produced will be the Hypophrygian mode; and the process of raising is similar in the case of the others. Thus they were not called *tropi* because the notes beginning at the bottom are transferred to the top and end up by recurring at the bottom, as Saint John [Damascene] believed. For there are some that do not begin in the lower regions but in the higher ones, as we will show a little later, treating them one by one.

28

◆

Pico della Mirandola
1463–1494

Giovanni Pico della Mirandola, the aristocratic *Wunderkind* of Florentine Neoplatonism, was only twenty-three when he went to Rome offering to debate his nine hundred Conclusions or Theses, all of which he claimed to be reconcilable with each other. (No one accepted his challenge.) The Theses are short statements, many of them extremely cryptic, summarizing points from Scholastic and earlier theology, Arabic philosophy, the Platonic schools, the Chaldaean Oracles, the Hermetic Corpus, Kabbalah, the Zoroastrian Magi, and the Orphic doctrines. Behind this endeavor lay Pico's own conviction that all were but different branches of the *prisca theologia* or primordial wisdom-religion of mankind (see Introduction to Ficino, no. 26).

For his Pythagorean theses, Pico drew on the mathematical works of Nicomachus, Iamblichus, and Proclus. I do not pretend to be able to explain them all. Music is here purely an intellectual concern: a branch of mathematics. We have no evidence that Pico shared Ficino's love for singing and hearing music—but see note 10 for his appreciation of music's therapeutic purpose.

Source: G. Pico della Mirandola, *Conclusiones sive Theses DCCC* [1486], edited by Bohdan Kieszowski (Geneva: Droz, 1973), pp. 48–49. Translated by the Editor.

Fourteen Conclusions after Pythagorean Mathematics

1. Unity, duality, and that which is,[1] are the causes of numbers: One, of unitary numbers; two, of generative ones; that which is, of substantial ones.[2]

2. In participated numbers some are species of numbers, others unions of species.

3. Where the unity of the point proceeds to the alterity of the binary, there the triangle first exists.[3]

4. Whoever knows the series of 1, 2, 3, 4, 5, 12, will possess precisely the distribution of providence.

5. By one, three, and seven we understand the unification of the separate in Pallas:[4] the causative and beatifying power of the intellect.

6. The threefold proportion—Arithmetical, Geometrical, and Harmonic—represents to us the three daughters of Themis, being the symbols of judgment, justice, and peace.[5]

7. By the secret of straight, reflected, and refracted lines in the science of perspective we are reminded of the triple nature: intellectual, animal, and corporeal.

8. Reason is in the proportion of an octave to the concupiscent [nature].[6]

9. The irascible [nature] is in the proportion of a fifth to the concupiscent.

10. Reason is in the proportion of a fourth to anger.

11. In music the judgment of the sense is not to be heeded: only that of the intellect.[7]

12. In numbering forms we should not exceed 40.

13. Any equilateral plane number may symbolize the soul.[8]

14. Any linear number may symbolize the gods.[9,10]

29

✦

Franchino Gafori
1451–1522

After wandering throughout Italy, Franchino Gafori (also known by his latinized name, Franchinus Gaffurius) settled in Milan in 1484 as canon and *maestro di capella* of the Cathedral. Here his friends included Leonardo da Vinci and the mathematician Fra Luca Pacioli. Gafori's creative life as a writer on music theory developed as he was able to take advantage of Greek sources, notably Ficino's translation of Plato (1484) and writings on Platonic philosophy, and the translations Gafori himself commissioned of the treatises by Ptolemy, Aristeides Quintilianus, Briennius, and others. Through these efforts, and by quoting these works extensively in his own books, Gafori opened up a new world of music theory for his successors.

De Harmonia Musicorum Instrumentorum Opus was written in 1500 and distributed in manuscript, to judge from several surviving copies, but not printed until 1518. In the present extract we hear many familiar voices, and indeed Gafori's own is not easily detected among them. His achievement is the establishment of a new canon of classical authorities, doing for music what his humanist contemporaries were doing for the other disciplines. Gafori's system of correspondences is beautifully summarized in the famous woodcut by Guillaume de Signerre, which he had already used as the frontispiece to his *Practica Musica* (1496).[1] In his early *Theoricum Opus* (1480), he had paraphrased Anselmi's correspondences of angelic orders with planets; now he borrows from Ramis de Pareja those of planets with tones, modes, and Muses. This is the ground on which so many Renaissance theorists met, whatever their arguments over practical music (and there were plenty of these between Gafori and Ramis) and with whatever attitude they approached the classical authorities in other matters.

Source: F. Gaffurius, *De Harmonia Musicorum Instrumentorum Opus*, translated by Clement A. Miller (American Institute of Musicology, 1977), pp. 197–204. Used by kind permission of Professor Miller and the publisher, Dr. Armen Carapetyan.

◆

Muses, Constellations, Modes, and Strings Belong to a Mutual Order

Book IV, Chapter 12. There are those who believe that muses follow the order of constellations and modes. Some enumerate only three muses and others list nine born of Zeus and Memoria. In *Metamorphoses*[2] Ovid calls them Mnemonides in this line where he said: O Muses, tarry, for he knew them. Diodorus Siculus in Book V said they were very celebrated and described the proper position of each as if of objects related to the art of music. Homer also honored them with extraordinary celebrity.[3] Saint Augustine in Book II of *De Doctrina Christiana* named nine muses. He refuted the idea that they were daughters of Zeus and Memoria and relied rather on the opinion of Varro that three workmen had each made three representations of the muses in the temple of Apollo, who were then given names by Hesiod in *Theogony*. Through the ancients it was desired to give the muses cognomens that were good and useful and that would teach men things that were unknown by the uneducated.

Some believe that the muses arose from the head of Apollo, as expressed by a poet in this verse: *Mentes Appolineae vis has movet undique musas.*[4] Others say that they were taught by Apollo; for this reason Apollo himself is called music, as we said more fully in Book I of *Theorica*. Many believe that Apollo's cithara is portrayed with ten strings, and others say seven, as if the seven essential strings which Virgil in

Book VI of the Aeneid[5] relates in this line: Also the long-robed Thracian priest plays and sings to a lyre with seven strings. They also have seven intervals: ditone, semiditone, fourth, fifth, major sixth, minor sixth, and octave. Number seven is arranged with a certain perfection since it and its aggregate parts equal number twenty-eight, [being the only number] between ten and one hundred [equal to the sum of its] aliquot parts.[6]

It is said[7] that Clio invented history, Melpomene tragedy, Thalia comedy, Euterpe pipes, Terpsichore the psaltery, Erato geometry, Calliope literature, Urania astronomy, and Polyhymnia rhetoric. But Anaximander Lampascenus and Zenophanes Heracleopolites in their books have interpreted the muses in another way, and others as the naturalist Pisander and Euximenses in the book *Theologumenon*. They compared the muses in order to the requisite instruments of the human voice. A tone sounds with four teeth placed in opposition at which the tongue strikes. To the upper teeth one muse was ascribed and another to the lower. But if one of them were lacking, they say a syllable rather than a tone would result. They also give two muses to the two lips as appropriate instruments of the words. They apply another muse to the tongue which as a curved plectrum striking a certain vowel forms the breath. A muse is given to the palate whose cavity produces sound. Another muse is given to the windpipe which in its round passage gives exit to the breath. They ascribed a ninth muse, fitting for quiet and silence, to the lungs, which like a goldsmith's bellows, produce and renew a tone.

Fulgentius in his interpretation of fables established nine muses of learning and science. Clio is chief of these, as if the first with an understanding of learning. For *fama* of Latin is called Clio in Greek, and so in Homer there is "we listened to fame alone" and also elsewhere. For great glory is a scepter, and no one looks to knowledge except to increase the worthiness of his own fame.[8] From this Clio is placed first, as if the thought of seeking knowledge. The second muse, Euterpe, is called well pleasing in Latin. Since the first is to seek knowledge, the second is to take pleasure in what is sought. The third muse, Melpomene, relates to persevering in contemplation, just as the first seeks, the second desires what is sought, and the third renews by studying what is chosen. The fourth is Thalia, capacious or germinal. Epicharmus says in *Comedia: Germina dum non viderit famem consumit.*[9] The fifth is Polyhymnia, as if having a memory of many things, for memory follows capacity. The sixth is Erato, which is found the same in Latin. It is proper that with the preceding capacity and memory one may find anything similar with his own ingenuity.

The seventh is Terpsichore, or pleasing instruction. Thus Hermes says in *Poimandre:*[10] without knowledge of food and nourishment and with an empty body. For after invention it is also necessary to understand and judge what you invent. The eighth is Urania or heavenly, for after a judgment you select what you affirm or what you reject. It is beneficial to choose the celestial in order to reject the noxious and transitory. The ninth is Calliope or fine-voiced, as Homer says: the voice of the songful goddess. Thus according to the plan of Fulgentius, knowledge and learning are sought as a gift of the muses so that in a requisite and successive way they are understood to agree with seeking, taking pleasure, persevering, grasping, keeping in mind, inventing, judging, selecting, and expressing.

Callimachus, the Greek writer, is a fine authority on the gifts of the Muses, as the poet shows in this epigram: Calliope invented the knowledge of heroic poetry, Clio, the sweet song of the noble chorus to the music of the cithara, Euterpe, the resounding tone of the tragic chorus. Melpomene gave the knowledge of the lyre to mortals for sweet sounds. Terpsichore, as a favor, skillfully constructed pipes. Erato invented the very delightful hymns of the gods. Learned Polyhymnus [sic] invented the delights of the dance. Polyhymnia gave harmony to all songs. Urania discovered the heavens and the celestial bodies. Thalia invented comedy and illustrious customs. We think, as many believe, that the muses fit the stars, modes, and strings, so that we ascribe them to single strings to which the beginnings of modes are given, placing one with each.

Subterranean Thalia is placed first, as if silent, as this verse says: In the beginning nocturnal silence germinates with song and Thalia lies silent in the bosom of earth. Marcus Tullius Cicero compared the earth to silence because it is immovable, just as the underground three-headed Cerberus is compared to the Apollonian feet. To the lowest added string is given the beginning of the Hypodorian (because it is the lowest of modes) and the moon, the home of Cancer[11] (as astronomers believe), since it is the lowest planet, and Clio, as in this poem: Persephone and Clio breathe and therefore the Hypodorian is born; here arises the origin of song. To the second string, *hypate hypaton*, is given the beginning of the Hypophrygian, and Mercury, home of Gemini and Virgo, and also Calliope herself, with this verse: The Hypo string produces a string connected to the Phrygian, which Calliope brings forth, and as an interpreter she produces divinity. The third string or *parhypate hypaton*, is given to Terpsichore, the Hypolydian, and Venus, Libra, and Taurus, as in the verse: The third string shows the beginning of the Hypolydian; Terpsichore stands opposite and Venus creates Paphos. The fourth string, *lychanos hypaton*, has the Sun, home of Leo, and Melpomene. It is given the Dorian mode in this verse: Melpomene and Titan form (I believe) the mode in the fourth place which is called Dorian.

On the fifth string or *hypate meson*, are Erato, Mars, Aries, Scorpio, and the Phrygian mode, as in the verse: Erato wanted to assign the fifth string to the Phrygian; also Mars is ever-loving of war and not peace. The sixth string, *parhypate meson*, has Jove, home of Pisces and Sagittarius, also Euterpe and the Lydian mode, as in this poem: the Lydian of Euterpe contains also the music of Jove; sounding sweetly, the sixth string rules because a goddess is present. On the seventh string, *lichanos meson*, is Saturn with Aquarius and Capricorn, also Polyhymnia and the Mixolydian, as in the verse: Saturn and also Polyhymnia govern the seventh string; from it the Mixolydian takes its beginning. The eighth string or *mese* is given to Urania, the Hypermixolydian, and the stellar orbit, as in this poem: When the Hypermixolydian sees the eighth string of Urania as a friend, it turns the heavens with skill. At the end of Book II of *Musica*, Aristides Quintilianus gave a somewhat contrary arrangement of muses and modes.[12] In Herodotus of Halicarnassus, who treats nine muses in his *Hysteria*, a different order can be found. The Greeks place the three Graces to the right of Apollo; they are called Charities or the attendant Graces of Venus. Aglaia is considered splendor, Thalia is freshness, and Euphrosine represents pleasure. But they appear in the subjoined diagram. (See figure 6.)

Among Heavenly Bodies Some Form Masculine Sounds, Some Feminine, and Some Are Common

Chapter 13. Aristides[13] has said that of the sounds which generate motion in celestial bodies, some are masculine in nature, some feminine, and others common, according to the property of each sphere. A masculine sound in a celestial body is sharp and firm, suitable for action and work; a feminine sound is weak and quiet, unsuitable for industry and labor. From these characteristics individually established or mixed a variety of sounds occur. For although the moon is weak and every source of corporeal movement emits a feminine sound, it is drawn for a little while to a masculine nature; since it receives the downward flow of other bodies its feminine nature is set free and it participates with the masculine, because the force of generating and nourishing bodies flows into lower bodies. Sacrificial priests and ministers believe this to indicate its masculine and feminine nature (I say more feminine) in invoking a goddess. As the orb of Mercury is mostly dry because of its proximity to the sun, if ever it is separated from it because of its size (although it has little humidity), it rarely delights in nocturnal appearances, but more often daytime [ones], and is believed to produce a mixture of a masculine and a feminine sound, with the masculine participation

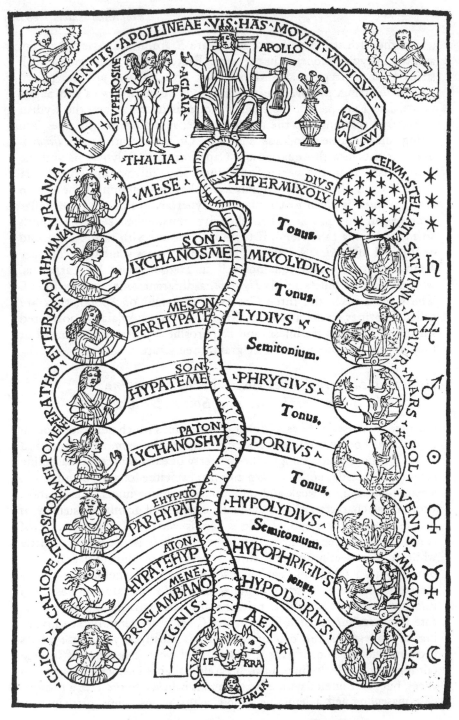

FIGURE 6: *Tones, Modes, Planets, and Muses*

greater than the feminine. Since the orb of Venus, delightful to view and mostly clear and pleasant, is humid, it is said to be pleasing at night and to emit a feminine sound. Because the orb of the sun is dry it mostly burns up in heat and energy; it is said to produce a masculine sound. The orb of Mars, warm and violent, takes pleasure in humid and nocturnal figures; it produces a sound that participates in both natures, but leaning more to the masculine. The orb of Jove, pleasant in all respects, is called the close rival of the orb of Venus; it is believed inferior in warmth to Mars and is thought to allay the coldness of Saturn. Like the orb of Venus it has a tempered mixture of both, since it is appropriate for producing the daily breath of life and procreating children. It is the author of marriage and is said to produce a feminine sound. Since Saturn is sharp, dry, and laborious, Aristides Quintilianus says it forms a masculine sound.

We do not think it incongruous to agree with the conception of Pythagoras and Plato, who said that celestial sounds are produced according to a certain order of instrumental sounds. Yet they are inaudible to us because we have ears that cannot hear the very great distance and rather confused mixture of heavenly bodies. There are some who accidentally have a weaker sense of hearing and who scarcely hear or do not hear at all a human tone. But those who are endowed with virtue and distinguished mores, and are removed from baser men (who live like beasts), can hear without difficulty celestial sounds with the uncorrupted sense of their nature. As we said in Book I of *Theorica*, Plato is the author of this subject. Therefore, just as those who reach the summit of virtue and a true understanding of knowledge and wisdom are able to see the presence of divine forms by avoiding all evil, so it is easy to hear harmony and universal sound. But those who are enveloped in unworthy vices find this difficult or impossible. For they know fleeting and earthly pleasures, adhering to brutal sensuality far from the first disposition of nature (I call it divine), as Ovid says: And when lower beings look at earth, and the face of man is able to see the sublime heaven, they withdraw with incredible loss. Vice is indeed found easily and without work, but virtue is prepared on a straight path for a long while and with great toil, as in the fine poem by Hesiod. For we can grasp vice at once without labor; it is an easy path and is never far away. But virtue must be prepared with long labor in remote places and must be sought on a straight path. You will not reach the heights easily at first; after that it is easy, although difficult at first. Thus Diogenes of Sinope cited those musicians who fitted the strings of the lyre to a suitable harmony and who brought them into pleasing and well ordered moral patterns.

The Ancients Understood Music More by Ratio than Aural Sense

Chapter 15. Thus far the things concerning harmonic capability have been treated. They are usually considered according to sense or according to ratio. Plato in *Timaeus* teaches that music perceived by the senses is much surpassed by music perceived by the intellect. By taking the available strings you can easily find the sounds of each by genus, some making a consonance by their outer numbers. From these, according to the outer numbers, the following consonances occur: duple, triple, quadruple, *hemiolia*, and *sesquitertius*. Others are contained between the outer numbers (an octave for example), as a ditone and trihemitone, or according to Ptolemy 5:4 and 6:5, or a major sixth or minor sixth; they are exceeded by the extremes of the octave consonance. Thus Pythagoras, in replacing death with life, recommended the use of a monochord to his pupils, as Aristides says, for he taught that the most sublime music is understood more by the intellect through numbers than by the sense of hearing. He suitably maintained this teaching as he also drew its beginning from the universe. But because this is mixed with corporeal material it falls from the exactness of numbers, since indeed in the place above us there is true and incorruptible music, as our holy church declares to exist in angelic choirs. Plato and the Pythagoreans also affirm this in the intervals and courses of heavenly bodies. From this we cannot have intervals divided into equal parts, and being encumbered by corporeal density we have imperfect consonances in the system.

30

Francesco Giorgi
1466-1540

Francesco Giorgi, or Zorzi, was a member of the Franciscan Order in Venice, where he enjoyed a high reputation for his learning. His knowledge of Platonic arithmology was called upon when he was engaged as adviser on the proportions for a new church of San Francesco della Vigna; his expertise in Hebrew law, when he was consulted on the legality of Henry VIII's divorce from Catherine of Aragon.[1] Living in Venice, where there was a strong Jewish colony, he was able to extend his Kabbalistic researches beyond those of Pico della Mirandola or Ficino, and it is this philosophy, in a christianized form, that underlies his enormous work on world harmony.

Giorgi is much more of a metaphorical than a literal harmonist. The sections of his book are called "Songs," "Modes," "Motets," etc. The pervasive theme is the harmony of all creation—elemental, celestial, and angelic—with the ten Sephiroth, archetypes directly emanating from the Divine Mind. To this stupendous picture of harmonious emanation Giorgi applies Platonic arithmology, writing in effect a vast commentary on the creation myth of the *Timaeus*. The scope of the work almost beggars description: not only is it full of the most encyclopedic learning, but every page shines with Giorgi's own Franciscan piety. For there is only one purpose behind his enterprise: to be reunited with God. Hence the friar's repeated emphasis, as found near the beginning of this extract, on the inviolable Unity which embraces all things and which the mystic may glimpse shining through them. Giorgi's book has only just begun to be read again, and it will probably be better understood as time goes on.

Source: Francesco Giorgi, *Harmonia Mundi* (Venice, 1525), fols. 89'–91', 94'–96. Translated by the Editor. Also consulted: French translation by Guy Le Fèvre de la Boderie, *L'Harmonie du Monde* (Paris, 1579).

The Consonance in All Things Results from the Unity of the First

Chapter 8. When all created things are separated and divided into many portions, they remain dissonant unless through due harmony they are brought into oneness. Consonance, as Boethius and Nicomachus define it, is the concord of equal and unequal voices reduced to one. Hence the consonance of this world-instrument is also a concord of equal and unequal things reduced to this first unity, such that all things are said to rejoice in this unity, as Plato argues learnedly and profoundly in the *Parmenides*. (Aristotle's contradiction argues more with the words than the sense, as Simplicius points out.) According to Orpheus, God is one through identity, whereas every other thing is called one through its diversity, dependence, participation and connection with the true One which is God. Now he creates them in their multiplicity of species: now he assembles them all into a single inventive Idea. And thus by enfolding them all with the principle from which they emanated, he drives and draws them to unity.

Hence Plato and the other Pythagoreans,[2] in order to demonstrate that this machine is unique through the supreme Unity and united through concord and conformity, collect the equal and unequal portions of things and their proportions and harmonic relationships, saying that they all proceed from unity and tend thither; and that the desire for unity is bestowed by the supreme One who gives it to all things that they may be one and aspire to oneness. Boethius says that everything has existence insofar as it is a single number. And all these things, says Proclus in his *Theological Theorems*, proceed from the One such that they hasten continually to return thither, agreeing with this One with the greatest concord when they have the greatest participation in it. One more thing is necessary: to know the end and goal in which everything is blessed or reposes. For Dionysius says that the Creator, as the cause of all, converts all things to himself, lest left to themselves they should fall short or be deranged.

The Notes of the Heptachord Correspond to the Soul

Chapter 9. Since Plato has described by this heptachord[3] both the human soul and the making of the world, we must investigate the way in which it is adapted to the human soul. First we assume that the soul, separated from all corporeal laws, is not [quantitative] number: hence it is not divided, nor multiplied by parts. But it is a substantial, uniform number, self-contained and rational, far surpassing anything corporeal or material. So its division is not after the fashion of matter, nor of base and gross things, but proceeds from the efficient cause (as Proclus says). The soul is divided by the perfect measure of all things which the Creator followed when he divided the universe into intelligible patterns. This kind of immaterial, intellectual and unadulterated division resides in the soul; it is the perfection of every animate substance, the generator of multiplicity which is in it and beneath it, bringing it to a single order through harmony. It joins sundered parts and is the cause of unsullied purity, giving to the soul a rational coherence of its multiplicity, so that the soul (as Timaeus and Plato teach) is both divisible and indivisible. For if, as Aristotle rightly says, there is something divisible in the indivisible things which contain them, so much the more must there be something indivisible which ever remains in things which have an undivided nature.[4]

For the Maker and eternal Fabricator of all made the soul a single whole before the division and otherness. But the eternal Producer, when he produced the one, did not entirely destroy the other: the totality was not destroyed by the production of parts, but remained and presides over those parts. Neither were the parts produced prior to the whole, but the whole before the parts, being not integral but essential. Thus the substance of the soul is a whole, having parts: one, and yet a multitude assembled in wondrous harmony. Whoever wishes to see it should not aim his industry toward the goal of mathematical speculations, but should arouse himself to the utmost toward the meeting-place of the living substance, weighing aright the triple whole. The first is prior to the parts; the second is constituted from those parts; the third is in each part. Plato has spoken of that which is prior to the parts, which is not consumed in the production of the parts, but remains. For that which makes the parts is good and does not dissolve them, that being the role of evil, not of good. So that whole, preceding the parts, remains; and that which is said to be of the parts dissolves; and the assembly of new parts made by harmonic ratio perishes in the dissolution of the parts. For that which is in the parts is what Plato signifies by the circles into which he divides the soul. The soul is therefore unique and multiple, having its unity in the Intellect, and its quality and multitude from below. And whatever is unique is divided when it descends to lower things, and, conversely, when it reverts to itself it is collected and re-united, according to the Dionysian and Apollonian way. Thus Proclus describes it, drawing on the ancient theologians Orpheus, Hesiod, Euripides, and Aeschylus, who clothed in fables that which they learned from divine men. For, as they teach, the divine Sun is the Archetype whence emanate all divers beauties, all adornments, all pleasant harmonies, and all life. On his right hand is piety and well-being; in his left is severe punishment.

But those gentile prophets, whether not perceiving the sacraments of God, or else desiring to conceal them, attributed all that draws its energy from the supreme Sun to that sun which we see with our bodily eyes. This they called Phoebus, Apollo, Sol, as we have said above, and said that he wielded graces in his right hand, and in his left a bow and arrows. They also called him Bacchus and Liber Pater. But leaving aside these names for the present discourse, they called him Dionysus by night, rending and dismembering, and himself broken in seven pieces; but Apollo by day, restoring what had been torn, and himself constituted of seven parts.

This septenary is so agreeable to him that he always proceeds by sevens; thus they call him the Lord of the Septenary. This results from unity, the binary, and the quaternary, whence is born the Disdiapason which is the most perfect harmony.[5] And to the septenary is attributed the authorship of life and its dissolution, being composed of the first odd number, 3, and the first even number, 4. (For the Pythagoreans did not reckon 2 a number, but a confusion of unity.) So from the even and the odd, like active and passive, are composed all things; and in the same way they are dissolved. Generation and life are from the supernal light and the uniting force, but dissolution is from the infernal regions and the dismembering night. That is why by night he is called the rending Dionysus, and by day the uniting Apollo, so called as dispeller of evils. Hence the Ancients said "Apello" instead of Apollo, and the Athenians "Alexikakos," i.e., averter of evils, while Homer calls him "Ulion," the author of health. In him resides this unity which makes all things perfectly consonant.

Even when it seems that division has taken place through obedience to inferior things, in fact it has not been divested of its state: but it assembles and reunites what is dispersed. This is perhaps the source of the royal Prophet's saying: "Thou takest away thy breath (the spirit of unity and unification) and all things abandoned to themselves fail. But if thou renewest that spirit of life and unification, all things are renewed and recreated" [Psalm 104:29]. And because it is the blessed God who does these things, he goes on: "Glory to God for ever; God be praised in his works," because he has led them to their due temperament and to union with the principle from which they proceeded, seeming to have separated from him by their divisible nature. Yet they are never abandoned by him, even though they remain in perpetual dissolution, for his unity is ever present in them. This unity assists them perpetually, preceding and following the parts and finally drawing all separate things to itself.

How the Three Novenaries[6] Are Found in Man

Chapter 10. Man, being the most perfect image of the world, contains all those things which are woven into the great world, and in the same proportion. First, the novenary of hundreds in his basest part, which is his twofold body (as Plato says), each part made from four elements: the corruptible from the four alterable elements, and the incorruptible, which will be after the Resurrection, from the nobler elements, not subject to corruption:[7] these two quaternaries making eight. To enable the body to cohere, and live, there is within it a single unique life, sometimes called the soul, to which the Hebrews give the name *nephes*, or else the vital spirit. The moderns call it the soul of reason, supplying all the energies of life. Adding this to the previous eight, the first novenary results, terminated by the ten, the highest proportion. The second novenary is contained in the soul, being the five exterior senses and the four interior ones, namely the imaginative, cogitative, fantastic, and common senses. These are terminated by the intellect as a tenth. And in this is another novenary, a purer one of single numbers, by which it corresponds to the nine orders of angels and to superior emanations. So that there is within it the whole image of the Angels and of God, as we shall treat of in its place. This novenary terminates in the ten which is the supreme source and parent of all, to whom all works and prayers are due. Thus the Son of God teaches when he says: "When you pray, say 'Our Father . . .'." By turning to him we shall be united with him.

Now the intervals of the enneads are full of operations, called secondary acts.[8] For man has a triple operation, and this consonant and harmonic: in nature, with the inferior world; in the senses, with the heavens; in the superior part, with the angels—that is, the rational or intellective and voluntary part. And if he were not consonant with all things, he could neither act on them nor understand them. And among living beings only man enjoys this honor of acting with all things and of having power over all things, even over his sovereign Creator, through intelligence and love.

But if we wish to assimilate man exactly to the scheme of Plato and the Pythagoreans, which is contained in seven principal parts made from the first even and the first odd number, or from active and passive, we must remember the common saying of the Peripatetics: that there is in man the passive intellect which receives all things, and the active intellect which produces all the conceptions drawn from the passive intellect. And in each of them, both the receptive and the productive, is this septenary mode from 3 and 4. For the ternary is suitable to the conception of supernal things, and the quaternary to that of composed things. Man is rather made of two parts: from a spiritual part, namely the mind, and from a gross part, the body; from whose mixture result the inferior senses, which are partly animal, partly intelligent, joining the two extremes.

If we may descend further than Plato into particulars, we will apply all these limits to man himself. For in him is unity: some portion of divinity, which understands God and operates around him. This is doubled into the active and passive intelligence; quadrupled into the fourfold elementary forces; made eightfold in the combined quadruplicities of the elements. On the other hand it comes to a trinity through the threefold part of the soul; to nine through the nine senses, the exterior corporeal ones and the interior ones; to twenty-seven, embracing all three enneads contained in man, in whom the intellect or supreme part is the leader and principle of life. And all the intervals between these limits are filled by the proportioned operations of all their powers and parts.[9]

What Can Be Said of the Fillings-in of the Intervals of the Three Enneads

Chapter 15. Now that we have treated the filling of the intervals [in Ch. 13, here omitted], we must survey all that can be said on the accomplishments of the first genera of 27, or the three Enneads. Certainly they are filled out according to superparticular[10] proportions, as the Pythagoreans say: "That which is made according to one portion alone dispenses to things." All the same, this unique life is not at all divided: but just as the Sun, unique in itself, is multiple and diverse in its rays and effects, and dispenses with the greatest efficacy its influence in one time and one place as in another, so the soul, being unique in the body, nevertheless has different effects. And to begin with the most noble, it understands through the intellect or the chief part, it wills in the heart, it sees through the eyes, it hears through the ears, with the hands it works, it walks with the feet, it digests in the entrails, and in short it goes so far as to expel the excrements. Nevertheless, however much it works through that which is indivisible, it does not act according to its wholeness but according to some part of its virtue. For if the virtue were always the same, its work would consequently be the same. Neither does the diversity of organs suffice to achieve such diverse effects, for otherwise when delivered from the body it would only be able to use one function. In fact, when separated from the body it wills, it understands, it changes its place, it uses the senses which are in the interior man. It is thus diverse in its effects and unique in itself, which partially limits it according to its organs and instruments, whence results a superparticular proportion, namely of the whole to that proportioned part.

Thus the life of the world is unique, whence it is said: "God, in whom all things live." And Saint Paul, celebrating the sentence of Aratus, says: "In him (that is to say in God) we live, we move, and we are;"[11] from the secret places of whose unity (as Orpheus says) proceeds the multitude of numbers. For this reason the Tree of Life planted in the midst of Paradise is unique, and yet is called *Hez chaijm* which signifies the tree of lives, because being unique it gives forth divers rays of life to other things, as Saint Paul says of the life of the mystical body: "One and the same spirit goes forth to each as it pleaseth him" [I Corinthians 12:11]. This spirit of life, being one and the same as the Tree of Lives, vivifies and invigorates all things, as we are apprised by the Oracle of Ezekiel, when he says "The spirit of life was in the wheels" [Ezekiel 1:20]: in the mundane wheels, I would say, or the celestial and terrestrial. For all things rejoicing in circular form imitate and follow the supreme and divine Sphere.

This life then, being all within itself, communicates to each a portion which is by no means separated from it: to one in the proportion of 1½, to another in the proportion of 1⅓, to another in sesquioctave or 1⅛, to another in the proportion of 1¹⁄₁₆. From these result the diapente, the diatessaron, the tone and the leimma. But none has hitherto demonstrated the particular genera and species in this or that proportion with which the first life, outspreading itself, disposes all things according to their numbers, weights, and measures. And it is unthinkable that it could be expressed or perceived except by those who are purified and filled with wisdom, by which they can know and embrace all things, as it is said in the Book of Wisdom: "He has given me the true knowledge of things which are, so that I know the disposition of the circle of the Earth, and the virtues of the Elements, the beginning, end and middle of the Ages, the changes of Fortune, the consummations of centuries, the mutations of customs, the disposition of the stars, the natures of animals, the furies of beasts, the force of the winds, the thoughts of men, the differences of plants, the virtues of roots: and, in short, I have learned all that which is hidden and not foreseen" [Wisdom 7:17-21].

That which is done by thought also gains by the outpouring of this chrism, of which Saint John speaks: "You have the unction of holiness, and know all things" [I John 2:20]. And again: "You have no need that any should instruct you, for this unction will teach you all things," I say the unction of which the Sovereign Truth speaks when he says, "He will make you remember all things and will teach you all things" [John 14:26]: that which is done by the Death of the Kiss,[12] as the most secret theologians teach us, of which it is said in the Psalms: "Most precious before God is the death of his saints" [Psalm 116:15]. Or, as Saint Paul says, by a transformation into the image itself; or according to the Platonists, by an essential contact with the Ideas of the first Intellect, capable of achieving all things within the recipient.

By this contact, as by an embrace, certain images are conceived and the species of things are grafted on: or rather the soul is reawakened, and by ambrosia and nectar purged of the contagion and pollution which it had drawn from the lethal river, just as Pythagoras, Empedocles, Heraclitus, and Socrates in the *Phaedrus* of Plato affirm, and which a great host of authors has sung after them. But the family of Peripatetics asserts that at the beginning the rational soul is like a clean tablet, and that it receives ever new forms. However, neither the one nor the other will prove their opinion, for God alone knows the truth of this matter, and those who are illumined by divine light. And Plotinus, speaking of the purgation which is required for receiving this wisdom, says that it reaches the point where the soul is denuded of all rust and foreign affections, and can take wing and conjoin to its first father, and enjoy the contemplation thereof.

Whence the Sovereign Truth admonishes us in a loud voice, saying: "Blessed are those who are pure in heart, for they will see God" [Matthew 5:3]. Living in the archive and fountain of Ideas, they can see distinctly and clearly all the degrees of things by this knowledge, which Saint Augustine calls matutinal, and then descend to the knowledge of things in themselves, which he calls vespertine.[13] Which not only the souls of the Blessed, but also those who are not yet fully detached from the body, enjoy in ecstasy, and by a certain elevation of thought beyond that which the human condition allows, obscured as it is by the darkness of this world. Or rather they are taught by the divine Oracles, such as we may conjecture of the great Patriarch Abraham through the Book of Formation[14] which is said to have been received from him; in which by the letters and characters of the names of things in the Hebrew tongue, and by the permutations and mixtures of these by the numbers they denote, their points and accents, he demonstrates precisely their proper essences and conformities by an admirable artifice, more one of contemplation than of explanation. So now we shall return to our discourse.

How All Things Are Established in Number, Weight, and Measure

Chapter 16. Since we cannot write fully of the conformities of things, we shall follow the covert course of translations and metaphors derived from things: which, though they may often seem diverse and variable, and their words judged different and foreign to each other, yet they often agree in sentiment. For what the Sage[15] demonstrates by number, weight, and measure, Plato describes through numbers, figures, and solid bodies. For by numbers he means to signify the special forms of things, which Aristotle also compares to numbers. And by measures he designates the virtues and properties given to each thing, of which also the Sovereign Truth speaks in the Gospel, saying: "He has given to each according to his proper measure" [Matthew 25:15], i.e., according to what has been meted out to each by the measures of the Archetype, which measures all things with a most exact balance. And by solid bodies are meant the weights of things, and the heavy and weighty elements which always drag us downward; or else the inclinations of things which drag us like weights and compel us either to good or to ill. Hence Saint Augustine calls Love a weight, saying: "My weight is my love, I am carried or I love."

Moreover, numbers signify the arithmetical mean, dimensions and measures the geometrical mean, and weight the musical mean, which is balanced proportion, and comprehends in similitude the lightness or heaviness of the movements and the high or low pitch of voices, and the inclinations of things both heavy and light. And just as pipes or flutes or any other sounding instruments need the temperaments drawn from weights, so in all things, and principally in man who is the bond of all things, it is necessary to temper the appetites, so that he may be prudent and his heart illumined, and must guide his natural inclinations toward the goal of reason, or rather of divine law grafted into the human mind. He must also temper according to their weight the food and all else which we convert to our use; so that in this great fabric the elements, the heavens, and all things rejoice in their proper temperaments, and an infinite number of species and forms, balanced in their due measure, concord with the greatest propriety.

The lower or subject bodies, though all composed of the four elements, seem to be innumerable; but this is no marvel, for if from the cube, which is made from three numbers, namely linear, plane, and solid, and which is formed of four surfaces, divers and almost infinite lines can be drawn from the center to the periphery, how much more so in the mundane cube, expanded as we have said above by the ternary up to 27, will innumerable forms and subjects be found? Although all things consist of the four elements, it is the superior numbers of their four principal surfaces by which the multiple forms answer to their superior subjects.

These things are ruled by the seven planets and by the seven intelligences that preside there, and by the seven divine names or *Sfires Sferales* whence the intelligences receive the influences which they give forth. And in truth this septenary resulting from the four surfaces and the three numbers is conducted to compounded things by the celestial septenary, angelic and divine, with the intervention also of the twelve [zodiacal] signs which are governed by the parts of the septenary. For thrice four, and four times three, engender the dodecad. Whence results the great consonance by which practical musicians following nature as a guide, are forced to set and adjust their harmonic instruments.[16] For the first instrument, of which Mercury is said to have been the inventor, contained the number of the elements, having only four strings. As Nicomachus bears witness, *hypate* was intended to designate earth, *parhypate* water, *nete* fire, and *paranete* air; and this number sufficed them until the time of Orpheus. And because invented things give easy issue to those who wish to alter them, those who came after easily surpassed them.[17] For Chorebe, king of the Lydians, added the fifth string, Hyagnis the Phrygian the sixth, and Terpander the Lesbian, inventing the seventh, furnished the number equal to that of the seven planets.

Well may they have attributed the lowest and slowest of these strings to Saturn, grave and tardy of movement, and followed by comparing each of the others to the remaining planets. But later they went even further, for Lycaon added the eighth string, Prophrastus of Pieria the ninth, to designate the harmony of the nine spheres or of the first ennead described above. But Estiacus of Colophon added the tenth, and Timotheus the Milesian the eleventh; and by interposing the semitone they reached as far as the fifteenth. Thus they made an instrument having twice a diapason. Finally, by mingling in the enharmonic and chromatic genera with the diatonic, they attained to 28: a number which contains the three enneads with the addition of one unit, namely the soul which is a single life filling all things, vivifying all things, and binding all things, so as to make a body of the whole machine and human fabric, as the Platonists affirm. And this is the true monochord of the three genera of creatures, angelic, celestial, and corruptible, resounding from a single spirit and a single life, of which Ezekiel said: "And the Spirit of Life was in the wheels."

Thus this world-machine, consisting of a simple number squared and cubed, leads to a sovereign concord, not only of the ternary to 27 but of the senary to 162 and from 384 to 10,368, as we have set them out above. And from each consonant number it extends in the same fashion along, sideways, and up and down, so that all the intervals filled with their due consonance show all the intermediate genera and all the consonant species, as in the whole world-instrument: species which are distinguished by numbers, formed by figures, and underpinned by solid bodies.[18]

31

◆

Heinrich Glarean
1488–1563

Heinrich Glarean, or Henricus Glareanus, was the foremost Swiss humanist of his time, a friend of Erasmus and Zwingli in Basel and of Lefèvre d'Etaples and Guillaume Budé in Paris. He was an esteemed poet and a keen educator, founding several schools on humanistic principles to teach children fluent Latin and the rudiments of Greek. At the Reformation, Glarean regretfully left the Protestant domain and many of his friends, taking up a post at the University of Freiburg im Breisgau. Here he taught poetry, theology, mathematics, and geography as well as music.

The *Dodecachordon* is a comprehensive work written during the 1530s out of a desire to revise and rationalize the modal system and to encourage the cultivation of plainsong in the Catholic Church. To this end, Glarean analyzes both plainsong and contemporary music, especially that of Josquin Desprez, and finds that the existing system of the eight church modes (two each on D, E, F, and G) is inadequate to describe actual practice. He therefore adds four more modes, two each on C and A, calling them "Ionian" (equivalent to the modern major) and "Aeolian" (the modern natural minor), bringing the total number of modes up to a dozen—hence the allusion of the Greek title to an instrument of twelve strings. His book is both perceptive and prophetic, for it was precisely these major and minor modes that were already dominating contrapuntal music, and would soon be used exclusively.

In his approach to the Harmony of the Spheres, Glarean is again a rationalist. While recognizing that the ancient speculations were intended "to help raise the human mind in every possible way to the contemplation of heavenly objects," he found them useful only in connection with his modal program. But his careful presentation of the sources, and his generous and sympathetic attitude to the ancient writers betoken the true humanist, while in the last resort, his motives are the Christian ones of using music to honor God and to lead souls to virtue and devotion.

Source: Heinrich Glarean, *Dodecachordon*, translated by Clement A. Miller (American Institute of Musicology, 1965), pp. 132–139. Used by kind permission of Professor Miller and the publisher, Dr. Armen Carapetyan.

Why the Number "Seven" Occurs So Frequently among Writers on Music

Chapter 12. This question is also worthy of consideration, namely, what the significance is of the number seven, named by so many writers and also so differently, when they happen to mention musical matters, as in Homer, the second hymn (to Mercury): "And he stretched seven concordant strings of sheep-gut."[1] And in Vergil on *Alexis*: "I have a reed-pipe made by joining seven unequal hemlock stalks."[2] And in book 6 of the *Aeneid*: "Orpheus accompanies different tones of the voices with seven strings."[3] And in Horace, book 3, *Carmina*, ode 11: "And thou, O lyre, skilled in resounding with thy seven strings."

Macrobius writes in book 1 of *Saturnalia* that Apollo's lyre had seven strings,[4] through which it may be preferable to understand the movements of as many heavenly spheres. I believe he may have taken spheres for planets, although this speculation is vain. For if the poets had brought forth eight strings, then an acute thinker would have immediately added that through them the movements of the same number of spheres could be understood, inasmuch as there are eight in all. Contrariwise, if they had mentioned nine strings, he undoubtedly would have said there were eight movements indeed, and also the same number of sounds arising from them, but that the ninth must be understood as a concord arising from the eight strings and therefore called Calliope, as Plato philosophizes in *De Republica* on the muses,[5] and not by chance, in the opinion of all learned men. The same treatment occurs concerning the *Somnium Scipionis* of Cicero, but more about this a little later. Yet Diodorus[6] says that Mercury was the inventor of the lyre, and arranged it with three strings, after the fashion of the three seasons, also choosing three tones, the low one from winter, the high one from summer, the middle one from spring, with great injustice to autumn, since it is banished most undeservedly, although a season laden with fruit. Some state that Orpheus constructed a lyre with four strings, either according to the same seasons, so that autumn, treated unjustly previously might return to its former rank, or according to the four elements, as others have written. These reasons brought forth by commentators as an explanation of this difficulty are useful to the matter only in that they explain who has added each string on the cithara until 15 strings have been reached. For the reader's mind is no less perplexed after he has read and understood all these statements. They are the subterfuges of our ignorance, which is excessively prolific whenever it cannot extricate itself. However, it must be explained why seven strings are mentioned so often by writers, and also what should be understood by them. Everyone passes over this point lightly, just as though it were well-known, and brings forth the spheres of seven planets, as if no one understood this.

Therefore I shall say briefly what I believe, namely, that by strings (*chordae, nervi*), or stalks, or differences of tones among the writers is meant those seven octave species about which we have said so much thus far, and which occur no less in the *trichord* of Mercury, the *tetrachord* of Orpheus, and the other remaining string instruments, than in the cithara of 24 strings. Every song is guided by them, it moves forward through them, and is enclosed by them, just as a poem by poetic feet. They do not fill out the double-octave, but lack a whole tone. And so the eighth octave-species of Ptolemy has been added, which is like the first, as we have shown extensively in chapter 2 of this book. For there is no song suitable either for the cithara or another instrument which does not fall into one of the seven octave-species. But if by chance this has seemed insufficiently clear to anyone, I beg him to remember how uneducated and unpolished our present age is, that among the highly learned, even among those teaching mathematics, not one in twenty has a clear conception of this matter who wishes to apply, or more truthfully, who is able to apply a healing hand to this study. Contrariwise, let him ponder how learned and instructed in all knowledge was the age of the ancients, both Greek and Roman. Thus there cannot be any doubt when Horace sang poems to the lyre and mentioned modes so frequently, that he really knew them in the best possible way, a matter certainly not very difficult if we could have had learned teachers from childhood onward in this study, as in other disciplines not so necessary to human life as music is unanimously said to be by all the ancients. But enough now of complaints; let us hasten to other things.

Two Opinions on Sound in the Heavens, and an Examination of Passages in Cicero and Pliny Concerning These Opinions

Chapter 13. Those who have believed that there was sound in the heavens seem to have arranged it according to the first octave-species, which is Hypodorian, as is evident in Boethius, book 1, chapter 27. Moreover, since at the same place he makes the sun the *mese*, he has turned to the opinion of the ancients, which does not disagree with ours. For in the diatonic genus the interval from *nete synemmenon* to *lichanos hypaton*, in which arrangement the sun is the *mese*, is the same as that from *mese* to *proslambanomenos* (as we arrange it), and is also the same as the octave-species which arises from the first fourth-species and the first fifth-species.

But writers have fallen into diverse ways in arranging the sounds according to their highness and lowness, as we also said in passing in the first book. Some have given the higher sounds to superior bodies, the lower to inferior bodies, because those which are moved more rapidly, as the superior, also seem to sound higher; this appears to have been Cicero's opinion in book 6 of *De Republica*. On the other hand, others have ascribed higher sounds to inferior bodies, lower or greater sounds to superior bodies, because greater bodies also produce a greater sound, smaller bodies a smaller sound. This opinion seems much more probable to me, if there are indeed bodies in the heavens of such a nature as those in this sentient world. But we decided to add a diagram of this matter which places both opinions under the reader's scrutiny.[7] (See figure 7.) For on the one side the greater celestial bodies have shorter strings, the smaller have longer strings, while on the other side the greater bodies have longer strings, the smaller have shorter strings. The matter is not very difficult if taught correctly.

Moreover, it does not appear likely to me that the intervals of celestial bodies in heaven also agree in their own relationship as they do in the octave of a string, regardless of the genus in which we may finally arrange them. Yet Pliny, transmitting the Pythagorean teaching, says in book 2, chapter 22: "But Pythagoras occasionally names a whole tone, in accordance with musical ratio, as the distance from the earth to the moon. From the moon to Mercury is half of this distance, and from Mercury to Venus is about the same. From Venus to the sun is one and

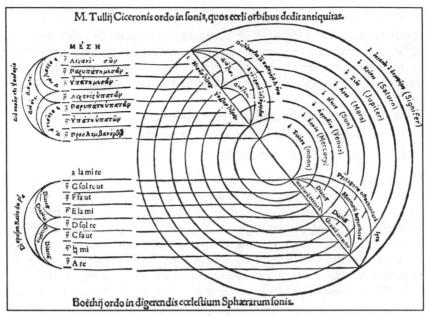

FIGURE 7: *Upward and Downward Planet Scales*

a half. From the sun to Mars is a tone, namely, as far as the moon is from the earth. From Mars to Jupiter is one half, from Jupiter to Saturn is one half, and from there to Signifer is one and a half. So six tones are produced, which they call an octave harmony, that is, a harmonious universe." So says Pliny. Not everyone interprets these words in the same way. Giorgia Valla says in book 1, chapter 2 of *Musices*,[8] that this "one and a half" in Pliny relates not to a whole tone, but to half of a tone, as he says there are three-fourths tones, which I understand as three fourths of a tone. But five whole tones and one half tone, not six whole tones are produced by this relationship. Others, relating the one and a half ratio to the whole tone, infer from Pliny's words that there are seven tones in the octave consonance, although it neither has this relationship nor has any musician proposed it anywhere, and all the old codices of Pliny have six, not seven tones. Besides, six is more nearly correct in the opinion of Aristoxenus, whom Pliny seems to have followed. In truth, however, the octave has almost six whole tones, for it lacks the comma, as we have shown in the previous book.

Yet Cicero has this in the sixth book of *De Republica*: "Therefore, the highest course of the starry heavens, whose revolution is more rapid, is moved with a high, powerful sound, yet that of the moon, which is the lowest, has the deepest sound. For the earth, the ninth body, remaining immovable, always adheres to the lowest place, and comprises the middle position of the world. But the eight paths in which the force of the two modes Mercury and Venus is the same, produce seven distinct sounds by their intervals, and this number seven is the crux of almost all problems." So says Cicero. He omits in this teaching the relationship of intervals that Pliny discusses, yet he has concluded that seven sounds are formed from eight paths by considering the extremes as the same sounds, just as it occurs in the octave. And so he voices the common saying of musicians: "Octaves must be considered the same."

However, to say what I think, this place in Cicero undoubtedly has been corrupted, and Macrobius did not understand it accurately although he wrote four very long chapters on it. When Theodore of Gaza, a man of distinguished learning, translated this passage into the Greek language, he omitted some things which are in the ordinary codices of Cicero, as these words: "Of the modes Mercury and Venus." He translated other places differently as, *hon ditte autois prosesti dunamis,* which is read frequently in some commonly known Latin codices as: *Illi autem octo cursus, quibus eadem vis est Modorum,* an interpretation differing widely from the words of Theodore. In these words of Cicero, "they produce seven distinct sounds by the intervals," some believe the word "seven" should be joined with "interval" according to the ablative case, in this sense: "These eight paths produce sounds distinguished by seven intervals." But this is undoubtedly forced, and is widely different from the phrasing of Cicero. I think this place in Cicero should be read so: "But those eight paths, in which the nature of the extremes is the same, produce seven intervals with separate sounds, and the number 'seven' is the crux of almost all problems." If these words are read in this way they are revealed according to their content, and are consistent with the tradition of musicians. For it so happens in nature, when eight strings are arranged according to a musical ratio and are measured out in a certain genus, that the eighth has the same nature as the first. And although there are eight strings, there are still only seven intervals; these, when changed about seven times in any genus, produce the seven octave-species which all the books of various authors mention very frequently, as we have discussed a little earlier. When Boethius, a true judge of this subject, saw these matters so remarkably varied among the ancients, and saw that Pliny was not afraid to speak in a pleasing way rather than with a necessary exactness, he tempered them in such a way that he could still place each opinion before the reader's eyes, as we have shown a little earlier.

Nevertheless, to indicate finally what we really think, we say frankly that this seemed to Aristotle to be an invention not without basis, and a saying more pleasant than actually true. In fact, if one should wish to link these subjects, so to speak, he will indeed find that the intervals of planets do not fit musical intervals at all, nor will he find present any definite relationship of sound either in the subject or in the effecting agent, as physicists state it. But this indulgence is allowed to antiquity, which has thought that the human mind must be raised in every possible way to the contemplation of heavenly objects. If I had not seen this discussed by great writers, I would have passed over in silence whatever concerns musical art, without any censure, I believe, but those conceptions which antiquity has made immortal with so great authority must certainly not be esteemed lightly; none the less, guided by truth, we must proceed with the matter at hand as it exists.

What Should Be Understood by "Nine Muses"

Chapter 14. But we must not pass over in silence the reason why Muses are spoken of by writers as nine Muses, since this relates somewhat to musical knowledge, as is apparent from the name. M. Terentius Varro has related that there were only three Muses among the ancients, and Saint Augustine mentions the same in the book of *Confessions*, undoubtedly because all sound is threefold, namely, produced either by singing, as from the larynx, or by blowing, as straight trumpets, or by striking, as the cithara. Some bring forth four Muses, some seven, from the seven octave-species, and this opinion would please me if it were generally accepted. But reckoning by nine prevailed among all the ancients, particularly Homer and Hesiod, the most distinguished writers on myths and on antiquity. Yet it refers to music in the same manner as numbering by seven does. Even so, I shall not relate now from whence the Muses were named, why they are thought of as young women, or what their surnames are, for these things are not pertinent to the matter. We must speak about the numbering by nine.

Plato relates in book 10 of *De Republica* that one Siren is appointed to each sphere, and through them he indicates the movement of the spheres, or rather the sound arising from this movement. On this account he calls these eight sounds *Muses*, and antiquity has not proposed more spheres of celestial bodies. And he calls the concord resulting from these the ninth Muse, which as some believe, was therefore called Calliope and considered by Hesiod the superior of all the others. Vergil, the best of the poets (as he was aware of every phase of philosophy), seems to have meant this in a line from book 9 of the Aeneid: "Thou, O Calliope, I entreat, be favorable to my singing,"[9] and he used for good reason the plural number *(vos)*, for a single Muse, as she included all the other Muses. He did not venture to say this in the seventh book concerning the Muse Erato.[10] Also, Christopher Landinus interprets this place from the 9th book of the Aeneid more correctly and learnedly than Servius, and in the same sense as we mentioned already.

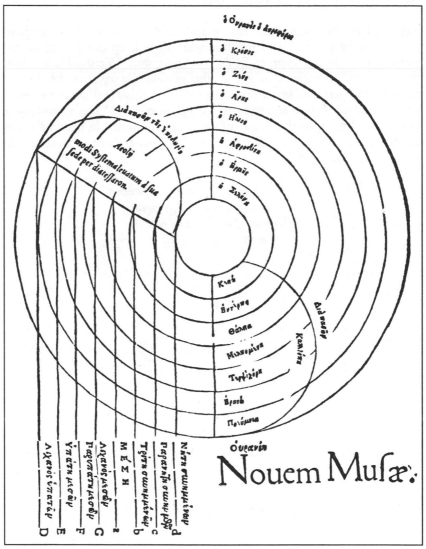

FIGURE 8: *Tones, Planets, and Muses*

But poets wished to signify by this enigma that it is allowable, as a poet, to be influenced by the upper worlds, namely, the stars, otherwise no art could be created. I am not certain myself, however, whether the names of Muses fit the sounds of the spheres or the spheres themselves in an ascending or descending order. Yet the order seems more suitable if one fits the names of the muses to these spheres in ascending order,[11] namely, that Clio is the moon; Euterpe, Mercury; Thalia, Venus; Melpomene, the sun; Terpsichore, Mars; Erato, Jupiter; Polyhymnia,

Saturn; Urania, the starry Heaven; and Calliope last of all. Grammarians will have a broad field for discussion if we keep this order, for the names of the Muses seem to fit the stars very well. We shall leave these problems for discussion by those who professedly assume the duty of interpreting writers, for it does not relate to this treatise. Now let us add a verse of Hesiod himself and a diagram of the Muses according to the celestial orbs, so that if any of these explanations can become useful to the reader, as we confidently trust, they also are not neglected here for the studious.

> *Clio, Euterpe, Thalia, Melpomene,*
> *Terpsichore, Erato, Polyhymnia,*
> *Urania, Calliope, who is truly*
> *the most excellent of all.*

32

◆

Gioseffo Zarlino
1517–1590

Zarlino, like Francesco Giorgi, was a Franciscan living in Venice, where he was *maestro di cappella* at Saint Mark's Basilica. Not for him, however, the mystical or symbolic dimensions of the *Harmonia Mundi*; like the humanists Glarean and Gafori, and indeed like Boethius himself, Zarlino was writing a treatise whose main subject was practical music. It would in fact be the most influential of all Renaissance treatises on the history of music theory. But once again, he prefaces the practical part with a consideration of other musics.

In Chapter 5, Zarlino has distinguished music into two parts: *musica animastica*, being "harmony born from the composition of various things joined together in one body," and *musica organica*, being "harmony born from various instruments," including the voice. The first of these is evidently the inaudible harmony that holds together the large world of the universe, as it does the small world of our own bodies. The following chapters break down *musica animastica* into the two traditional categories, of which the present extract treats *musica mundana*. *Music, Mysticism and Magic* contains a translation of its companion, Chapter 7 on *musica humana*.

Zarlino's account of *musica mundana* is the most succinct and balanced of his time, taking in the major classical authorities and the few relevant biblical verses, and extracting from them what is of philosophic value, rather than analyzing controversial and perhaps meaningless details of correspondences. In his very lack of fascination with the latter, he would set the tone for future music theorists, enabling them to dispense with the subject altogether, leaving it to cultivation by amateurs like Fludd, Kepler, and Kircher, or to eccentrics like Werckmeister and Tartini.

Source: Gioseffo Zarlino, *Institutioni Harmoniche* (Venice, 1558), Part I, Chapter 6. Translated by the Editor, using the edition of Venice, 1573, reprinted Gregg Press, 1966, pp. 16–21.

On Musica Mundana

Chapter 6. Returning now to the soul's music *[musica animastica]*, we will say that it is of two kinds: heavenly and human. The heavenly sort is not only that harmony which is known to exist among things seen and known in the heavens, but is also included in the linkings of the elements and in the changing of the seasons. It is, I say, seen and known in the heavens from the revolutions, distances, and placements of the heavenly spheres, as well as from the aspects, nature, and position of the seven planets: the Moon, Mercury, Venus, the Sun, Mars, Jupiter, and Saturn. For it is the opinion of many ancient philosophers, notably of Pythagoras, that the revolution of so vast a machine at such speed could not possibly take place without giving forth some sound. This opinion, although rejected by Aristotle, was nonetheless favored by Cicero, in Book 6 of his Republic, when the elder Scipio Africanus replies to the younger. The latter asks: "What is this sound, so loud and beautiful, which reaches my ears?" The elder replies, "It is caused by the impulse and movement of the spheres themselves, joined by unequal intervals but nonetheless distinguished by definite proportions. The high sounds mixed with low ones make different harmonies: for so great a motion could not take place in silence. And Nature has arranged that the extremes of one end should sound low, those of the other end high. Therefore the highest circuit, that of the starry heaven, which has the fastest revolution moves with a higher and louder sound: and the lowest, lunar, one with the deepest sound." Thus Tullius writes, following the opinion of Plato, who, to show that harmony is born from this revolution, imagines that on each sphere there sits a Siren. By a Siren he means nothing less than one who sings to God. Hesiod suggests the same in his *Theogony*, calling *Ourania* the eighth Muse who belongs to the eighth sphere, that of *Ouranos:* the name by which the Greeks called Heaven. And to show that the ninth sphere is the one which produces the grand and concordant unity of sounds, they called it *Kalliope*, which comes to signify the eighth tone. By this they meant to show the harmony which results from all the other spheres, as the poet indicates with the words: "You, O Calliope, I beseech with uplifted voice." He invokes Calliope alone among their number as the principal one, at whose will all the others move and turn.

So firmly did the Ancients believe this that at their sacrifices they would use musical instruments and sing some hymns composed of verses set to music. These contained two parts, one called *strophe* and the other *antistrophe*, to show the various gyrations made by the celestial spheres: by one they indicated the movement which the sphere of the fixed stars makes from East to West, and by the other the various movements made by the other spheres, those of the planets, moving in the contrary direction from West to East. And with similar instruments they would also accompany the bodies of their dead on the way to burial, because they thought that after death the soul will return to the place whence music derives its beauty, namely to heaven. The ancient Hebrews also observed such customs on the death of their kinsfolk, as is clearly testified by the Gospel [Matthew 9] in which is described the raising of the daughter of the ruler of the Synagogue: musical instruments were present, whose players our Lord told to cease playing. This they did, as Saint Ambrose says,[1] in observation of the custom of their ancestors, who would thus invite those standing around to mourn with them. Again, many believed that in this life every soul could be conquered by music, and that, although trapped in the prison of the body, by remembering and being aware of the music of heaven it might bear every hard and wearisome pain.

This does not seem strange in view of the testimony of the Holy Scriptures to the harmony of heaven, as when the Lord speaks to Job, saying: "Who will tell of your ordinances, O voices of the heavens? And who will make their music sleep?"[2] And if I were asked why it happens that so great and sweet a sound is not heard by us, I could not reply otherwise than Cicero does, in the place cited above: that our ears, filled with so great a harmony, are deafened. The same happens, for example, to those who inhabit the regions called Catadupa, where the Nile rushes down from great mountains: on account of the enormous noise, they lack the sense of hearing. Or, again, as our eye cannot stare at the light of the Sun, our sight being overcome by his rays, thus our ears cannot make out the sweetness of the celestial harmony on account of its excellence and grandeur.

But no reasoning will better convince us that the world is established by harmony, than that its very soul is harmony, as Plato says; also that the heavens are turned around by their intelligences with harmony, as one can see from their revolutions which are proportionately slower or faster with regard to each other. Again, this harmony is known from the distances of the celestial spheres, since the distances between them (as many believe) are in harmonic proportion: not measurable by the senses, but nonetheless measurable by reasoning. For the Pythagoreans, as Pliny shows, measuring the distance of the heavens and the intervals between them, made out the distance from the Earth to the first, lunar sphere to be no less than 126,000 stadia, and this they called the interval of a tone. (But it seems to me completely irrational to attribute a sound to the Earth, knowing that things which are by nature immobile, as the Earth is, cannot possibly generate harmony. As Boethius says, sounds have their origin in movement.) Continuing, they place between the sphere of the Moon and that of Mercury the interval of a major semitone; between Mercury and Venus a minor one; and between Venus and the Sun a tone plus a minor semitone. They say that the Sun is three tones and a semitone distant from the Earth: the interval called a fifth. And from the Moon to the Sun they reckon the distance of two tones and a semitone, which constitutes the interval of a fourth. Returning now to the principal sequence, they say that the distance from the Sun to Mars is the same as that from the Moon to the Earth; from Mars to Jupiter is the interval of a minor semitone; from Jupiter to Saturn is the distance of a major semitone; and from Saturn to the last heaven, where the zodiacal signs are, they reckon the distance of a minor semitone. Consequently the distance or interval from the last heaven to the sphere of the Sun is five tones and two minor semitones, i.e., the octave.

Whoever will examine the heavens in detail, as Ptolemy did with such great diligence, will find by comparison of the twelve portions of the Zodiac, in which are the twelve heavenly signs, various musical consonances: the fourth, fifth, octave, and others in their turn. He may know that the deepest tones are associated with the movements made from East to West, and the highest ones with the midheaven. In their elevation he may find the diatonic, chromatic, and enharmonic genera. Similarly in the horizontal dimension he may find the tropi or modes, which we will name, and in the face of the Moon, according to its various aspects with the Sun, the conjunctions of the tetrachords.

Not only in the things mentioned may one find such harmony, but also in the various aspects of the seven planets, in their nature, and in their positions or sites. First, from the aspects which they make with inferior things, such as Trine, Square, Sextile, Conjunction, and Opposition, and according to their good or bad influences, comes such a diverse harmony of things that it is impossible to describe it. Then, as to their nature, there are some (as the astrologers say) of an unhappy and malignant nature, which come to the good and benign ones to be tempered. From this results harmony, and great convenience and advantage to mortals. One finds the same in their sites or positions, because they are related among themselves in the same way as the Virtues and Vices. Thus, just as those who go to extremes can be restored to virtuous conduct by a suitable mediator, so these naturally malignant planets can be reduced to temperance by means of another, benign planet placed in their midst. Suppose, then, that Saturn and Mars, malignant by nature, are placed in an elevated position: if Jupiter is placed between them, or if the Sun is beneath Mars, then that malignity is tempered by a certain harmony, and its evil influence is not permitted to operate with such ill effect on lower things as would be the case without this intervention. Their influences have so great a power over inferior bodies that at the time of year when the two planets first named have the rulership, the harmony of the four elements is disturbed: for they corrupt the air in such a way that it generates universal pestilence in the world.

They say, moreover, that the two greater luminaries, the Sun and Moon, cause a corresponding harmony of benevolence among men: for instance between two people born, the one with the Sun in Sagittarius and the Moon in Aries, and the other with the Sun in Aries and the Moon in Sagittarius. A similar harmony is struck, they say, when they have had the same sign, or similar ones, at birth; or the same planet, or similar ones, on the ascendant; or if two benign planets are in the same aspect to the rising cusp. It can also occur if Venus is in the same house of their horoscopes, or in the same degree.

It was after considering all these opinions, and the world (in the words of Mercurius Trismegistus) as the instrument of God,[3] that I defined the music of the spheres above as the harmony which exists among things seen and known in the heavens; and that it includes in addition the linkings of the elements. Because these states, as well as all other ones, were created by God, the great Architect, by "number, weight, and measure" [Wisdom 11:21], we may understand such harmony in each of these three things. It occurs in Number, first, on account of the qualities found in the elements, which are four and no more: dryness, coldness, moisture, and heat. There is one quality most appropriate to each of them: to Earth dryness, to Water coldness, to Air moisture, to Fire heat. Dryness is also attributed as a secondary quality to Fire, heat to Air, moisture to Water, and coldness to Earth. Hence, although there are elements which are contrary to one another, they are nevertheless concordant and united in one quality through a mediating element. Although with each one of them, as we have seen, two others are incompatible, by this means they are marvelously joined together. Just as when two square numbers meet in a mean, proportional number, so two of these elements are joined through one intervening. The squares 4 and 9 meet in the number 6, which exceeds 4 by the same proportion as it is exceeded by 9. In such a manner Fire and Water, which are contrary in two qualities, are joined through an intervening element. Fire is naturally hot and dry, Water cold and moist; through warm and moist Air they are miraculously united in fit proportion. For while repelled by Water on account of its heat, it is united by its moisture. And if the moisture of Water repels the dryness of Earth, their coldness nevertheless brings them together. Thus they are mutually joined in so marvelous an arrangement that no more disparity remains among them than exists between two mean proportional numbers, joined by means of two cubic numbers—as can be seen clearly in the figure below.[4] (See figure 9.)

Boethius explains this harmonious linking, saying:

> Thou temp'rest elements, making cold mixed with flame
> And dry things join with moist, lest fire away should fly,
> Or earth, opprest with weight, buried too low should lie.
> Thou in consenting parts fitly disposed hast
> Th' all-moving soul in midst of threefold nature placed.[5]

And, in another place:

> This sweet consent
> In equal bands doth tie
> The nature of each element,
> So that the moist things yield unto the dry.
> The piercing cold
> With flames doth friendship keep,
> The trembling fire the highest place doth hold
> And the gross earth sinks down into the deep.[6]

Those who wish to understand the heavenly harmony through Weight will find it there, too. For each of the elements is relatively heavier or lighter; and they are mutually enchained and bound so that the circumferences of each are distant from the center of the world in proportions which make a certain harmony. We can see that those which are naturally heavy are drawn upward by those which are naturally light, and the heavy ones draw the light ones downward in such a way that none of them can stray outside its own place. In this manner they remain

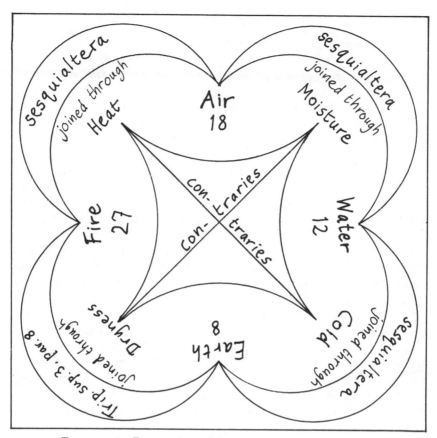

FIGURE 9: *Proportions Between the Four Elements*

forever united and compacted together, so that the vacuum, utterly abhorred by Nature, may nowhere occur among them, not even momentarily. Their mode of conjunction is such that Earth, which is simply heavy by nature, and Fire, which is simply light, possess the extreme positions. Earth is at the bottom because every weighty thing tends downward; Fire is at the top, since every light thing tends thither. But since the middle elements reflect the nature of the extreme ones, the Creator has well arranged it so that Water and Air, in certain respects both heavy and light, should hold the middle place: Water, being heavier, accompanies Earth; Air, being lighter, is next to Fire. So each accompanies the one most similar to it in nature. This order and linking is charmingly expressed by Ovid, who says:

> *The fire most pure and bright,*
> *The substance of the heaven itself, because it was so light*

Did mount aloft, and set itself in highest place of all.
The second realm of right to air, for lightness did befall.
The earth more gross drew down with it each weighty kind of
matter,
And set itself in lowest place. Again, the waving water
Did lastly challenge for his place, the utmost coast and bound,
Of all the compass of the earth, to close the steadfast ground.[7]

But should we wish to examine the matter more subtly still, we will find the celestial harmony in their measure and quantity, through the transmutation of parts which occurs from one to another. The Philosopher[8] shows that just as one part of earth is transmuted into water, one part of air is transmuted into fire; and that as one part of fire is transmuted into air, and one part of air into water, so one part of water is transmuted into earth. When the earth is transmuted into water, this transmutation takes place in decuple proportion, in the following way: when one handful of earth is transmuted into water, according to the Philosophers, ten handfuls of water are generated; and when that amount of water is transmuted into air, it makes a hundred handfuls of air. Finally, when all this is transmuted into the highest element, it is multiplied into a thousand handfuls of fire. Contrariwise, a thousand handfuls of fire are converted into a hundred of air, those into ten of water, and ten of water into one of earth. Because of their rarefaction and density there is more in one than in another. The closer they are to heaven and the further from the center of the world, the more rarified they are; the further from heaven and the nearer to the center of the world, the more dense they are.

If therefore one wished to judge their measure, it could be said that the quantity of fire is in decuple proportion to that of air; the quantity of air, similarly, in decuple proportion to that of water; and the quantity of water in the same proportion to the entire quantity of earth. Now the elements are all bodies of the same type: the whole with all its parts concords in a single nature and follows the same law. So one can also say that the proportion found between the sphere of fire and the whole mass of earth is the same as between the number 1000 and unity.

From the movement, distances, and parts of heaven; from the aspects, nature, and positions of the seven planets; and from the number, weight, and measure of the four elements, we arrive at knowledge of the heavenly harmony. This concord and harmony divide the harmony of the seasons, seen primarily in the year through the changing of Spring into Summer, Summer to Autumn, Autumn to Winter, and Winter to Spring. It is also found in the months, in the regular waxing and waning of the Moon, and lastly in the day, as light and darkness appear alternately. From this harmony arises the diversity of flowers and fruits; which is why Ovid says in this regard:

Autumn gives fruits; Summer is fair with harvest;
Spring provides flowers; Winter is cheered by fire.[9]

When hot and cold, dry and moist, unite proportionately, as Plato says,[10] there results from the harmony of these qualities the year, so serviceable to every living thing, full of divers sorts of fragrant flowers and excellent fruits; and no other kind of plant, nor any animal suffers hurt. But if the opposite should befall, discord and distemper will engender pestilence, sterility, infirmity, and everything hurtful to men, beasts, and plants. Truly Nature has followed a fair and excellent rule in arranging that whatever Winter shrinks and seizes up, Spring should open and make fruitful: what Summer dries, Autumn finally ripens. One season thus renders assistance to another, and of the four seasons harmoniously arranged a single entity is made.

Mercury and Terpander would have understood this harmony very well. One of them, having discovered the lyre, or cithara, put upon it four strings in imitation of the heavenly music (as Boethius and Macrobius say), which is perceived in the four elements, or in the variety of the four seasons of the year. The other strung it with seven strings after the likeness of the seven planets. Henceforth the one with four strings was called Quadrichordo or Tetrachordo, meaning the four-stringed one, and the one of seven strings Heptachordo, the seven-stringed one. But the first one was accepted and embraced by musicians, so that the fifteen notes of the Greater Perfect System might be built up according to the number of the strings of the said Tetrachord— as we shall see; also so that the distances from one to another might accord with different proportions. And this will suffice for the exposition of *musica mundana.*

33

✦

Jean Bodin
1523 or 1530–1596

An enigmatic figure at an enigmatic time, Bodin moves through the religious and political currents of late sixteenth-century France, always in touch with people of influence, always at the center of controversy, often working to heal the wounds of religious division, but sometimes causing further trouble. His denunciation of witchcraft, *De la démonomanie des sorciers* (1580) unleashed a spate of persecution and witch-hunting, while also doing its best to blacken the reputations of the Christian Kabbalists Pico della Mirandola and Cornelius Agrippa. It was not that Bodin disbelieved in the efficacy of Kabbalistic magic, but rather that he regarded its use of Hebrew as a desecration. His views were in the end more Jewish than Christian.

While in some respects Bodin was extremely bigoted, in others he showed an openness to religious plurality rare for his time. The *Colloquium Heptaplomeres*, completed in 1588 but not published until 1857,[1] has the form of a conversation between representatives of seven faiths: Catholic, Lutheran, Calvinist, pagan, Jewish, Muslim, and Naturalist. They are all searching for a truth on which they can agree and be united. Bodin's vision of a religious harmony which recognizes different ways to God is explicitly inspired by his belief in a universal harmony of numbers and of Nature; it is also close to Pico's optimistic synthesis. At the close of each day's conversation, the participants are entertained by boys performing sacred music with voices and instruments, in a ceremony that restores harmony between the seven friends, no matter what their temporary disagreements may have been. The entire Colloquium is like a work of musical polyphony in which the different voices combine, both in concords and discords, to make a harmonious whole that is true to Nature itself.

Source: Jean Bodin, *Colloquium of the Seven about Secrets of the Sublime*, translated by Marion Leathers Daniels Kuntz (Princeton, N.J.: Princeton University Press, 1975), pp. 144–147. Copyright © 1975 by Princeton University Press. Excerpt reprinted with permission of Princeton University Press.

A Conversation on Harmony

On the following day Octavius brought Coronaeus[2] a tragedy that he had written about the parricide of three children of Prince Solimannus; Coronaeus asked me to read it because of the erudition of the author and because of the merit of the theme. After I had completed the reading at lunch, all congratulated Octavius. Coronaeus praised him profusely and said that the tragedy had been elegantly composed in regard to the choice of words, the seriousness of the sentences, the arrangement of the topic and the variety of verses.

When they had given thanks to God according to their custom, they sang hymns of praise to their soul's delight. Coronaeus said: Often I have wondered why there is such sweetness in a tone that has the full octave, the fifth and the fourth blended at the same time; just now you have heard the sweetest harmony with the full system of the highest tone blended with the lowest, with the fourth and fifth interspersed; although the highest tone is opposite to the lowest, why is it that harmonies in unison, in which no tone is opposite, are not pleasing to the trained ear?

Fridericus:[3] Many think that the harmony is more pleasing when the ratios of numbers correspond.

Curtius: I am amazed that the most learned men approve of this, since no ratios seem to combine more aptly than geometric progressions; the last members accord with the first, the middle with each, all with all, and also positions and orders are related, as 2, 4, 8, 16. Still in these systems that most pleasing harmony fails. When the numbers are arranged in this manner, 2, 3, 4, 6, and the ratios have been separated, we delight in this harmony. Indeed, what is the reason that the interval of the pure fifth (3/2) is most pleasing, but the apotome (9/8) is heavily offensive?

Octavius: I think harmony is produced when many sounds can be blended; but when they cannot be blended, one conquers the other as the sound enters the ears, and the dissonance offends the delicate senses of wiser men.

Senamus:[4] I do not think a ratio of numbers or a blending of tones produce this sweetness, since a variety of colors presented to the eyes is more pleasing than if all are mingled simultaneously. Likewise, the flavor of fresh oil and vinegar is very pleasing, but it cannot be mingled by any force. Also the most dissimilar songs of birds, blended by no ratio, produce a most pleasing delight for the ears. Plato thought it strange that no dissonance is perceived in the song of birds, however much it is joined with men's voices or lyres.

Toralba:[5] Indeed I think that pleasing delight of colors, tastes, odors, and harmonies depends on the harmony of the nature of each, a harmony which depends on the blended union of opposites. For example, something too hot or too cold offends the touch; likewise too much brightness or too much darkness offends the sight, and too much sweetness and too much bitterness offends the taste. But if these are blended by nature or art, they seem most pleasing. I find it hard to agree with Seneca's opinion,[6] based on the Stoics, that nothing bad can touch a good man, since, he says, opposites are not blended. If boiling water is mixed with the coldest and driest dust, the greatest blending of opposites exists, tempered by art and pleasing to touch. We also see elementary bodies, which are joined together in nature herself, blended from opposing qualities and elements which Galen[7] thought could not be united by any craft. And so we can defend Seneca and vindicate him from censure if we say that he spoke about substances, not about qualities and accidents. Indeed nothing can be so easily united as water with wine; still they are not blended among themselves as things which are mingled by nature, because wine is separated from water by soaking it in a sponge which has been saturated with oil. Likewise gold is blended by design with silver, and bronze infused with silver; they are separated by gold water, yet they would never be drawn apart if they had been blended by nature herself. An example is amber, which nature herself tempered with equal portions of gold and silver.

Senamus: If there is no contrariety in substances, how is it possible that contrary substances are blended?

Toralba: Aristotle believed that nothing was contrary to substance.[8] However, since there is opposition of form with form, as the form of fire to the form of water, and since the contrariety of accidents, for example, extreme dryness and humidity, severest heat and cold, could happen only from the contrariety of the forms of fire and water, which differ in their whole nature, who can doubt that forms of each, that is, substances are contrary among themselves? Indeed that most certain decree in nature is proclaimed, not as plainly as appropriately for the question. On account of this one each thing is of such a kind, and that the more. If accidents of fire and water are opposite because of the power of forms, forms must be much more contrary among themselves. Therefore, things which are contrary to each other in nature herself cannot be mingled by design, but only blended, joined, or united so that they seem to be one. For example, oxymel is very pleasing to taste; it is made from vinegar and honey; and from the bitter with a light burning of sweet meats it becomes bittersweet,[9] a flavor most pleasing to the palate.

Fridericus: In musical modulations that contrariety does not seem to be destroyed, but extreme opposites are brought together by intermingling of the middle tones. For the simple sound of the *hypaton* produces a sweet harmony with the highest sound of the *hyperbolaion* or the whole octave since from opposites it is united in the whole diameter. If you shall join the middle voice to these, with one it will produce a full octave, with another a fifth, from which comes the most pleasing harmony of all modulation from the proper union of opposites.[10]

Toralba: This is very apparent in all of nature. Opposites when united by the interpolation of certain middle links present a remarkable harmony of the whole which otherwise would perish completely if this whole world were fire or moisture. In like manner tones in unison would take away all sweetness of harmony.

Fridericus: Surely heavenly revolutions, though contrary and moving in an agitated pattern, are held in check, and the contrary force of Mars and Saturn is restrained by the intermediate light of Jupiter.

Senamus: How then does it happen that we see that salutary mean always confounded by some hindrance, as health with sickness, pleasure with pain, peace of mind with anxiety?

Curtius: That impediment is as useful as a drainage ditch is in a city. The poisonous toad in the garden or the spider in the house are as necessary for gathering poisons as the hangman in the state. Even that keenest sweetness of harmony which we have heard most eagerly just now would not have been so pleasing unless the musician had contrived some dissonant or harsh note for our sensitive ears, since the pleasure is not perceived without a pain that precedes it and produces boredom when continued too long. We have attempted to imitate in these verses this contrariety in all things which has been tempered by immortal God with remarkable wisdom:[11]

Creator of the world three times greatest of all,
Three times best parent of the heaven,
Who tempers the changes of the world,
Giving proper weight to all things,
And who measures each thing from His own ladle
In number, ratio, time,
Who with eternal chain joins with remarkable wisdom two things
  opposite in every way, preparing protection for each,
Who, moderating melody with different sounds and voices yet most
  satisfying to sensitive ears, heals sickness, has mingled cold with
  heat and moisture with dryness,
The rough with the smooth, sweetness with pain, shadows with
  light, quiet with motion, tribulation with prosperity,
Who directs the fixed courses of the heavenly stars from east to
  west,
West to east with contrary revolutions,
Who joins hatred with agreement,
A friend to hateful enemies.
This greatest harmony of the universe though discordant contains
  our safety.

IV

BAROQUE

34

Johannes Kepler
1571–1630

Kepler stands out among all the authors in this collection, because while many of them talk happily about the Harmony of the Spheres, he alone takes the trouble, and has the expertise, to demonstrate it. Living at the most exciting era astronomy has ever seen, Kepler readily accepted the Copernican scheme of a heliocentric universe. But he went further than Copernicus and even Galileo in breaking with the dogma, common to Aristotelians and Platonists alike, that the heavenly bodies had to move in perfect circles. This enabled him to establish that the true orbits of the planets in our solar system are not circular, but elliptic, with one focus occupied by the sun.

Such a wrench away from both popular and scientific opinion necessitated both courage and a strong element of self-persuasion. Kepler had to show how the elliptical orbits did not impair the perfection of God's design, but revealed an even more delightful perfection: that of geometry ruled by musical harmony. The following long extract from his masterwork, *Harmonices Mundi*, enables his argument to be followed in its every step. Preceding that is an earlier venture into the musical explanation of the cosmos, from *Mysterium Cosmographicum*. I would make a connection from this chapter on the division of the Zodiac to the excerpts from Censorinus (no. 6): in both cases, some common rules or beliefs based on the twelve signs are shown to have a musical rationale.

Kepler's obsession with musical harmonies, like Newton's with alchemy, seems to the conventional mind to be a lapse on the part of a great scientist. Yet what makes them "great" might be their freedom from such conventional views, and their openness to other dimensions of their subject. The recent researches of Rudolf Haase (included in my collection *Cosmic Music*) show that the predominance of harmonious intervals in the solar system, as discovered by Kepler, not only far exceeds random expectation, but is reinforced by measurement of the outer planets that were not yet discovered in Kepler's day. So Kepler was right, and it remains for us to draw conclusions appropriate to our own time and convictions.

Source: Johannes Kepler, *Mysterium Cosmographicum* (Tübingen, 1596), Chapter 12, and *Harmonices Mundi Libri V* (Linz, 1619), Book V, Chapters 5–7. In Kepler's *Gesammelte Werke*, edited by Max Caspar, vol. I (Munich: C. H. Beck, 1938), pp. 39–43, and vol. VI (1940), pp. 317–323. Translated by the Editor. Also consulted: Caspar's German translations, *Mysterium Cosmographicum: Das Weltgeheimnis* (Augsburg: Filser, 1923), and *Harmonices Mundi: Weltharmonik* (Munich & Berlin: Oldenbourg, 1939). These translations were previously published in *Cosmic Music*.

◆

From Mysterium Cosmographicum:
The Division of the Zodiac and the Aspects

Chapter 12. Many people consider the division of the Zodiac into exactly twelve signs a human invention, unsupported by any natural phenomenon. They believe that these divisions do not differ in natural force or influence, but that they are merely made because the number is suitable for calculation. While I do not entirely disagree with them, in order not to reject something out of hand I would suggest, on the grounds of this division, a reason why the Creator may have disposed these qualities (if indeed they do have distinct ones).

We have seen above what the object of numbers is. To be sure, without quantity, or something that is like quantity and endowed with a certain power, nothing in the whole universe can be numbered except

God, who is the Holy Trinity itself. Now, therefore, we have sectioned all bodies by means of the Zodiac. We shall see what the Zodiac itself has attained or suffered by this sectioning. By sectioning in the aforesaid way the cube and octahedron, a square will result; from the pyramid a triangle; and from the other two figures a decagon. 4 x 3 x 10 makes 120. Therefore when a square, a triangle, and a decagon are inscribed in a circle, starting from the same point, they will mark off various arcs on the circumference, all of which are measured by a portion no greater than the 120th part of the whole circle. Thus the natural division of the Zodiac into 120 arises from the regular placement of the solids between the orbits. Since three times this is 360, we see that this division is in no way irrational. Now if we inscribe a square and a triangle separately, starting from the same point, the smallest arc will be $^1/_{12}$ of the circumference, namely one sign. And it is remarkable that both the monthly movement of the sun and moon and the great conjunctions of the outer planets[1] so nearly fit the arcs which are determined by the same solids through the triangle and the square.

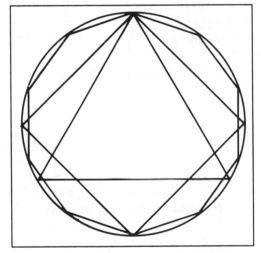

FIGURE 10: *The Astrological Aspects*

Furthermore, you can see from another example how highly nature prizes this twelvefold division; for although the cause is not known, it gives occasion to learn more about these five figures. Take a string tuned to G. As many notes as are consonant with G between it and the higher octave (g), so many divisions can one make in the string in which the parts are consonant both with each other and with the whole. The ear will tell how many such notes there are. I will give them in notation and numbers:

FIGURE 11: *The Intervals Corresponding to Aspects*

You will now see both those harmonies and the proportions of string lengths in numbers: the lower note represents the note of the entire string; the upper one, the shorter portion; the middle one, the longer portion. The largest number indicates in how many parts the string is to be divided; the others, the length of the portions.

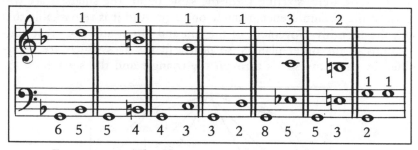

FIGURE 12: *The Harmonious Divisions of a String*

And these notes seem to me the only natural ones, because they have definite numbers. The other notes cannot be expressed in exact (*certa*) proportion to the ones already given. For the note F is different, depending on whether you reach it from C above or from B-flat below, although these both seem to be perfect fifths. But to resume: the first and second consonances are, as it were, neighbors: thus, too, the fifth and sixth. For they are all imperfect,[2] and so always unite in pairs—one major and one minor—as if to resemble the single and perfect ones. Their ratios are also not very different. For 1:6 and 1:5 are to one another as 5:30 and 6:30, differing by only 1:30. Similarly, 3:8 and 2:5 are to one another as 15:40 to 16:40, and therefore differ by only 1:40. And thus properly speaking we have only five consonances in music, the same as the number of solids.[3] Now if one seeks the lowest common multiple of 6, 5, 4, 3, 8, 5, 2, one will again find 120, as we did above when treating the division of the Zodiac; but the lowest multiple of the perfect consonances is again 12. It is just as though the perfect consonances came from the square and triangle of the cube, tetrahedron and octahedron, the imperfect from the decagon[4] of the other two solids. This is the second correspondence of the solids with the musical consonances. But since we do not know the causes of this correspondence, it is difficult to accommodate the individual intervals to individual solids.

We see, indeed, two orders of chords, three simple and perfect, and two duplex and imperfect; likewise three primary solids and two secondary ones. But since the rest does not agree, we must give up this relationship and try another. Just as the dodecahedron and icosahedron above increased the 12 to 120 through their decagon, so here the imperfect harmonies do likewise.

Let therefore the perfect consonances correspond with the cube, pyramid, and octahedron, and the imperfect to the dodecahedron and isosahedron. Then something else occurs which, to be sure, points a finger at the deeply hidden cause of these things (which we will discuss in the next chapter): for there are two treasures in geometry: one, the ratio of the hypotenuse to the side of a right-angled triangle; the other, the Golden Section. The construction of the cube, pyramid, and octahedron derives from the former, that of the dodecahedron and icosahedron from the latter. That is why it is so easy and regular to inscribe a pyramid in a cube, an octahedron in either, or a dodecahedron in an icosahedron. But it is not so simple to accommodate the individual intervals to the individual solids. It is only clear that the pyramid should be the interval they call a fifth (our no. 4), because in it the lesser part is $1/3$ the greater, just as the triangle's side subtends $1/3$ of the circle. Many things will confirm this when we treat the aspects below; but to understand it here we must imagine that the string is not a straight line but a circle. The said interval will therefore be given by a triangle whose angle opposes the side as the corner of a pyramid opposes the surface. For the cube and octahedron, then, the octave and fourth remain (our nos. 3 and 7). But which of them supports which interval? Or should we say that the secondary figures contain the intervals which lead to straight lines, the primary ones those which lead to figures? Then the cube would be the fourth. For if you make a circle of the string, and draw a line from one quarter to another until you return to the same point, you will make a square—which also arises from the cube. To the octahedron will correspond the octave, which is half the string. For if one divides the circle in two and joins the points of division, one will only have a line. So the dodecahedron will be the first double imperfect consonance. For if a circle is divided in five and six, the pentagon and hexagon will result. The icosahedron remains as the second double imperfect consonance, since only lines result when one joins points of $2/5$ of the circle apart until one returns to the starting point. It is the same when joining points $3/8$ of the circumference apart. Or would we rather give the octahedron the fourth, since its edges quarter the circle twelve times, as no cube's edge does? Thus the cube

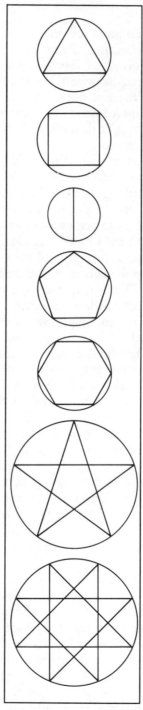

FIGURE 13: *The Regular Divisions of a Circle*

would be left for the octave, the most perfect interval, just as it is the most perfect solid. Perhaps it is also more accurate to give the icosahedron the first imperfect consonance on account of the hexagon, which is closer to its triangular surfaces than the pentagon, but to give the dodecahedron the eightfold division because of the cubic number 8, since the cube can be inscribed in the dodecahedron. All this is in the balance, until someone can find the causes.

Let us now look at the aspects.[5] Since we have already made a circle of the string, it is easy to see how the three perfect consonances can be compared beautifully with the three perfect aspects; opposition, trine and square. The first imperfect consonance, B-flat, is very close to the sextile, which is said to be the weakest aspect.

Thus we have a reason (such as Ptolemy did not give) why planets separated by one or five signs are not counted as being in aspect. For as we have seen, nature knows of no such concords in tones. Since in the other cases the ratios of influence and interval are the same, it is credible that it is the same here. Without a doubt both have the same cause, to be sought in the five solids: but I leave the seeking to others. Since, then, all four intervals concord with their aspects, while in fact three more intervals remain in music, I once suspected that in judging a nativity one should not neglect planets 72°, 144°, or 135° apart, especially since I see that one of the imperfect consonances has its aspect. It will soon be obvious to any observer of meteorological phenomena whether these three rays contain any power, since constant experiments confirm the changes of the atmosphere under the other aspects. The causes which one may plausibly adduce for the fact that $^3/_8$, $^1/_5$, and $^2/_5$ sound on the string but do not operate in the Zodiac may be as follows:

1. One opposition, two squares, a trine plus a sextile each make a semicircle: but these three aspects have no kinship with this function which music would not absolutely reject.[6]

2. The remaining aspects have a simple rational from the diameter, while the one subtending one or two sides of a pentagon or three of an octagon are one step more remote, and irrational.

3. Another cause: a trine plus a sextile, a square plus a square, make a right angle: the other aspects no such angle with any accepted line.

4. The imperfect consonance of B-flat is in a way perfect, since it possesses the same division as the perfect ones, and is half a fifth. Therefore it is not surprising that among the imperfect consonances it alone corresponds to any aspect, namely the sextile, which itself is half the trine. The others do not fit the twelvefold division, nor are they parts of any perfect consonance.

5. Finally, six angles of a trine, four of a square, three of a sextile, and

the space enclosed in two semicircles fill the whole area of the plane. But three angles of a pentagon are less than four right angles, while four are more. So it is also clear from this why neither the aspect of an octagon, nor that of a decagon, nor any other is efficacious.

And here I would separate the causes of aspects from those of harmony. Certainly a reason drawn from angles is true of aspects, since they work on account of the angle made on a point on the earth's surface at which they meet, not because of their configuration on the circle of the Zodiac, which exists rather in the imagination than in reality. Now the division of a string is not made on a circle, nor does it use angles, but is done on a plane with a straight line. Nevertheless, concords and aspects may have something in common, since, as we have said, they both have the same cause. I leave it to the industry of others to investigate them.[7]

From Harmonices Mundi:

In the Proportions of the Apparent Planetary Motions (As If for Observers on the Sun) Are Expressed the Steps of the System of the Notes of the Musical Scale, as well as the Major and Minor Modes

Book V, Chapter 5. Since therefore harmonic proportions exist between these twelve termini or motions of the six planets circling the sun upward and downward, (or at least intervals that approach these by an imperceptible difference), we have proved what was posited in numbers on the one hand by Astronomy, on the other by Harmonics. We first elicited in the third book the single harmonic proportions one by one in Chapter 1, then in Chapter 2 compiled every one of them into a common system or musical scale, or rather divided one of them, the octave, which embraces in its power the rest, into its degrees or steps by means of the others, so that a scale arose.[8] Now having formed the harmonies which God himself has incorporated into the world, we shall see whether they stand singular and separate, none having a relationship with the others, or whether in truth they all fit together. Certainly it is simple to conclude, without further investigation, that these harmonies are fitted one to another with the greatest ingenuity, as if they were parts of a single assemblage, so that none oppresses another. In all our manifold comparison of the same terms we see that harmonies never fail to arise. But unless all things were fitted together in one scale, it could easily happen (and here and there it necessarily has happened) that many dissonances should exist. If, for example, one were to put a major sixth between the first and second terms, and between the

FIGURE 14: *Inharmonious Intervals*

second and third a major third, without respect to the first, one would admit the dissonant and unmelodious interval of 12:25 between the first and the third.

Let us see whether what we have now learned by reasoning is found in reality. We would, however, issue some cautions so as not to impede our progress. First, we must for the present ignore those excesses or defects which are smaller than a semitone: what their cause is, we shall see afterward.[9] Next, we shall reduce everything to the system of one octave by means of the doubling and halving of motions,[10] because of the identical nature of every octave.

The numbers which express all the steps or notes of the octave system are shown in a table in Book III, Chapter 8, folio 47:[11] these numbers are to be understood as lengths of two strings. Consequently the speeds of the motions will be in inverse proportion to one another.[12]

Let us now consider the movements of the planets in the fractions which arise after repeated halving. They are as follows:

[Diurnal] Motion				
of Mercury at Perihelion	divided by		2^7 or 128	3' 0"
of Mercury at Aphelion	"	"	2^6 or 64	2'34"-
of Venus at Perihelion	"	"	2^5 or 32	3' 3"+
of Venus at Aphelion	"	"	2^5 or 32	2'58"-
of Earth at Perihelion	"	"	2^5 or 32	1'55"-
of Earth at Aphelion	"	"	2^5 or 32	1'47"-
of Mars at Perihelion	"	"	2^4 or 16	2'23"-
of Mars at Aphelion	"	"	2^3 or 8	3'17"-
of Jupiter at Perihelion	"	"	2	2'45"
of Jupiter at Aphelion	"	"	2	2'15"
of Saturn at Perihelion				2'15"
of Saturn at Aphelion				1'46"

Now let the slowest motion at the aphelion of the slowest planet, Saturn, represent the lowest note G of the system, within the number 1'46". The same step will also represent the aphelion of Earth, only five octaves higher (for who would argue about 1" difference from the motion of Saturn's aphelion? The difference would be no more than 106:107, which is less than a comma). If one adds to this 1'47" a quarter, i.e., 27", it makes 2'14", while the perihelion of Saturn is 2'15" and Jupiter's aphelion the same, but an octave higher. Therefore these two motions represent the note B, or a very little higher. If one takes from 1'47" a third, i.e., 36"-, and adds it to the whole, it will make 2'23"-for the whole note C: and here is the perihelion of Mars, the same magnitude except four octaves higher. Add to the same 1'47" its half, 54"-, making 2'41"- for the note D, and just here is the perihelion of Jupiter, only an octave higher: it is very close in value, namely 2'45". If one adds two thirds, i.e., 1'11"+, they make 2'58"-: and look, the aphelion of Venus is 2'58"-. So this represents the step or note E, but five octaves higher; and the perihelion of Mercury is not much more, being 3'0", but seven octaves higher. Lastly, divided twice 1'47", i.e., 3'34", by nine, and subtract one part of 24" from the whole; leaving 3'10"+ for the note F, which represents nearly the aphelion of Mars, 3'17", but three octaves higher; this number is a little larger, approaching the note F#. For $\frac{1}{16}$ of 3'34", namely 13½", subtracted from 3'34" leaves 3'20½", which is very close to 3'17". In fact in music the note F# is often used in place of F, as one can see everywhere.

Thus all the notes of the major mode are represented within one octave (except for the note A, which was also not represented by harmonic divisions in Book III, Ch. 2)[13] by all the extreme motions of the planets except for the perihelions of Venus and Earth and the aphelion of Mercury, whose value of 2'34" approaches the note C#. For taking from D, whose number is 2'41", the sixteenth part, 10"+, there remain 2'30", the note C#: therefore only the perihelions of Venus and Earth are absent from this scale, as can be seen in this table.

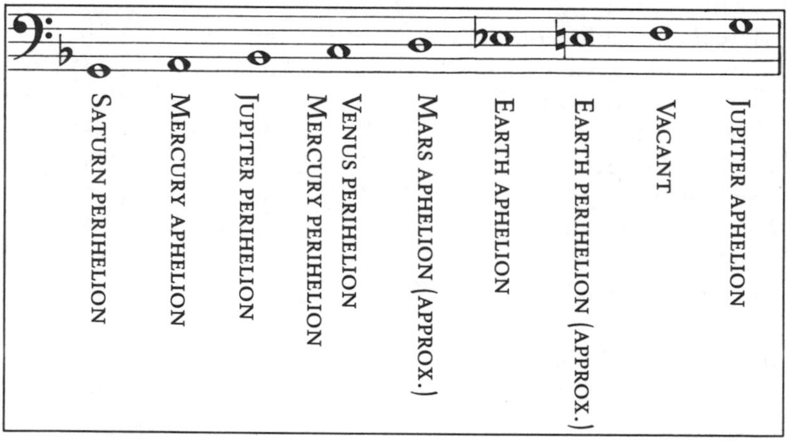

FIGURE 15: *The Minor Scale of Planetary Tones*

If, on the other hand, one begins the scale with the motion of Saturn at perihelion, 2'15", and makes that represent the note G, then the note A will be 2'32", which closely approaches the aphelion of Mercury; the note B-flat will be 2'42", which is very nearly the perihelion of Jupiter, following the equivalency of octaves; the note C will be 3'0", nearly the perihelion of Mercury and Venus; the note D will be 3'23"—which is only a little deeper than the aphelion of Mars, i.e., 3'18", so that its number here is as much less than its note as it was formerly more than it. The note E♭ will be 3'36", which approaches closely the aphelion of Earth; the note E will be 3'50", while the perihelion of Earth is 3'49". Jupiter in aphelion takes G again.

In this way all the notes within one octave of the minor mode, except F, are expressed by most of the planetary motions in aphelion and perihelion, especially those which were formerly left out, as can be seen in this table.

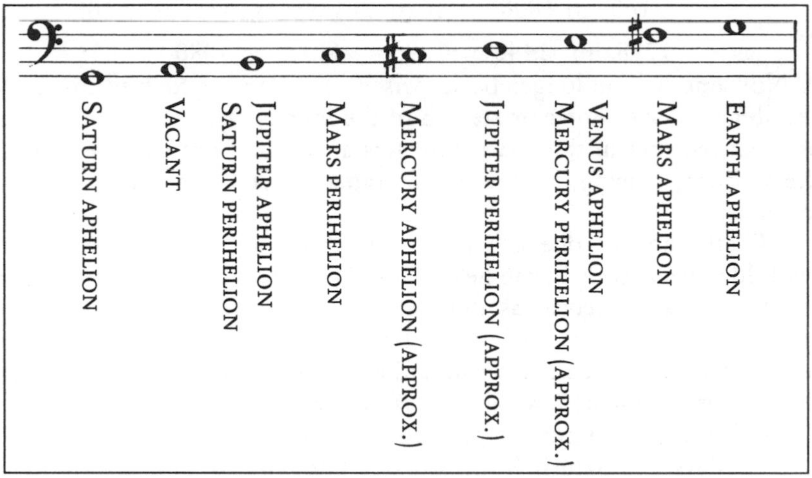

FIGURE 16: *The Major Scale of Planetary Tones*

Previously F# was specified, A left out. Now A is specified, but F# left out; and in the harmonic divisions of Book III, Chapter 2, the note F was also left out.

The musical scale or the system of a single octave is thus expressed in heaven in a twofold way, and as it were in the two melodic modes, with all the steps through which a natural [diatonic] melody moves in music. There is only one difference: that in our harmonic divisions both ways begin with one and the same G, whereas here in the planetary motions what was formerly B now becomes, in the minor mode, G.

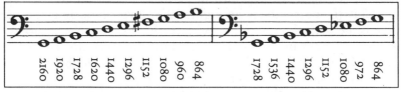

FIGURE 17: *The Twofold Scale of the Heavens*

FIGURE 18: *The Songs of the Planetary Motions*

For just as in music 2160 is to 1800 as 6 to 5, even so in the system which represents the heavens 1728 is to 1440 as 6 to 5, and so with many others:

	2160	to	1800	:	1620	:	1440	:	1350	:	1080
as	1728	to	1440	:	1296	:	1152	:	1080	:	1864

Now one will no longer be surprised that man has formed this most excellent order of notes or steps into the musical system or scale, since one can see that in this matter he acts as nothing but the ape of God the Creator, playing, as it were, a drama about the order of celestial motions.

There remains one other way in which we can understand the twofold musical scale in the heavens.[14] The system here is the same, but the tuning is conceived as double: one to the motions of Venus at aphelion, the other at perihelion. For the motions of this planet vary the least, being contained within a diesis, the smallest of intervals. And the tuning at aphelion, as above, has the aphelion motions of Saturn, Earth, Venus, and Jupiter (approximately) as G, E, and B; the perihelion of Mars and Saturn (approximately) and, as appears at first sight, also Mercury, as C, E, and B. The perihelion tuning, on the other hand, gives the aphelions of Mars, Mercury, and Jupiter (approximately), the perihelions of Jupiter, Venus, and Saturn (approximately); in a certain respect also the Earth's, and doubtless Mercury's too. For now that it is not the aphelion of Venus but its perihelion of 3'3" that occupies the step of E, the perihelion of Mercury at 3'0" approaches it most closely, two octaves away, as was observed toward the end of chapter 4. And subtracting from this perihelion of Venus, 3'3", a tenth, 18", we are left with 2'45", the perihelion of Jupiter, providing the step D. And adding a fifteenth, 12", makes 3'15", nearly the aphelion of Mars, providing the step F. Similarly, at the step B of this tuning there follow nearly the movements of Saturn in perihelion and of Jupiter in aphelion. But if one subtracts an eighth, 23", five times, it gives 1'55" which is the perihelion of Earth. Admittedly, this does not square with the previous ones in the same scale (since this does not contain the intervals 5:8 below E and 24:25 above G). But if, outside this order, one gives the perihelion of Venus and also the aphelion of Mercury the step E♭ instead of E, then this perihelion of Earth will take the step G. The aphelion movement of Mercury will also agree with this, since a third of 3'3", 1'1", multiplied by five, makes 5'5", whose half, 2'32"+, comes very close to the aphelion of Mercury, which in this extraordinary arrangement will take the step C. Thus all these motions relate within one and the same tuning. But the perihelion of Venus divides the scale in another way; the first three (or five) are in the same mode as in the aphelion tuning, i.e., in the major, but with the last two motions it divides the scale in another way, not into different intervals but into a different order of intervals: one proper to the minor mode.

This chapter has sufficed to make the matter in question visible. But the reasons why everything has been made thus, and what are the causes not only of so much agreement but also of disagreement in details, will appear with the clearest proof in chapter 9.[15]

The Musical Modes or Tones Are Somehow Expressed in the Extreme Planetary Motions

Chapter 6. This follows from the above, and requires few words. The individual planets in a way represent, by their movement at perihelion,[16] the individual steps of the system, in that it is given to each one to traverse a certain interval in the musical scale, lying between certain notes or steps of the system. Each begins from the note or step which in the previous chapter was assigned to aphelion motion: Saturn and Earth had G, Jupiter B, which may be transposed higher to G, Mars F#, Venus E, Mercury A in a higher system. (See the individual planets in the usual notation.) They do not form the intermediate steps which one sees here filled out as notes in an articulate way, like extremes, for they move from one extreme to the other not by leaps or intervals but by a continuous rising or falling, touching on all the intermediate notes, potentially infinite in number.[17] I could express this in no other way than by a continuous series of intermediate notes. Venus remains almost in unison, since the range of its rising is not equal even to the smallest of melodic intervals. (See figure 19.)

Now the specification of two of the common system of notes, and the formation of an octave framework by setting up a certain harmonic interval, is in a way the first means of distinguishing scales or modes. Thus the modes of music are distributed among the planets. (I know of course that more things are required for the formation and definition of the different modes, such as are proper for human melodies proceeding by intervals; hence I used the term "in a way.")

The harmonist will be free to form his own opinion as to which planet best expresses which mode by the extremes here assigned to it. Of the modes commonly used I would assign Saturn the 7th or 8th,

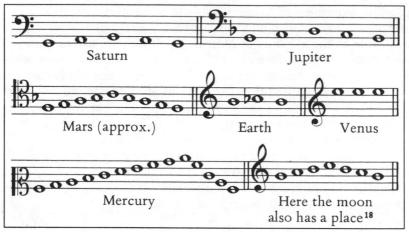

FIGURE 19: *The Melodies of the Planets*

because if one puts G as its keynote, its motion at perihelion rises to B; Jupiter the 1st or 2nd, because if its motion at aphelion is aligned with G, its perihelion reaches B-flat;[19] Mars the 5th or 6th, not only because it covers about a fifth (for this interval is common to all the modes) but mainly because when reduced to a uniform system with all the others its motion at perihelion reaches C, at aphelion approaches F, the keynote of the 5th and 6th tone or mode. To Earth I would give the 3rd or 4th, since its movement oscillates within a semitone, and the first interval of these modes is a semitone. Nearly all the tones and modes would suit Mercury equally, because of the width of its interval; Venus, because of the narrowness of its interval, would suit none, unless it were the 3rd and 4th of the common system, because in respect to the other planets it occupies the step E. (The Earth sings Mi-Fa-Mi, from which syllables you can infer that MIsery and FAmine obtain in this our home.)

There Are Universal Harmonies of All Six Planets, Similar to Common Four-Part Counterpart

Chapter 7. Now, Urania, let it sound louder, as I ascend to the heights by the harmonic scale of the celestial motions, where the tones and secret archetype of the world's making are preserved. Follow me, you musicians of today, and judge the matter by your arts, unknown to antiquity. In these last centuries, after two thousand years in the womb, ever-prodigal nature has finally produced you to give the first true image of the universe.[20] Through your counterpoints of many voices and through your ears she has suggested what exists in her innermost bosom to the human intellect, most beloved child of God the Creator.

I have shown above which harmonic proportions obtain between the extreme motions of two adjacent planets. It is certainly a very rare thing for two of them, especially the slowest ones, to coincide with their extreme intervals. The conjunctions [*apsides*] of Saturn and Jupiter, for example, occur about 81° apart.[21] Therefore 800 years must elapse before this aspect has run through the whole Zodiac in its leaps of twenty years; and even then, the leap which concludes the eighth century is not precisely on the original point of conjunction. If it misses by a short distance, it will take another 800 years to see if a happier leap is to be expected from one's reckoning; and this will be repeated so many times that the quantity of aberration will amount to half a leap in length. Similar periods, though not so long, also occur in the various other planet pairs. In the meantime, however, other harmonies have occured between the two planets: not between their extreme motions but with one or both of them in intermediate position, as if in different tunings. For Saturn is tuned from G to B and a little higher, and Jupiter from B to D and a little more: so between Saturn and Jupiter there can occur both kinds of thirds and a fourth (all plus an octave). The thirds can each be sounded over a range equal to the other,[22] but the fourth only over a whole tone, i.e., from the G of Saturn and the C of Jupiter to the A of Saturn to the D of Jupiter, and in between at all pitches from G to A and from C to D. But the octave and the fifth only occur at the extreme points. Mars, whose own interval is higher, has the property of making an octave with the higher planets within a certain tuning range. Mercury has received such a wide interval that in one of its revolutions, lasting no more than three months, it usually makes all the harmonies with all of the others. Earth, on the other hand, and Venus even more so, have narrow intervals which restrict them in their harmonies not only with the other planets but especially with each other, to a remarkably limited degree.[23]

If three planets are to coincide with the same harmony, one must wait for many revolutions. These harmonies, however, are very numerous, and occur all the more easily as one follows closely on its neighbor. It seems, too, that three-part harmonies occur somewhat more often between Mars, Earth, and Mercury. Four-part harmonies of the planets already begin to be scattered over centuries; five-part ones over myriads of years. The cases of all six harmonizing are separated by immensely long periods. I even think it may be impossible for this to have occurred twice by precise evolution: such a harmony may rather indicate the beginning of time, from which the whole age of the world proceeds.

If there were only a single six-part harmony, or one outstanding among others, we could doubtless regard it as the constellation of the Creation. The question is, therefore, how many ways the motions of all six planets can be reduced to a single mutual harmony.[24]

35

◆

Robert Fludd
1574-1637

Fludd was an amateur musician as well as being a successful London doctor versed in the new, Paracelsian medicine. In his earlier life he spent some years elaborating a system of universal knowledge, inspired by Hermetic and alchemical principles.

This great work, the *Utriusque Cosmi Historia*, though never completed, devoted hundreds of pages to music in both its heavenly and earthly modes. For one of the most powerful symbols in Fludd's imagination was that of the Divine Monochord, stretched between heaven and earth, on which all the levels of being, spheres, elements, and creatures are as it were different notes, tuned by the hand of God. The powerful image of this monochord, which was engraved for Fludd's book by Johann Theodor de Bry, has often been reproduced, but never before accompanied by Fludd's own explanation of it.

In order to understand Fludd's musical cosmology, it is necessary to grasp his dualistic view of all things as compounded of the Dark and the Light. Again resorting to graphic mode, he depicted this as two intersecting pyramids. Consideration of the actual proportions of light and dark present at any given level led him naturally to think in terms of musical ratios, just as Boethius had suggested that the elements were meted out according to the laws of harmony. The application of the scheme to the human body and psyche follows from Fludd's assumption that the Microcosm reflects the Macrocosm.

Source: Robert Fludd, *Utriusque Cosmi Maioris scilicet et Minoris Metaphysica, Physica atque Technica Historia*[1] (Oppenheim, 1617), Vol. I, i, Book 3 ("De Musica Mundana"), Chapter 3; Ibid., Vol. II, i, 1 (*De Supernaturali, naturali, praeternaturali et contranaturali Microcosmi historia*, Oppenheim, 1619), Book 13, Chapter 5. Translated by the Editor. Also consulted: Spanish translations by Luis Robledo in Robert Fludd, *Escritos sobre Música.*

◆

On the World-Monochord and on the Discovery of Its Consonances, Both Simple and Complex, Which Bring about the World-Harmony

Chapter 3. Since all these proportions on either one of the said pyramids[2] are incapable of producing musical consonances unless the two are joined together (just as a lute without a player, or a player without a lute, cannot make musical consonances), in order to create the harmony of the world it is necessary for the two pyramids to meet. These are the material pyramid, which takes the place of the musical instrument, the string of the monochord, or the singing of instrument or voice; and the formal one, which assumes the office of the soul that plays or sings and produces the notes. The increase of formal substance therefore produces a higher and more subtle air, and consequently makes finer and more excellent world-harmonies, just as the more intense spirit of man or the tighter string of an instrument makes higher sounds and subtler air with its more violent vibrations.[3] Its decrease, on the other hand, makes denser air, and consequently a lower harmony, in no wise differently from a more relaxed human voice or a loose monochord string which produces milder sounds: and this occurs on account of the lesser density of the air. Light therefore acts on the world-matter in the same way as the human spirit upon the air. (See figure 20.)

It is clear from this that a single nature, whether material or formal, is incapable of anything without the other, and consequently that the music of the world comes from both pyramids in at least some degree of combination. For without the presence of light the humid matter would not have taken pyramidal form, nor would the created form have

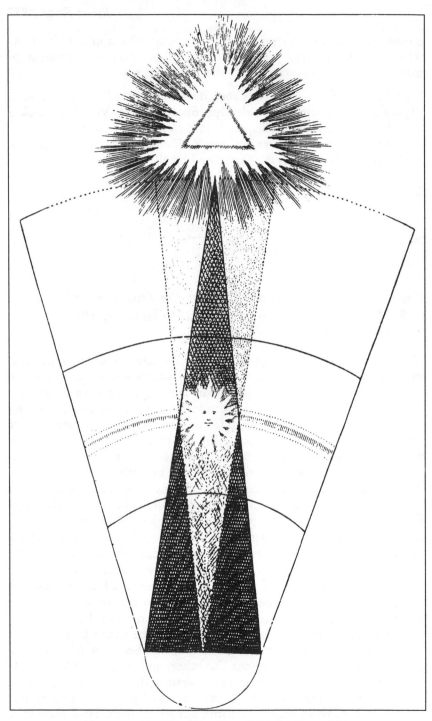

FIGURE 20: *The Dark and Light Pyramids*

taken its [shape] without the humid matter. Thus they must certainly be mingled in larger or smaller quantities according to the nature of the portions of the respective pyramids which move in opposite directions in the Macrocosm and correspond to one another.

Knowing this, therefore, let us begin with the material of the world, which we have represented as the string of the monochord (whose great instrument is the Macrocosm itself), so that we may conveniently compare to musical intervals, both simple and compound, the degree by which any difference of place between the centre and periphery of the world-instrument is to be discerned. For this one must know that the string of an instrument is customarily divided by intervals proportionally measured upward from G. I have therefore divided both its matter and its form in quantitative degrees, and separated them by proportions similar to those which make up the musical intervals. Then, if you will imagine a monochord stretching from the heights of the Empyrean heaven to the very bottom of the Earth, constituting the ray of either pyramid, let us see it divided into the parts which constitute the consonances. If it were stopped halfway, it would give the consonance of the octave, which also occurs, as experience shows, on the instrumental monochord.

But we must consider that these simple and compound consonances (and, similarly, the intervals which measure them) cannot be marked on the world-monochord except in the way that we divide the instrumental monochord in proportional parts. For the operation of cold on the Earth, and consequently the very matter of the Earth, naturally relates in density and weight to the cold and matter of the lowest region (in which there is just one fourth part of light and natural heat) as 4:3, which is sesquitertia proportion, as we have said above: the proportion from which the consonance of the fourth arises, compounded of three intervals, namely from water, air and fire. For the earth's place in the world-music is as Gamma in music, unity in arithmetic, and the point in geometry: it is as it were the terminus from which the proportional ratio of matter is derived, since it is the base and foundation of the material pyramid. Water will therefore occupy the place of one tone, and air the next interval of a tone. But the sphere of fire has the place of the minor semitone, because it is reckoned to be only the summit of the airy region.

Two portions of this matter are raised up in a pyramid as far as the midpoint of heaven, withstanding the action of the supernatural heat; and as many portions of light descend pyramidally, acting on those two portions of matter. These together compose the sphere of the Sun which is naturally given the name "[sphere of] equality." The sesquialtera proportion is produced in it insofar as three parts of spirit, or the lower nature of the middle heaven, are related to those two parts of the solar sphere, consequently producing the interval of a fifth. That is in fact the difference between the Moon and the Sun, for between the convexity of this heaven and the middle of the solar sphere there are four intervals: the whole spheres of the Moon, Mercury, and Venus compared to whole-tones, and half of the solar sphere which we have made a semitone.

Now since the consonance of the octave consists of a fourth plus a fifth, an octave must necessarily be produced therein. And this, the more perfect material consonance, cannot possibly achieve perfection unless it satisfy its appetite with solar form. Besides, it is in the very heart of this middle heaven, namely the Sun, that the more perfect material consonance ceases and the movement to the formal octave begins. Within its concavity, however, there sounds only the consonance of a fifth, either above the sphere of equality or beneath it. And this is more appropriate to this place than any of the other consonances, because just as it is less perfect (being in between the perfect and imperfect ones), so this heaven, however perfect and free from corruption, is said to be less perfect in comparison to the higher heavens, and holds a median position between the two heavens (namely the perfect and imperfect). Moreover, the exact point of equality which is the exact intersection of the two pyramids is the true terminus of the more perfect consonance. The union and embrace of matter and form here is such that there could never be any separation or division. But in truth the parts of the pyramids in the ethereal region, both above the sphere of intersection and below it, are sufficiently equal for perfection: for the material ether outside the sphere of the soul has sufficient form to fill its appetite—though not to the degree of perfection of the solar sphere. On this account the proportion joining the intervals is called a perfect consonance, though less complete than that of the Sun, which the best musicians call *Diapende* (fifth). And this was the reason why some philosophers distinguished its substance by the name of Quintessence, seeing that its composition in respect to either of the outer heavens participates more of the consonance of a fifth. For it contains the sesquialtera proportion downward from the solar sphere to the lowest heaven, while the same proportion is found going upward to the boundaries of the spiritual heaven. The higher consonance, however, differs from the lower one, because one is related to form, the other to matter. Insofar as form is more excellent than matter, even so the natures of [formal] consonances are more eminent than those of matter.

Thus from the center [of the Earth] to the sphere of the Sun an octave can be produced which is material, just as the other octave which raises its summit from the top of the material one to the height of the Empyrean heaven is spiritual. For the part of the middle heaven above the sphere where the triangles intersect at the Sun's center, up as far as the crystalline region, is more spiritual than the part beneath it, on account of its greater propinquity to the Empyrean heaven: therefore the consonance produced in it is also spiritual. And as the two portions of light of this superior part of the heaven compare to the three portions which inform the matter of the Empyrean heaven, we will find their relationship determined by sesquialtera proportion.

From the part of the material pyramid contained between the dwelling-place of the Sun and the boundaries of the Empyrean heaven, four intervals emerge because of the relation and repetition of the portions of the light to one another: and from them the spiritual fifth is made and perfected. From one portion of the solar sphere, namely the upper one which extends from the intersection of the pyramids to the top of the solar sphere, there is an interval comparable to a minor semitone; so just as the lower part of the same solar sphere occupies the place of the minor semitone in the material fifth, the upper part serves as the minor semitone in the formal fifth. Surely the amount by which two major semitones exceed two minor ones, namely two commas, concurs with the exact composition of the sphere of equality, in which the centre of the solar body, the vehicle of the world-soul, is suspended by equal weights. Next the sphere of Mars, that of Jupiter, and that of Saturn above them are counted as the other three intervals of the spiritual fifth. We may compare them to whole-tones because each of these orbs has its distinct nature and operation both on the middle heaven and on the lower regions.

Above this less perfect spiritual consonance there extends that most perfect consonance of all, which is called the spiritual octave. And in this place a divine mystery will reveal itself to those who diligently contemplate it. For as the formal octave is superior to the material one, so its foundation is more excellent, and far nobler both in the proportion of its harmony and in its base. And as the Earth is less noble than the solar body, so the material octave and the consonances contained therein are far less estimable than the spiritual ones. For the Earth is the base of the material fourth and fifth, constituting the material octave, while the Sun itself is the foundation upon which the spiritual fifth, fourth, and octave are raised. Besides, the fundamental consonance of the material octave is the fourth, the most imperfect of all consonances; but the fundamental concord of the formal octave is the fifth, an indisputably perfect concord.

Here, then, the degrees of all material things can be seen as in a mirror, and hence the spiritual scale and order are manifested to the eyes of the intellect. All generation occurs beneath the Sun and receives its energy from the lower minor semitone of the solar sphere, which we have called the material one. But regeneration, on the other hand, arises from the superior semitone of the solar sphere, from which sublimation takes place up to the peak of the spiritual. So those things which do not attain the summit of the solar sphere cannot obtain the perfection of regeneration, nor be turned from corporeal into spiritual. Blessed therefore are those bodies which are permitted to reach such heights and dignity.

Returning to our theme, those three parts of the light that informs the subtlest matter are related in sesquitertia proportion to those four formal parts constituting the integral form in which there is nothing material (as is the nature of the Uncreated Light), and constitutes the spiritual consonance of a fifth. Its most subtle spirit, as it were transformed and liberated from corporeal substance, is divided into three final intervals to which the three Hierarchies are attributed.[4] Their two lower circles are likened to whole-tones, the upper one to a semitone, seeing that the orders of the *Epiphania* and particularly the Seraphim, who are said to be the very attendants of God, extend beyond the limits of matter and are thought of as utterly formal on account of the ineffable splendor which they receive from the Divine Presence.

From the union of these concords, therefore, arises the spiritual octave, whose perfection cannot be comprehended by human understanding, since it is confined in the triangular nature of purest form. It is the sum of all perfection, the peak of universal purity, beyond which nothing exists but God, single and alone. And this very consonance is produced both from the material and the formal part: for if the two parts of matter which give substance to the ethereal heaven are compared to the one bodily part within the Empyrean heaven, there will result the double proportion which makes the consonance of an octave. From this it is clear that there is a quadruple proportion between the Earth and the convex surface of the spiritual world, since its radius is made by the double octave which we have imagined as the string of our macrocosmic monochord in the following demonstration. And it is evident from this that just as the musical instrument known as a monochord, stopped at the midpoint between the two extremities, will sound an octave between the middle and either end, even so the action in the middle elevation of the world, or the middle of the radius from its center to its circumference, eminently displayed in the solar body, is understood to produce an octave.

It is quite obvious, then, that the perfection of the spiritual heaven extends to the center of the Sun, and that the Earth asks and receives from that solar limit everything that has any quality or quantity of perfection. The movement of the light, therefore, also tends downward so as to fill lower things with perfection, and its motion terminates in the Sun; so the solar body, which collects the rays of light emitted, projects them in the same proportion on to the surface of the Earth. Thus the formal heaven bestows form, whereupon matter avidly seizes on what is given.

We conclude, therefore, that by virtue of the spiritual harmony, through the formal octave ordered proportionally with its intervals, the Sun (the God of Nature, but only a created one) receives all formal and lucid virtue from the greatest God of all, the supernatural and uncreated Creator. The Earth, on the other hand, received the influences of the same God through the material octave, whereby it corresponds to the Sun as the latter to God Most High: for which reason the royal Psalmist said: "He hath placed his tabernacle in the Sun."[5]

We have demonstrated in this music of ours that both consonances of a fifth are found in the middle part of the world, such that philosophers call its very substance "Quintessence"; both the consonances of a fourth are found in the exterior parts of the world, the octave in the Sun, and the double octave in God himself.

This, then, is the natural harmony of the universal machine, which no one up to now, so far as I know, has explained so succinctly or lucidly.

In this picture[6] (see figure 21) we have represented the world-monochord in more detail, with its proportions, consonances and intervals, whose Mover is outside the world.

The upper fourth agrees with the material fourth in this way: the lower fourth, though more perfect with respect to the Earth beneath it (seeing that it has more form informing it than the latter), is still found imperfect in relation to higher things, i.e., to the fifth of the same nature, and can in no way be compared to the matter of the ethereal heaven, any more than the consonance of the fourth can be compared with that of the fifth. So it is in every respect with the formal fourth: for if it is compared to the nature of the ethereal heaven, it is more perfect than the latter beneath, even than its consonance of the formal fifth, because this interval is more material, the fourth more spiritual. But if it is compared to the unsullied form above, the intervals of this consonance are now imperfect. So although this consonance is found (as our intellect sees it) in the region which seems to us the purest and most perfect of all regions in the world, yet even this is a vast distance away from the perfection of the Heaven of the Trinity.[7]

The Mutual Harmony of Soul and Body

Marvelous is the love and fellowship of the flesh and the soul, of the spirit of life and the mud of the earth: for the whole man may be said to be formed from these two conjoined. For thus it is written: "God

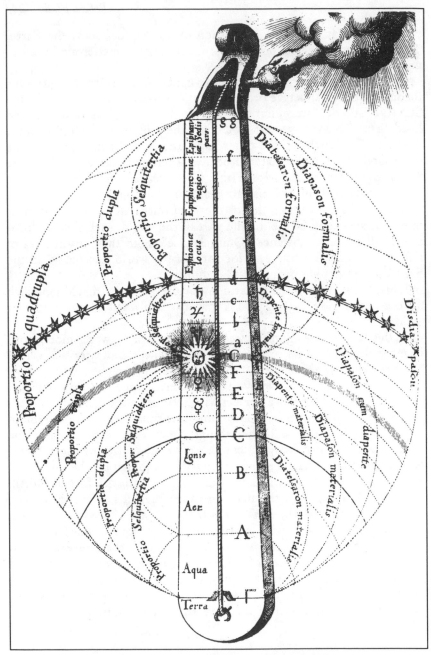

FIGURE 21: *The World-Monochord*

made man from the mud of the earth, and breathed the breath of life
into his face "[Genesis 2:7], giving him sense and intellect, so that
through sense he might vivify the clay associated with him, and through

intellect rule it; that likewise he might enter inwardly through the intellect and contemplate the wisdom of God, and outwardly through the sense behold the works of his wisdom. God illuminated the intellect from within but adorned the sense without, so that the whole man might find recreation in both, namely felicity within and enjoyment without. But since outward things cannot last long, man is bidden to turn from the things without to the things within and to ascend from the things within to the things above, that is to say from sense to imagination, from imagination to reason, from reason to intellect, thence to mind or intelligence and thus to God.

It was marvelous, I say, ineffable, and indeed fully miraculous, that things so diverse, divided and contrary in nature as the body and soul could have been joined together; and it was no less admirable that God should have contracted himself into our mud, so that God and mud might be coupled together: such sublimity, and such vileness—for there is nothing sublimer than God, nor viler than mud. The first conjunction was marvelous, the second was likewise, and no less marvelous will be the third when man, Angel, and God shall be one spirit: for man is good by the same goodness as the Angel, and by the same goodness both are blest. But if both can desire the same with an identical will and an identical spirit, and if God can join such disparate natures as those of flesh and soul in a single federation and friendship, it will certainly not be impossible for the Rational Spirit, humbled during its fellowship with the earthly body, to be exalted with that same body when it is glorified by the fellowship of blessed spirits which remain in their purity, and to rise to participation in its glory. To this end the Most High created him, not through necessity but through charity, that he might make him a participator in his blessedness. If therefore there is so much joy and happiness in this temporal life which consists of the presence of the spirit in that corruptible body, what joy and happiness there will be in eternal life which consists of the immediate existence of the Deity in the Rational Spirit! The body therefore should be subject to the soul [*animus*], and the soul to God, and become one spirit with him, if only the spirit remain in humility and give thanks to its Creator, by which it is glorified and exalted.

Man is thus made from body and soul, and each has its goodness in which it rejoices and exults. The goodness of the soul is God, with the abundance of his sweetness; that of the flesh is the world, with the abundance of its happiness. But the world is exterior, God interior; and there is nothing more interior or more present than he. He is inside everything, since all things are in him, and outside everything, since he is above all things. We should therefore pass beyond ourselves, ascending from this world to God as if struggling up from below. For to ascend to God is to enter into oneself, and not only to enter oneself but to pass, in an ineffable way, through one's own center. Hence Mercurius Trismegistus says: "Whoever transcends by passing inward and penetrating within himself, he truly ascends to God." For indeed this is done by drawing our heart away from the distractions of this world, and recalling it to internal joys. And if we cannot always keep our heart on these, at least we should restrain it from wicked and vain thoughts so that we may at some time hold it fast in the light of divine contemplation. For the repose of our heart is when it is fixed through desire on the love of God; its life, when it contemplates its God and is tenderly refreshed in that very contemplation, which is always sweet to behold, to love and to praise. For nothing seems more efficacious for a blessed life than to turn every feeling [*affectum*] within oneself, as if one were outside the flesh and the world, with the carnal senses closed; [to turn] unsuitable affection away from the desires for mortal things into oneself alone, and to converse with God.

Since therefore the soul is incorporeal, it administers to the body through the latter's subtler nature, that is through fire and air (which are the more excellent bodies in the world itself and therefore more familiar to spirit). The latter, being closer to the incorporeal nature than fluid and earth, are the first to receive the impulse of the vivifying soul, so that the whole mass is administered by their direct government. For [water and earth] are incapable of sense in the body, and of voluntary bodily movement in the soul. Fire and air, since they are lighter, move earth and water, which are heavier. On this account we will see that a body scarcely moves after the soul has departed, because after the separation of the soul the fire and air, which are held by the soul's presence in the earthly, humid body so that there should be a proper mixture of all [the elements], have fled to higher places and freed themselves.

As for the composition of the body, it should be understood that it consists of members with distinct duties; these consist of similar parts; those parts of humors; the humors of nutritive substances, and the latter of elements. Hence we can see that the soul is none of these, but acts upon them as if on organs naturally destined for itself, and through these supports the body and the life by which man is made a living soul. So then, if all these are tempered and in order, they collaborate in giving life and the soul never departs; but if they become distempered and confused, the soul unwillingly withdraws, taking everything with it: sense, imagination, reason, intellect and intelligence, concupiscence and irasciblity; and according to the merits of these it is affected by pleasure or pain. Thereupon the body which before was whole, like an instrument tempered and prepared to play a melody and keep in rhythm, now lies broken and mutilated. The soul, on the other hand, once the elemental parts have returned to their home, having nowhere to exercise its energies, rests from all the motions by which it used to move the body in time and space. For the instrument may perish, but the melody does not perish, nor that which set the instrument in motion.

From this, therefore, we can gaze as it were with open eyes on that admirable harmony which the two extremes of the most precious and the most vile make together, and how they concord with one another; and we see how the intermediate spirit of the world, the vehicle of the soul, is the nexus maintaining them in peaceful concord and harmony; and that God is he who gives breath to the human music or plays the monochord string, or is the internal principle producing as if from the center the consonant movements of all things and the vital activities in the Microcosm. The string which by its vibration spreads through the Macrocosm and the Microcosm the lucid effects of its inspirer like the accents and sounds of love is the limpid spirit which on account of its site and position naturally participates in both extremes and connects the extremes with one another. In the same way it makes the degrees or notes of the human system by which the soul's descent from on high into the body takes place, and conversely also its ascent to higher regions—the most noble place—after the death of the body—the vilest place—and the dissolution of the vital bonds.

Now we show thus the arrangement of this human harmony compared with that of the world, according to the difference of each of its regions correlated to the world. (See figure 22.)[8]

We see here the whole human being represented at his full length, in whose middle and along whose longitude we have traced a straight line representing the diameter of the Microcosm. For the rest, the radius of this string A C measures the tones and semitones of the monochord and gives musical proportions similar to those produced in the world-music. So on the left-hand side of it we have explained the music appropriate to the Microcosm, while on the right-hand side we have delineated well enough the symphonious aspect of the soul descending from heaven to earth and its returning ascent from earth, or body, to heaven, or spirit. On the left, therefore, we have demonstrated that the proportion between intellect and life is a duple proportion on the monochord, containing the spiritual octave. Similarly, we have shown that the interval on the monochord which stretches from the sphere of the heart (or Sun) to the Earth (or genitals) sounds in the same proportion a corporeal octave, uniting life with the sense or carnal nature which dwells in the lower belly. In the same way, therefore, as the string of the musical monochord, stopped at its halfway point, sounds one octave above and another below, so the heart situated in the middle of the human radius (no differently from the Sun in the middle of the world) relates by one octave-consonance to

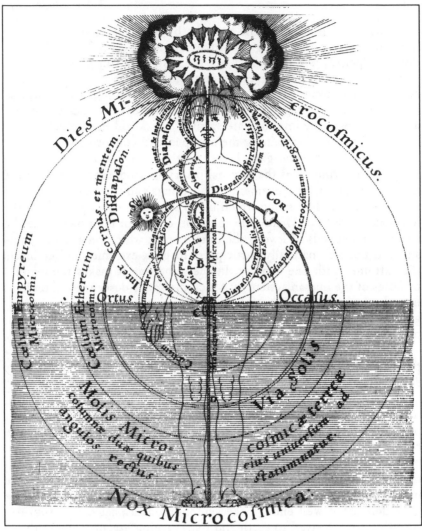

FIGURE 22: *The Human Monochord*

lower things, i.e., to the nature of the lower region, and by another to
the highest heaven of the Microcosm in which mind, intellect and
reason hold sway.

On the right-hand side of the Microcosm we have set an octave
between intellect and imagination, a fourth between intellect and reason, and a fifth between reason and imagination; the latter two joined
together make the spiritual and invisible octave. Similarly, from the
sphere of the heart downward we find the same proportion between
imagination and body (or the center of the Earth), since the distance

traversed from imagination to sense corresponds to a fourth and that between sense and flesh or body to a fifth; joined together, these compose the corporeal and as it were visible octave. The first breath of life gives motion to all of these, breathed directly by God into the face of man. Its second voluntary motion, like a microcosmic *primum mobile*, sets the faculties of the lower regions in involuntary movement. Thus we see that the animated action contained in the higher region is voluntary, whereas the others common to the lower regions, namely the vital and natural, are said to be involuntary.

We have also explained in the preceding presentation the day and night of the Microcosm:[9] how the fountains of rational, vital, and natural light always ascend and rise up above the human earth or center of the living man, whereas beneath his earth nothing is to be found but his two supports or columns, sustaining and underpinning the universal mass of earth like Atlantides. Therefore in the upper hemisphere of the Microcosm there always shines the light of the first heaven—of the mind—illuminated with divine splendor; in the center the heart, filled with solar light; and in the depths the liver, abounding in fire and the heat of the aerial element.

And thus I have explained the nature of the whole man, both internal and external, deliberately not as elegantly as you might wish, but as diligently as I could; and as a son of light, not of darkness, I believe [I have done so] lucidly enough, and not without great and most grateful thanks to God my Creator; spurning, ignoring, and having no care for the bitter maledictions of Momus, the worst calumnies of the malevolent, and the impious derision of the envious; washing away and expelling the virulent bites of such viper-tongues with the balsam of patience, the antidote of a clear conscience, and expunging them with my own floods; and to heal wounds inflicted in private I have sent out and published them in public. *May the GOD JEHOVA be my helper, and look down on me from His holy cloud, or maintains, or hill of Sion, and keep me safe from the injuries of the wicked under His blessing wings. Amen.*

36

✦

Marin Mersenne
1588–1648

Kepler, Fludd, and Mersenne are the three outstanding luminaries of our subject in the early seventeenth century. Each develops speculative music further than ever before, and each in a way that differs markedly from the others. Mersenne, a member of the Franciscan Order of Minimes, was at the center of the early scientific revolution. Enjoying an enormous number of friends and correspondents, he acted as a clearinghouse for discoveries by savants throughout Europe.[1] Like Fludd, he sought to grasp the world through encyclopedic knowledge. Like Kepler, he was entranced by the possibilities of quantitative science. And like both, he saw the universe as God's harmonious creation, to which the study of music offered a precious key.

His very breadth of knowledge and of approach led Mersenne into some difficulties. He loved the Florentine Platonists, but was embarrassed by their too-pagan enthusiasms. He learned much from Giorgi, but felt obliged to reject the Venetian's Kabbalism. The Platonic and Pythagorean heavenly harmony was for Mersenne merely a matter of proportions, found in everything in the world that has relations with anything else.[2] To some thinkers, though, such as Jacques de Liège (see no. 22), this had been the best of reasons to regard the whole world as musical, and Mersenne comes close to doing so when writing of universal harmony. One also senses some uneasiness in this friar's great love for music and learning, living as he did under the command to fix his mind solely on the love of God and of his neighbor. Such a dilemma, suffered by erudite Christians from Saint Augustine onward, led Mersenne at times to cynicism, as in his indecisive and skeptical *Questions Harmoniques*. Here, in reply to the question "Is music pleasant?" he replies that cards, dice, and tennis all pass the time agreeably, and with much less trouble than music, which only breaks the silence so conducive to the contemplation of celestial bodies and prayer!

Our extract presents what is essentially the same point of view, though in a more serious vein. Coming at the beginning of a large work on every aspect of theoretical and practical music, it calls into question the whole enterprise. If the single tone, sounded in unison, is the most sacred form of music, what need for anything else? But then it would have been spiritual pride for Mersenne, or anyone else, to claim to have attained the state of genuinely preferring the monotone to any other music. What Mersenne shows us, as does no other author, is the musical absolute against which the fictive drama of harmony is played out. The unison is to all other music as God is to his creation, and its contemplation invites Christian and Platonist alike to turn their attention, in the end, from the Many to the One.

Source: Marin Mersenne, *Harmonie Universelle* (Paris, 1637), pp. 10–19. Translated by the Editor.

To Determine Whether the Unison Is a Consonance, and Whether It Is Sweeter and More Agreeable than the Octave

Book I, Proposition IV. Those who contend that the unison is among consonances as unity is among numbers, say that it should not be called a consonance, because it contains no variety of sounds as to high or low. On the other hand, those who believe the unison to be the Queen of Consonances are of the contrary opinion, believing that in order to make a consonance it is enough for the sounds to differ in number, and that since the union of sounds is the formal cause of these consonances, the one which unites them so perfectly that they seem to the ear but a single sound cannot be denied the name granted to the others. One can confirm this by the names we give to God when we call him Being, the Good, the Beautiful, etc., because despite the fact that God does not possess being, goodness, or beauty as we have them, and although he possesses these perfections to an infinitely more perfect degree, we are still permitted to speak of them in this way, being thus taught by the Holy Scriptures. The unison is the exemplary cause and the goal of the consonances, for they all tend toward it, having it as their origin, just as unequal relationships have theirs in the relationship of equality; and consequently one may not unreasonably hold it to be the first consonance.

There are two types of unison to be considered. The first continues with the same tone, that is to say on the same pitch, as is done when they sing without raising or lowering the voice in those monastic choirs which do not use plainchant. One can call this singing in *Isson*, which means equal; the notes following are like the first, and all the parts are unisons.

The other type is that of plainchant which uses all sorts of degrees to rise or fall. It has more variety than the first, which is like a voice holding fast to a single note and having no other distinction than that which comes from the different syllables or from various interruptions, pauses, and rests for catching the breath, inhaling, and relieving the voice and diaphragm.

These two types of unison differ in that the first has only a single sort of note or sound while the other has as great a number of them as there are differences of high and low pitch. This is why the first unison is simpler than the second, yet both are consonances, since they make a union of two or more sounds which is agreeable to the ear, although they have no other difference but their particular and individual nature—the least difference that can exist between substances.

This difference is sufficient to establish the relationship of equality, which is distinct from the simpler one of identity, although Aristotle says in Problem 39 of Section 19[3] that the octave is more agreeable than the unison because the latter is only a single sound. And in Book 2, Chapter 5 of the *Politics* he says that one who includes the unison among consonances is like one who would introduce the community of all things into states, or who confuses lines with the [metrical] foot.[4] But Jean de Muris, in Book 2, Chapter 10 of *The Mirror of Music*,[5] maintains that it is a consonance: with which all who follow reason and experience rather than authority are in agreement. And as for the community of states, to which Aristotle was opposed in order to contradict his Master,[6] it is very desirable, though not to be hoped for so long as diversity is preferred to equality. All the best things lead us toward this equality and communion of goods: in nature the earth, the air, and the sky are made equally for everyone, and in the state of grace there is only one faith, one hope, the same commandments and the same law; in the state of glory there is only one God who will be all in all, *omnia in omnibus*,[7] when all things are made subject to him and have left behind diversity, which is the source of corruption. In the same way one may say that Plato, whose spirit seems to have reached the very summit of the light of nature, was contemplating the beauty of the eternal Ideas when he proposed the happy communion of goods. His disciple ought to accept this since he avows that friends should have their goods in common. Now all men should be friends, because they are brothers and children of the same father, and because the true religion teaches us that the faithful should be one body and one spirit, since they all have as their final goal the love and glory of God. Hence the whole of Holy Scripture has no other purpose but to make us embrace the community of goods both spiritual and bodily, and to unite us to God forever, in order that the Unison, less esteemed here than it should be, may triumph over diversity, the source of error, and enjoy eternally the prerogatives of which people want to deprive it through the differences of rhythm and movement which they currently employ.

As for the second part of the proposition, namely whether the unison is sweeter and more agreeable than the octave, I will say first that there is no doubt of its being sweeter, since it unites its notes more often and more easily. Unison is one against one, the vibrations of the air uniting at each stroke,[8] whereas the vibrations of the octave only unite at every second stroke, and one finds invariably, in the operations of every sense, that whatever unites most easily is the sweetest. But it does not follow that it is the most agreeable: for though sugar and honey are extremely sweet, they are not the most agreeable to those who prefer sour and bitter things. That is why we must see whether the unison is more agreeable than the octave.

Second I will say that it seems that the unison is more agreeable than the octave because it tickles the ear more, and is more easily comprehended by the imagination which is the principal seat of pleasure.

If one wishes to use comparisons to confirm this truth, nature provides them in all the disciplines; for the great pleasure of Algebra consists in finding all sorts of equations which are solved by means of equality; the science of Mechanics has its basis in equilibrium, which is a certain species of unison; and Medicine, it seems, has no more exalted end or speculation than the temperament of the body reduced to the equality of humors. If we may be permitted to ascend higher, we will find an eternal Unison in the Deity, for the three Persons are all of the same nature and all have a single will, power, and virtue, although they are in fact distinct. Perhaps this is the reason why the Blessed sing perpetually in unison, so that their song may confirm to the equality of the three Persons, and to the state of equality which traces its origin to that eternal beatitude which is not susceptible to any alteration, and which, being so simple, requires the simplest songs. And nothing can be simpler, when many sing together, than for them to sing in unison.

One can also confirm the same thing by the beginning and ending of pieces which are almost always in unison, this being the goal of music: for one finds that all the consonances tend toward the unison, as I show elsewhere. And if one compares the power of the unison of plainchant with that of the consonances of music one will find it more effective, making a stronger impression on the spirit. No longer distracted by the variety of consonances and dissonances, it begins to sense the Music of the Blessed when it hears the unison, reminding it of its origin and of the bliss for which it hopes and waits.

The power of the unison imposes its effects not only on the spirit and on souls, but also on inanimate bodies; for whenever one plucks a string of a lute, viol, or any other instrument, it sets in motion and causes to vibrate the other strings tuned in unison to it. Consequently this can serve to set in motion all sorts of machines, and to touch off cannons. In this way one can besiege and take towns by means of the unison, just as Orpheus is said to have built them with the sound of his harp.[9] But this discourse must be kept for the treatise on the sounds used in warfare.

One of the strongest reasons which argues that the unison is more agreeable and natural than the octave comes from the experience which shows that one tires much sooner of hearing singing at the octave than at the unison, which one can happily listen to in church for several hours. And although children naturally sing an octave above men, their intention is nevertheless to sing in unison: the goal toward which all voices tend when supported and strengthened by their equals. For resemblance is the source of love, and the conservation of the being and nature of every object, which is better preserved by conformity than by deformity. Now the movements which our spirits receive from the unison are perfectly uniform and equal, while those of the octave are unequal, since some are twice as fast as others.

If we compare sounds to the objects of touch we will find that the ear receives as much pleasure from hearing unison singing as the sense of touch does from handling objects that are smooth, soft, and mild, such as satins and a thousand other such things. Hence dissonances are called hard and rough, because their sounds resemble hard, rough, and uneven bodies which hurt the hand and destroy the spirits which serve the sense of touch.

Nevertheless, many believe that the octave and other consonances are more agreeable than the unison because they have some variety in their union, and nature rejoices in diversity (as we have proved in a special proposition). And if one thinks about the harmonies which charm the spirit during concerts, one will have to admit that one sometimes encounters passages which ravish the listener more powerfully than is ever achieved by the unison.

Moreover, the different voices of music which are doubled are so many unisons enriched and elevated by the diversity of consonances, so that while they are good and agreeable on their own, they will be better and more agreeable still when they are joined to those harmonies.

As for the great equality and unity of sounds brought about by the unison, it seems to be too simple to give pleasure. For one finds in many things, and especially in visible ones, that whatever is too simple and whatever is not made from several parts, is not thought agreeable. A straight line or a circle, taken alone, is neither beautiful nor agreeable. So the unison is similar to isolated lines, just as the note is similar to the point and to unity.

One may add to this that a perfect unison cannot be distinguished from a single sound; when, that is, the voices which make the unison are perfectly equal. Consequently it is no more agreeable than a single voice, since it is heard in exactly the same way as if there were only one.

Here one may refer to all the arguments I employed elsewhere to prove that diversity is pleasing to the senses, and to the comparison which Zarlino uses in Book 3, Chapter 8 of his *Institutions*, where he holds that the unison and octave are comparable to the extreme colors of white and black, while the other consonances in between (the fifth, fourth, and thirds) correspond to the colors in between, green, red, and blue; and that consequently the unison and octave are not so agreeable as the other consonances, for white and black are less agreeable than the mixed colors in between. Nevertheless, this does not prevent one from concluding that the unison would be better pleasing to all kinds of men if they were in the state of perfection: a state, it seems, which rejects a diversity whose pleasure only bears witness to our poverty and imperfection. For since all music is only for the sake of the unison at the end, why not take that rather than all the other chords? But those who take more pleasure in the other chords than in the unison are like people who prefer the weather gloomy and overcast in broad daylight to the pure light of the sun; who prefer the colors that participate in darkness, like the median ones, to white which is the image of light, serving as the color of the garments which angels wear to appear to men, and which our Savior used at the Transfiguration (for his garments were white as snow and shone like the light). In truth, those who make more of green and the other mixed colors than of white, and more of the imperfect consonances than of the unison, are like those whose eyes cannot bear the light, and who are more content to behold partial truths than the universal truth which is in God; who prefer to rejoice in creatures and passing pleasures than in the Creator and eternal joys.

As for those who have risen above everything created, and who have felt a thousand times the distaste which one has for all the truths of Mathematics and Physics as soon as they are discovered (from which one gains almost no satisfaction but that of the toil one suffers while searching for them), they get no enjoyment from concerts, preferring singing at the unison to that in parts: for the unison represents for them the state of the Blessed and the perfect union of the three Divine Persons who are in the unison of perfect equality.

And because the unisons we make here are not perfect, those who rise above everything corporeal and begin to unite in ardent love with God receive no satisfaction from unisons except when they are using some reading to be rapt in contemplation of the Sovereign Being; and they are happier to hear no singing at all, so as not to be distracted from their thoughts of the uncreated Unity, on which they are so fixed that nothing in the world can separate them from it.

Therefore I reckon that the unison is more agreeable than the consonances, and that one must have compassion on the fragility and inconstancy of men who do not share this sentiment, making more of diversity and inequality than of unity and equality: the more so since they judge things not by what they contain of simplicity and excellence, but by what best agrees with their appetites and fancies.

One can further confirm this truth by a persuasive consideration of all that renders things agreeable, namely of what gives them their being, their faculties and action; for there should be no doubt that what renders things agreeable cannot be more agreeable than the things themselves, since the former have nothing but what they have borrowed from the latter, and are only agreeable by virtue of their borrowing from the source that is their own origin.

Lines, figures, and solids possess nothing that they have not borrowed from the point, because the line is none other than the movement of a point, and figures and solids only the movements of lines and planes. If one takes away all points, nothing remains; just as if one contemplates the beauty and perfection of the point, one will confess that it has the beauty of lines and figures in eminence and perfection.

All creatures depend no less on God than lines on the point; they have no beauty nor anything agreeable except that which they receive from the presence of God, who creates them perpetually, just as nothing is perfect in creatures except for God. Hence the greater the beauty in creatures, the more assistance God gives to them, and the greater the quantity of light which he dispenses to them and with which he illuminates them; just as numbers are greater according to the greater multitude of rays which unity sends out to them, and to which it communicates itself more fully; just as one can say that all possible numbers are nothing other than unity made common, or the love, perfection, and communication of unity without which no number can exist.

Now consonances depend on the unison as lines on the point, numbers on unity, and creatures on God: this is why they are sweeter as they approach closer to it, for they have nothing sweet or agreeable but what they borrow from the union of their sounds, which is the greater as it tends more toward the unison. Many, it is true, do not get so much pleasure from it as from the other consonances, as their spirit is not strong or lofty enough to contemplate the point and the unity in their simplicity, nor to fix themselves on the divine presence alone, considered devoid of any relation with visible things. The spirit of most of mankind is so shut up in the body and limited by phantasms that it cannot rise above the senses; and if they should happen to raise themselves up to the center of the Divinity which the Cabalists call the dark Aleph and the Ensoph,[10] they would be totally lost in the darkness which would overwhelm their understanding; for the phantasms which gave them some semblance of light would no longer accompany them, forcing them to fall back into the false light which eclipses the rays of the intelligible Sun, robbing us of true beauty in order to feed us a false one which brings no solid and lasting pleasure. This is what Saint Augustine has said in Chapter 3 of his book on the knowledge one needs to have of the true life, in which after showing the power of dialectics he adds: "Dialectics is the power of differentiating, capable also of explaining dubious matters, shaking up and pulling to pieces all the scriptures, and annihilating all human wisdom when it pretends to divinity; then, thrown back by such light of majesty, it bows its head in fear and trembling, and fleeing, hides away in the secret places of worldly wisdom, and is struck dumb when the links of syllogisms are broken."

Thus no wisdom and no power of human understanding can ever reveal to us the light of the principal truth, nor make us acknowledge that perfect contentment consists in perfect simplicity, which is never rightly experienced except when one contemplates it in itself, and abandons diversity altogether in order to embrace the divine unity. Thither the Royal Prophet aspired when he sung these words: "I will be satisfied when thy glory appears" [Psalm 16:15].

Nevertheless, when one knows the art and practice of meditation on true pleasure, one will soon discover that the eternal ideas are its only true object, and consequently that we err in believing that Beauty has her seat in the existence of creatures distinct or separate from the existence of the Creator. For beauty, and that which we call agreeable in sensible or intelligible things, depends on the uncreated Being, just as numbers depend on unity, lines on the point, time on the moment, movement on the motionless, and the consonances on the unison.

Now numbers have nothing in themselves but unity, which renders them smaller or larger in proportion to its greater or lesser communication of itself. The number 1000, for example, is ten times as great as the number 100, because unity communicates itself to 1000 ten times more than to 100.[11] But unity has an infinite power which is so much its own that it cannot communicate it, for it cannot make number infinite, just as God cannot communicate his infinity nor his independence. Hence one may conclude that one must behold the Creator in his creatures, like unity in numbers and the unison in consonances.

One finds by experiment, in fact, that consonances are twice, thrice, or four times better and more excellent according to whether unison communicates itself to them twice, thrice, four, or more times,[12] as I will show in a separate discourse. When it communicates two degrees of union to them, they are twice as good as when it communicates only a single degree, and so on until they are reduced to unison by the subtraction of degrees of variety which determine the material of consonances, as union determines their form. If inequality and diversity act as the body of consonances, equality and union are their soul and spirit, as we shall see in the treatises on the Divisions and Superpositions of each consonance. There I demonstrate that of all the divisions of each consonance, the sweetest and most agreeable is the one which unites the notes most perfectly; and that of two or more superpositions of a single consonance or of several, whether upward or downward, the best and most natural is the one whose union is greatest.

When we have stripped creatures of their differences and diversity, and when the veil of exterior and finite appearances is lifted from them, we will behold the divine spirit which makes them move; and then we will be one spirit with God, following the beautiful words of the Apostle — "One who cleaves unto God is one spirit with him" [1 Corinthians 6:17]. For as soon as we see that there is no goodness or beauty in creatures but the divine goodness and beauty, our spirit will cleave so powerfully to that object which entrances the Blessed Ones that it will seem to be one and the same thing with it, as objects known are one and the same as the knowledge of them in the Peripatetic school.

But while we suffer patiently in the imperfection of our present state, let our ears be assailed by the variety of consonances as we await the abode where we shall be ravished by the perfect unison, whose beauty we cannot perfectly understand so long as we have need of diversity for our preservation. Neither shall we be able to know the divine beauty and excellence until it takes the scales from our eyes, explains the enigma which hides it from us and robs us of the sight of it as mists and clouds rob us of the Sun's light. For the same reason, the nourishment provided by thin soups and jelly is not so strong, useful, and agreeable to healthy people as that of bread, meat, and other foods which have not been deprived of their various imperfections: for the human body has many different parts, each of which requires a different food. Thus potable gold, or the elixir of which the alchemists and cabalists boast, is not suitable for food because it is too simple and pure.

Moreover, experience shows us that we cannot last long here without a variety of different actions and passions, each of which wearies and displeases us after a while. For instance, if one is tired one is happy to sit down; but after having sat for two or three hours one feels as weary as before and, rather than remain seated, prefers to resume work until the next bout of weariness. This proves clearly that man's pleasure cannot last without variety so long as he is in a variable state, and consequently that the continuation of unison cannot be as agreeable to him as when it is interrupted by other chords, even dissonances. This diversity, however, in no way prevents the unison from being more agreeable than the other consonances when it is used in the places where the rules of the art require or permit it.

This state of variety in which we are is the reason why the unison is avoided as much as possible: it is too sweet and excellent for this life. This is why pieces end more often with the octave, fifth, third, or their inversions than with the unison, and when they do end with it accompany it with other chords. For while the spirit is subject to matter which subjects it to delusions, obscurations, and error, one dare not, as it were, raise oneself up to the perfection of a unity which is entirely free from the variety and inequality found in the other chords. Here is evidence that the unison is as it were outside music, as God is beyond one's reach, and that when one hears the unison one should remember that the least pleasure of divine harmony is more excellent than the perfect knowledge of harmony as we use it. One may infer this from a beautiful saying of Saint Augustine: "To know even the tiniest fragment of God with a devout mind surpasses in its incomparable felicity the knowledge of the whole universe of created things" (*Genesis ad litteram* V, 16). Hence one may conclude that the pleasure which the spirit receives from the unison, when it is free from error and delusion, surpasses proportionately all the delights that come from other chords, for it is the image of the divine harmony and the source of those pleasures.

But an objection may still be raised which seems to deprive the unison of the prerogative we are giving it, namely that the spirit has more pleasure in conceiving things which increase its knowledge. This is why Nature rejoices in diversity, as I have proved through the same argument in a separate discourse. Now one learns nothing by considering the unison, because it contains no intervals and all its notes are one and the same, whereas in the other consonances one perceives the difference between low and high notes and the pleasure that comes from their mixture. Consequently the unison is the poorest chord and the least agreeable of all consonances, since it affords us no new knowledge whatsoever.

What is more, if the greatest unity of sounds were the cause of the greatest satisfaction, it would follow that there is more satisfaction in seeing a room whose walls touch, or nearly so, and a little house, than in seeing a great Louvre, because the parts of the little house are more united than those of a great palace. One could say the same of all that is grand and magnificent, and which may be abridged and curtailed, because abridgments cause the parts of the things in question to become more united than when they had a greater scope.

Finally, the differences of all creatures will be entirely preserved in Heaven, where it will please the Saints better than if they were all one and the same thing and there were no difference between them. For it seems that the whole pleasure of the knowledge of creatures consists in the relationships and comparisons that one makes between them and God, and between one and another.

It is easy to answer these objections because they presuppose the imperfect state of mankind, whose knowledge will be more perfect by far when they see clearly the great unity of all creatures and recognize that the diversity of objects exercises a great tyranny over our spirits. It diverts them from contemplation and thoughts that lead us to unity, which one cannot attain except by stripping creatures of their diversity in order to recognize the unity which has absolute rule over them, and to see in them only the root of being and the center of the sovereign intelligence. It is just as one sees only the radical terms of harmonic, arithmetical and geometrical ratios when one strips the larger numbers of what is superfluous and useless in them; and as one perceives only the spirits and quintessence of compounds when one has discarded the earthly and all that made them subject to corruption and to different alterations.

As for the greater knowledge that comes from the other consonances, one could compare it to the light of many little candles, or glowworms, whereas the light of unity and the unison is like the sunlight which dims all others by its presence, as the grace and excellence of the unison makes that of the other consonances vanish. For while we may not enjoy here the whole pleasure which can come from the unison, being distracted by the diversity of consonances, nevertheless the little attention which we do pay to considering its excellence gives us a far nobler and higher knowledge than that of the others, just as the little knowledge that we have of Heaven is far more excellent than what we have of the elements, however greater or more certain that may be.

Neither does it follow that a cramped room is more agreeable than a great hall, or a little house more beautiful than a great palace. One does not measure the beauty of buildings by the unity but by the proportions and symmetry of their parts, as one measures that of consonances by the union of their sounds. And as for the diversity of bodies and spirits in Paradise, it will be so much tempered by union that, according to some, all the bodies of the Blessed will be embraced by the humanity of Jesus Christ, and all their spirits engulfed by his divinity. So God will be all in all and reign absolutely in the being of all creatures, who cannot attain to a higher degree of perfection than to enter into the perfect unison of the created being with the uncreated, which consists of having no knowledge or love but that of the Divinity.

One can prove further that the unison is more excellent than the other consonances through Astrology, which sees the consonances in the aspects of the planets. The conjunction is the most powerful and excellent of all aspects, but many deny that it merits the name of aspect, just as they deny that the unison is one of the consonances. In fact, if the conjunction of planets represents the unison—as they hold that the opposition represents the octaves, the trine the fifth, the square the fourth, and the sextile the thirds and sixths—and if the conjunction is more powerful than the other aspects, one may well say that it has a close correspondence with the unison. But I will explain the planetary aspects in the first book on stringed instruments;[13] for the present it will suffice to consider that all things act with as much affection and inclination to union as they do for their own preservation.

This is why man does all he can to unite himself to all sorts of possessions by which he hopes to increase his comfort and to preserve and enlarge his being. And the greatest possession which can enter the spirit of man, namely eternal glory, consists of the union which man will have with God in spirit, and in body with the humanity of our Savior. Thus Saint Paul teaches in his Epistle to the Ephesians, Chapter 4, consoling them in the hope that all Christians should have of the transformation of their bodies, presently subject to all kinds of change, into another, spiritual body in which we will encounter Jesus Christ. Using years as an image, we will then be in the springtime of the most agreeable and perfect age: "Until we come to a perfect man, unto the measure of the age of the fulness of Christ" [Ephesians 4:13]. All these considerations, then, bring us to the recognition that the unison is the most perfect and agreeable consonance in music because it participates most abundantly in that which makes it sweet and agreeable, and that there is nothing but imperfection in the variety which preoccupies us and makes us prefer that which is more like our own fragility and misery, which cannot subsist here without diversity, the mother of corruption, albeit we aspire to unison and to unity. This is what is represented by those excellent words of the Gospel, "One thing alone is needful" [Luke 10:42].

If music serves for anything in this world, it should be used in particular to recall the memory of some of these considerations, so that it will not be said in eternity that men who profess to be rational, and who should have employed recreations and speculations to the end destined for them by God, have abused the chaste and rational pleasure of music and emulated certain musicians who rise no higher than the passion and action of the senses and the pleasure of the ear. But the latter should serve merely as a channel to give easy access to the contemplation of eternal things, and to the pleasure which comes from thinking of the final end, with which true Philosophers should always be occupied. And now it is time to speak of other difficulties which are met with in the unison, whose definition is explained in the following Proposition after the five Corollaries which I add to forestall many of the difficulties and objections based on the preoccupation of musicians and other persons who imagine things which do not exist. They may also influence the spirits of singers or music lovers to use harmony to lift themselves up to God, and to contemplate the grandeur of his goodness and the sweetness of his blessings and mercy, enjoyed by all those of whom the Royal Prophet speaks in the first verse of Psalm 72: "Truly God is good to Israel, even unto such as are of a pure heart."

◆

37

Athanasius Kircher
1601 or 1602–1680

Kircher's *Musurgia Universalis* of 1650 was the last great summation of speculative music, before interest in this subject waned and treatises became more and more practical. Living in the Jesuit College of Rome, and assured on all sides that he was the most learned man alive, Kircher was encapsulated in an archaic worldview: that of the Renaissance man who took all the world for his subject, and of the Christian Hermetist to whom the esoteric keys of knowledge gave perfect mastery.

One can see this view vividly illustrated in the following excerpts from the last part of his book on music. Keeping in mind that music was only one of half a dozen subjects on which Kircher wrote huge tomes, his scope, ambition, and authority are awe-inspiring. Having chosen the passages which apply to our subject, we find everything falling neatly into place in a scheme of universal harmony, whose author is God and whose method of operation is through the law of correspondences. *Music, Mysticism and Magic* contained one of Kircher's elaborate tables of correspondences, in which nine notes of the musical scale went with the ninefold division of the angels, of planets, metals, plants, animals, and much else; the idea was that as one "string" is plucked, the resonances are taken up by the member of each class tuned to that note. Kircher gave this quite seriously as an explanation of natural magic, in which the invocation of a planet, for example, through a talisman made of the appropriate metal at the correct hour and day, twangs the string of that planet and sends its influences radiating down to us.

The passages here serve to amplify that idea, first through Kircher's explanation of the workings of music on our "spirits" or intangible, etheric bodies (an essential factor in Renaissance acoustics, as in medicine), then through an ascent from the harmonies of the elements to those of the solar system. Kircher was not permitted to embrace publicly the heliocentric cosmology of Copernicus and Galileo, but he worked with the compromise of Tycho Brahe, the official view of the Society of Jesus. While keeping the earth static, this allowed certain innovations of which the most fertile, in this instance, was the "choir" of Jupiterian moons.

Kircher makes a surprisingly original response to the challenge of recent astronomy. At other times he simply borrows, or steals, as he can be seen doing here from Kepler, Fludd and Giorgi. Kepler he distrusted as a Protestant; Fludd, worse yet, as a Rosicrucian; Giorgi's book was on the Index. But all were grist to the Kircherian mill, and in the end he could display his orthodoxy in the magnificent vistas of a Dantesque heaven, dancing to the music of God the conductor. (See *Music, Mysticism and Magic* for the conclusion of the book that follows immediately on this excerpt.)

With Kepler, Fludd, and Kircher we complete the trio of late Renaissance theorists, unsurpassed before or since in their efforts to rationalize the intuitions of Pythagoras. True, Kepler quarreled with the irascible Fludd, and Kircher censures them both; but three centuries later, they can all be seen as representatives of the worldview which, as fate would have it, lost the battle for the mind of Western civilization: the view of an ensouled cosmos, resonant with meaning and redolent at every level of the divine intelligence.

Source: Athanasius Kircher, *Musurgia Universalis* (Rome, 1650), vol. II, 203–205, 370–373, 381–388, 458–461. Translated by the Editor.

On the Nature and Production of Consonance and Dissonance

Everyone wonders why the soul is so much affected by consonances and so much recoils from dissonances. In order to understand their causes and reasons, let us first observe that there are two things to be consid-

ered in consonances: the collision of bodies which occurs through the
motion of sounds, and proportion. These are the two principles, respectively physical and mathematical, by which all consonances are
caused. Physics considers the motion, mathematics the quantity, number, weight, measure, and every proportion of one sound to another.
Now since every rising and falling of notes and sounds derives from
faster or slower motion, it is necessary that the note or sound should
be higher, tighter, and denser when the motion is faster: lower and
looser when it is slower. All this was learnedly considered by Boethius,
De Musica, Book 1, Chapter 3, where he proves that sounds will be
lower when made by slower or less frequent movements, and higher if
the motions are more rapid or dense, so that the same string will sound
higher if it is stretched, lower if it is loosened. For when it is more
tense, it makes a faster pulse, it returns more quickly and frequently,
and impels the air more densely. When it is loose, it makes feebler
pulses, and these weak strokes vibrate more briefly and with lesser
effect. In *Musurgia Organica*[1] and Book 7 we demonstrated how tubes
and pipes sound higher or lower according to whether their apertures
are open or closed, and according to whether the air is compressed or
released. This is seen both in organ pipes and in the larynx or human
voice-box. The longer the tongue or epiglottis, or else the cleft through
which the air escapes to cause sound, and the wider the pipe, the lower
will be the sound: when it is shorter and narrower, a correspondingly
higher sound will emerge. Consequently, a person with a broader and
longer larynx will normally sing with a lower, bass voice, and someone
whose larynx is narrow and short will have a high voice. Much on this
appeared in Book I.

Secondly, it is to be noted that the sound caused by the motion of
a string or emitted by a voice is not continuous, but made of discrete
and interrupted movements, as shown in our *Chordosophia*[2] above. But
our ears are scarcely able to perceive its intervals, just as the eyes cannot discern whether a burning torch which someone whirls rapidly
around is present in the whole circle, or whether the whole circle is
burning. Boethius gives the example of a top spun by children: if red
stripes are painted up and down its cone, it will seem when spun to be
altogether red in color. Of course it is not so, but the speed unites the
red-colored parts, however small, and does not allow the others to
appear. Similarly, a tighter string will rebound or vibrate with more
strokes, a looser one with less. Whenever a string is struck, even if only
once, there comes forth not one sound alone, but as many sounds as the
trembling string conveys to the air. The addition of movements will
raise the pitch, while the subtraction of them will lower it, since high
pitch is caused by many movements. We have dealt fully in our *Algebra
Harmonica*[3] with the way in which one can divide any number or mag-

nitude such that the divisions are in a given harmonic proportion.

From all this I infer that the marvelous power which music has for moving the emotions does not proceed directly from the soul, for that, being immortal and immaterial, neither gives proportion to notes and sounds, nor can it be altered by them:[4] it comes rather from the spirit [spiritus], which is the instrument of the soul, the chief point of conjunction by which it is annexed to the body.

This spirit is a certain very subtle sanguine vapor, so mobile and tenuous that it can easily be aroused harmonically by the air. Now when the soul feels this movement, the various impulses of the spirit induce in it corresponding effects: by the faster or stronger harmonic motions of the spirit it is excited or even shaken up. From this agitation comes a certain rarefaction causing the spirit to expand, and joy and gladness follow. The emotions felt will be the stronger as the music is more in accord and proportion with the natural complexion and constitution of man. Hence when we hear a perfectly crafted harmony or a very beautiful melody we will feel a kind of tickling in our heartstrings, as if we are seized and absorbed by the emotion. These various effects are best promoted by the different modes or tones of music which we have already discussed thoroughly, especially by diminutions of small notes running up and down, and by the skillful combination of dissonances mingled with consonances.

MUSICAL EXPERIMENTS

In order to show how the spirit is moved in this way and in no other, I make the following experiment:[5]

Take a glass tumbler of any size and fill it with pure, transparent water. Moisten the index finger, and with it rub the rim of the glass in a circle. You will hear an extraordinary sound like ringing metal, which will excite the water to agitation as if blown by the wind. Now fill your glass to the midpoint, and you will again hear a sound, but twice as low, i.e., a perfect octave below the first, and will notice the water somewhat less agitated. If now you divide the same tumbler into five parts, and fill three with water, leaving two empty, the glass will resonate with a note accompanied by smaller ripples. If you divide it in seven parts, filling four and leaving three empty, you will find the note followed by even less agitation of the water.

From this it is obvious that our humors are moved in the same way, and especially our spirit, which from its seat in our heart has the most powerful dominion over all our emotions. Hence, when we hear any violent sound like thunder or the discharge of a cannon, we are seized with fright because our spirit suffers a blow and is dissipated: hence the fear and terror. We cannot bear to hear the noise of a knife grating on iron, for its harshness influences us ill by impinging on certain muscles leading to the teeth and brain. All of these things also apply to animals: there are some which enjoy and are tamed by music, which could not be the case if the power of music proceeded directly from the rational soul. But man enjoys music in an infinity of ways denied to animals, because he understands it better and more fully.

The Symphony of the Four Elements

Different people explain the harmony of the four elements in different ways. The alchemists reproduce it through this experiment: they extract the spirit from wine, then the oil which is innate within the spirit, and thirdly the phlegm, after which there remain the faeces which they call the *caput mortuum*. Sealing this in a hermetic vessel, they identify five regions therein:[6] the lowest, black one they call Earth; the extracted liquid, Water; the subtle and spirituous region, Air; the inflammable oil, Fire; above all of which floats the Quintessence, far separated from all impurity and from the faeces of earthly contagion. This latter they call Heaven, which rises by degrees, marking off the single mass as it were into harmonic proportions. They maintain that the world itself is created analogically to this example of art and nature, as will be explained at greater length below.

Others compare the whole world harmonically to a monochord,[7] into which the spirit or soul of the world—or rather God himself—introduces all the consonances from outside, producing higher notes in the nearer creations and lower ones in those which are further from him. The lowest sound of all may be assigned to the Earth: it is rather a silence and a darkness, giving place to the horrors of materiality, on account of the great density of the air and its resistance to spirit. They say that light acts on the substance of the world in the same way as the spirit of man on the air; that, moreover, the material of the world is a string—a megacosmic monochord, in fact—in which the harmonic steps reproduce perfectly the disposition of an harmonic scale. Thus Robert Fludd attributes to the Earth cold, density, and weight, proportionate to those of the lower region up to the Moon in the ratio of 4:3. This is sesquitertia proportion, because in this latter region there is a fourth part of light and color. Out of the Water, Air, and Fire is made the interval of a fourth, the earth being the gamut of the monochord analogous to arithmetical unity or the geometrical point.

In order to explain these proportions better, he invents a double pyramid, one light or formal, the other dark or material, intersecting such that the light one has its base in the Empyrean heaven and its apex touching the Earth. The dark one has its base on Earth and its apex fixed to the Empyrean. The two pyramids intersect at the sphere of the

Sun, where there is a remarkably equal mixture of light and dark, and continue their harmonic progression thus: from Earth to Water is one tone, from Water to Air another, from Air to Fire (which is nothing but the highest summit of Air) a semitone; from Fire to the Sun three more tones—to the Moon, to Mercury, and to Venus—and from Venus to the Sun a semitone. Thus there are in all five tones and two semitones between the Earth and the Sun, and the sphere of equality divides the string exactly in two, combining the fourth and fifth into an octave.

From the Sun to the Empyrean Fludd makes another, spiritual octave, so that the whole universe spans a double octave, the formal and material accommodating each other to make this perfect scale system of fifteen notes. When played by the hand of God, a marvelous concert arises in both worlds, whose depiction can be seen in our last Register.[8]

The Ancients, believing the elementary world to be made of concord and discord alike, and observing a certain admirable mixture of divers qualities, compared the elements to a tetrachord. For Orpheus, as Briennius[9] relates, fashioned a lyre according to their example, tuning all its strings to the interval of a fourth, as shown in this diagram:

$$
\begin{array}{ll}
\text{Nete} & \text{Fire} \\
\text{Paranete} & \text{Air} \\
\text{Parhypate} & \text{Water} \\
\text{Hypate} & \text{Earth}
\end{array}
$$

And just as two of the elements are heavy and two light, and of each pair one is heavier and one lighter, so in this tetrachord there are two low sounds, two high sounds, and each pair has one lower and one higher. Again, as the four elements mutate in one way and another, taking on various characteristics according to variations of density and temperature (whence arise the different species of things), so in this tetrachord the mutations from one string to another produce variations of sound. Moreover, between the four elements there are three intervals, whence are born the mutations and distinctions of physical forms. Just as in the tetrachord various forms of consonance can arise from the four strings, the same occurs in the year, divided into four seasons; and this elemental tetrachord, played by Apollo's plectrum, gives forth the concert and symphony which we admire yearly in the regeneration of things and the production of other effects. And though this tetrachord be in itself dissonant and devoid of harmony, so soon as the radiant fingers of Apollo play upon it, it performs the most excellent symphony, reconciling consonance and dissonance so that things most disparate and contrary join in friendship and concord to make a single piece.

Pythagoras[10] investigated a little deeper into this secret harmony, and discovered what nature has cunningly concealed in the five regular solids. The Earth's harmony he found in the cube: for a cube has 6 faces, 8 solid angles, 24 plane angles, and 12 edges. 12:6 is the octave, 8:12 the fifth, 6:8 the fourth, 24:6 the double octave, and 24:8 the twelfth: all the perfect consonances contained in a single solid. Yet this should not surprise anyone, for the Earth is the foundation and basis, the proslambanomenos, in which lie the harmonic seeds of all generated things (as we shall explain); and as in the polychord all the strings acquire a certain relationship when sounded with the first, so do the other elements with the Earth.

Pythagoras also found Fire in the pyramid [tetrahedron] which consists of 4 faces, 4 solid angles, and 12 plane angles, from which comes the simpler nature we observe in Fire. The ratio 4:4 gives the unison and 4:12 the twelfth, from which we can understand the sharpness and fineness of this consonance. The harmony of Air he found in the octahedron, with its 8 solid angles, 8 sides, and 24 plane angles. Lastly in the icosahedron, with 20 sides, 12 solid and 36 plane angles, he found the nature of water, as we shall explain in what follows. The harmonic proportions of these bodies are thus interrelated: the octahedron and pyramid have their sides in duple proportion, their solid angles in sesquialtera, their plane angles in duple, from which emerge the duple and sesquialtera proportion, i.e., the octave and fifth. The pyramid's proportions compared to the cube's give the sides in sesquialtera, the angles in duple, making again the octave and fifth. The cube and the icosahedron compared have triple proportion in their sides, sesquitertia in their bases, sesquialtera in their angles; from which the twelfth and fourth emerge in the bases, and in the angles of a fifth. You see, therefore, that the more they approach Earth, the greater harmony is manifested in their bodies. Thus you can regard the cube and icosahedron, which answer to Earth and Water, masculine and feminine, swelling with harmonic seed, as the marriage bed of nymphs, most fitted for the generation of things (whence the icosahedron is inscribed in the cube); for Water and Earth must be properly wedded in the most fertile manner for generation.

We will illustrate the harmony of the elements with an analogy of the proportions which illuminate them. Now also must be revealed the sensible symphony of the elements in which they are truly consonant with one another. The agitation of various elements and their physical impacts certainly cause a perceptible harmony. For what is the Earth but a sort of organ, furnished with hidden channels or pipes, by whose passages the igneous spirit is diffused through its whole body? When a violent shock occurs it is as when the wind in our artificial organs makes high and low sounds, according to the width or narrowness of the openings. I first observed music of this kind in the amazing earthquakes of Calabria, in 1638,[11] when the sides of the mountains shook from the air trapped within; and as it forced its way out, we straightway heard a sound now like a trumpet, now like the murmur of receding waters; sometimes like thunder, or like the noise of people wailing and the soughing of wind. Waters in the belly of the mountains are also responsible for the prodigious music of the Nile Cataracts near Syene, which as mentioned above is said to be audible through certain clefts in the mountains, and which we will explain fully in our *Subterranean World*.

The unceasing motions of earth and water give rise to a further kind of harmony by perpetually pounding on the seashore, creating through the numerous caves and channels the harmonies of which historians have told. Pausanias says that the beaches of the Aegean Sea have the sound of a lyre. The shores of certain islands make a sound like organs or wind instruments of every kind; those of the Botnic Sea, according to Olaus, like people wailing. Now as we have shown before, the concavities of caves, hollows, and beaches can utter manifold intervals: if their volumes are related as 1:2, they will sound the octave when struck by the waves and torrents of water as by a plectrum. If their relation is 2:3, then they will sound the fifth; if 3:4, the fourth; if 8:9, the tone, and so forth. So it is no marvel that such a roaring or crashing noise should be pleasant, seeing that it contains the hidden seeds of harmony within itself.

Next the Air, like Earth and Water, gives birth to divers harmonious sounds when it is aroused by violent agitation in mountains, rocks, caverns, alleyways, trees, and torrents. It, too, can make a pleasant concert of concords and discords, of high and low notes. In the mountains, where the wind blasts through narrow channels of different sizes, it displays an amazing variety of sounds. When a southeasterly wind blows in the crevasses and caverns of Etna, one hears a ceaseless harmonious whispering just like strings tuned to the fifth, third, and octave. I remember observing the same sounds when the wind blows through trees of different sizes. For if one is twice the height of another, as may often occur in cypresses and poplars, they will whisper an octave apart; if their heights are 2:3, then a fifth apart, etc. The same must obtain with the proportions of the sea's waves. Lastly, in the case of parchment windows, a perpetual music of this kind can be created by the movement of the air alone.

Draw on a parchment window a number of circles having their diameters in the same proportion as the pipes of an organ, viz. 1:2, 2:3, 3:4, 4:5, 1:3, 1:4, etc. Now carefully cut round these circles, but leave them still joined at some point; they should be like lips, which act like a plectrum or a tongue, forming a sound when struck by the wind. Now your instrument is ready. When the wind blows on this window, the lips within the circles will soon vibrate with a quavering sound, and when all the circles are singing together they will make a marvelous harmony with each other, to the admiration of all listeners. The instrument with strings played by the wind, which we have described earlier, is also pertinent to this. The air forced by the heat of a fire through the variously tuned strings of an Aeolian harp[12] also makes an unusual and harmonious sound. I could relate here innumerable other tales of this elemental symphony, if I had not already done so throughout the work.

Certainly if our ear, placed as it is, can project the ability to hear distinctly the noise of wind and ocean crashing on the shores of the lands, I would say that it can no less perceive the perfect symphony of all numbers; and this to the end that it might be aroused thereby tirelessly to hymn the Creator's praises, just as we behold the very elements joining their voices like a fourfold chorus of soprano, alto, tenor, and bass[13] in perpetual hymns to GOD, the omnipotent creator.

That There Is a Music of the Heavenly Bodies, and in What It Consists

That there is some harmonious concord in heavenly bodies worthy of that eternal Harmost[14] of infinite beauty, both Holy Writ and all the theologians and philosophers agree. But it is scarcely explicable in numbers, for this indescribable concord of the heavenly bodies, "eye hath not seen, nor ear heard, neither hath it entered into the heart of man" [1 Corinthians 2:9]. It is reserved by the eternal, supramundane Organist to beautify the eternal life of his Elect alone. As much as it can be grasped by the feebleness of our human understanding and the dimness of our mental eyes, it is not to be understood as the sensible impact of heavenly bodies, but solely in their admirable disposition and in the ineffable proportion which unites them. For these heavenly bodies are so linked one to another, that if one were moved or changed, the harmony of all would be marred.

The Sun wields the imperium in this planetary economy subject to him, like a choirmaster arousing all into harmony with the plectrum of his beams. This harmony consists, as we have said, in the admirable disposition and proportion of the heavenly bodies, each ensphered by another, and in the most exact analogy of their mass and size, each appropriate to its own ends. Thus the distances between Sun, Moon, and Earth and their sizes are such of necessity, without which the cosmos would perish. Owing to these intervals, the worlds are able perfectly to aid and preserve one another by their mutual influences. Consider as an example how igneous bodies (of which the Sun is one, being altogether fiery by nature) heat other bodies less and less as the distance from them increases: not only nature but also the empirical art clearly teaches us that. We have already seen how the Sun's rays are reflected in various ways upon the Earth, having one effect in the frigid zone, another in the temperate, and another in the torrid zone (as we have demonstrated fully in our *Anacamptical Art*);[15] and this admirable variety of diverse effects is harmonious in such a way that without it the Earth could not exist, and things would be in perpetual dissonance. Again, the distance from the Earth to the Sun and the size of their bodies produces a certain disposition of the Earth, a certain tempera- ture, a certain harmonic variety, and no other. Again, this distance and size of the Earth cause exactly equal portions of light and darkness to fall in every place (as demonstrated in many ways in the *Anacamptical Art*). If the distance were greater or less, the whole harmony would be reduced to the most confused of discords. In the same way, strings of such and such a length are concordant, but if they should diverge in length, in place of melody there will occur I know not what concord and discord. Thus if the Earth, without changing size, were to come closer to the Sun, or to go further from it than it in fact is: the nearer it went, the more strongly would the Sun's rays strike it; the further off, the weaker they would be. Apart from the inequality in light and dark- ness that would ensue, the temperature would be not only unpleasant but incompatible with human life. The same would follow, were the distance to remain the same but the bodily mass of either Sun or Earth to be greater or less. For just as with burning glasses a definite focus and a certain distance are required to effect combustion, so God's eter- nal wisdom has placed between the Earth and the Sun the perfect distance for the augmentation of heat: one, moreover, that is harmonic. Different effects are felt on the Earth from the different degrees of heat caused by the continuous movement of the Sun between the tropics: sometimes its rays fall directly, sometimes obliquely, sometimes ob- tusely on any given place.

The empirical art shows that the different applications of fire simi- larly cause varying degrees of heat. By their application in distilling, a simple substance will change in a marvelous way from one state to another, symbolizing with its elements a wondrous harmony. What was at first compounded in a confused and homogeneous mass is exhibited through the operation of fire in its various discrete natures, its various bodies of differing qualities, by fermentation, coagulation, fixation, dis- solution, composition, softening, and hardening.

Since the most fiery Sun and the most humid Moon could not instill a perfect harmony of humors, or a perpetual octave, in the Earth's globe without the intervention of other bodies, the Divine Providence has placed between the Moon and the Sun two other heavenly bodies, Venus and Mercury, powerfully endowed with different virtues.[16] Through them the vehemence of the Sun's rays is removed or blunted, and the Moon's humidity augmented as is necessary to bless the Earth with greater variety. The fecundating power of Venus is communicated to the influx of the Moon, and enriches her humidity, while Mercury, the divider of consonance from dissonance, tempers that which is noxious and superfluous both in the Sun and in Venus, and thus pours upon the lower world the necessary kind of harmony. This exerts on all things ever-changing combinations of effects, through the successively different aspects and eccentric positions of the said planets. Yet since no instrument is so perfectly tuned that in the course of time it will not become dissonant through the slackening or tightening of its strings, so in the inferior world it is inevitable that the retreats and advances of the other planets will from time to time cause a hiatus [*discrasia*]. Then it is just as when the body, in the course of time, suffers a congestion of the humors or an attack of fever: when the humidity retreats it is evacuated and restored to health.

To this end, God has placed in heaven by his natural art two dissonant bodies, Mars and Saturn, from whose pestiferous vapor all ills in the sublunary world have their origin. But lest their evil should bring disaster to the undefended economy of the whole lower world, he placed in between them the benign planet Jupiter, whose healthful influences restrain and arrest the deadly and utterly pernicious forces of the other two. Lest the unprotected Earth should suffer ruin from Mars' virulence, Nature has given the latter the most eccentric orbit of all. So, coerced from below by the Sun and Venus and from above by Jupiter, it desists somewhat in its evil.

But Mother Nature should not be accused, from the conduct of such pernicious bodies, of being rather a stepmother (as she well might be, by the highest and the lowest counsel, for thus disposing matters). No: the world could not be maintained without them. When the greater world is aggravated by a great mass of evil humors, the same occurs as in the microcosm: Cantharides, a medicine powerful in caustic virtue, attracts the morbid material and, dissipating it from the center to the circumference, releases the endangered person. So there is nothing evil in the nature of things which does not eventually contribute to the good of all and the conservation of the universe.

What else are Mars and Saturn but dissonances? Tied and syncopated by Jupiter in perfect consonance, they give to music not sweetness, exactly, but a great embellishment. What else is Mercury than a dissonance between the Moon and Venus, tied like a syncopation between two consonances lest, set at liberty and bound to none, it should pervert the Earth from the benign influences of the Sun, Venus, and the Moon. Certainly whoever will consider this a little more deeply will find that the seven planets sing with the Earth a perfect four-part harmony, in which dissonance is combined so artistically with consonance that it gives forth the sweetest chords in the world. That the curious reader may have some example of the celestial four-part harmony, we give here an idea of our speculation in notation.[17]

FIGURE 23: *The Harmony of the Planets*

Saturn, Jupiter, and Mars sing the highest voice, in whose notes Jupiter (the consonance) always ties and weakens Mars and Saturn (the dissonances). The Sun proceeds singing the middle part in the most perfect consonances, always an octave or a twelfth from the bass (Earth). Venus, Mercury, and the Moon sing the Hypatodon, Venus and the Moon binding Mercury, the intermediate dissonance, bringing him into friendship, and resolving him straightway to consonance, as can be seen. The Earth, lowest of all, receives the consonances and dissonances perfectly mingled, thus forming, as we may imagine, a perfect music with the planets.

Nature loves variety; and in order that the planets in the course of time should never offer an identical harmony to the lower world, they turn now around their own centers, then make eccentricities by going nearer and farther from each other, enter into various positions relative to the Zodiac, and ever and again create new kinds of harmony—yet always under the laws and within the limits set by Nature. In their eccentric orbits, which are as it were the limits of planetary motion, they gradually alternate ascending and descending tones or modes, now from Dorian to Phrygian, then to Lydian, next to Aeolian; and so proceeding to the rest until they return to their first, and the whole circulation begins again. Since certain parts of the globes or planetary bodies are heterogeneous, and their qualities, properties, and forces vary with respect to the surrounding planets, Nature has made them revolve around their own centers so that they are always turning different parts of themselves to Earth. Thus celestial music, like artificial music, shows a mixture of dissonances with consonances. You will understand, then that the planets are so placed that if only one were to stir beyond its determined limits, or its harmonic interval, the whole harmony would be destroyed thereby.

COROLLARY

From this it is plain that the Earth is like the bass above which the rest of the harmony of the universe progresses. If it were removed, the other voices would remain in an unpleasant and imperfect harmony of many dissonances. Believing sufficient to have been said on the universal symphony of the planets, we will now turn to the particulars of their symphony.

On the Particular Symphony of the Planets

Lest the unceasing noise of numerous voices should displease the ear, composers have been recommended to distribute their forces in several choirs, dividing the various instuments or groups into alternating choruses so that a greater variety and pleasure may obtain for the listeners. Not for this reason does the Harmost Nature distribute those celestial songsters [*Phonasci*] in various choirs, which although different in sound conspire in a consonant-dissonant union, ornamenting the world with their diversity and witnessing to the ineffable wisdom of that supermundane Harmost. Thus I will begin with Saturn, the highest choir of the planetary music. The Lyncean School of astronomers[18] has recently discovered Saturn's remarkable constitution, as it has pleased GOD, Best and Greatest, to arouse in us awe and love of his wisdom by opening a new register of the celestial organ. At first it was observed to have three bodies, then two of these were ascertained to be in perpetual rotation, like attendants pressing in upon their leader. Now the Sun, the leader of the chorus, is so distant that the plectrum of his light cannot reach Saturn: but in order that the latter should not remain dissonant, he has most wisely been provided with two companions who follow him constantly like delegates of the Sun, and foster him with the mutual exchange of energies. By this admirable cooperation the same harmony as the Sun, Moon, and Earth have with one another is established in the Orb of Saturn (as we shall henceforth call this planet), the planet having two satellites; and these attendant singers (whom we might well call the Saturnian Sun and Moon) work like a distinct choir. And as it moves this way and that on its own axis, the middle Orb of Saturn is pleased to turn once in 29 days and 10 hours—the same number of days as it takes of years to complete its great circle: a most amazing thing, and only recently discovered. And if God had chosen to place us on the Orb of Saturn, our single days would last not 24 but 706 hours and 38 minutes! Lest the subpolar regions of Saturn should be condemned to perpetual darkness, it is made to incline a little to each satellite so that the whole globe is imbued with uniform light. Probably these followers also rotate both around their own axes and around the Orb of Saturn, in a period agreeable to their bodies; and for the greater variety of influence it is likely that the two satellites of Saturn are of a heterogeneous nature, imbuing it with different qualities from the different points of their revolution. Their rotation would thus introduce more of a mixture into Saturn's globe, mingling concords with discords so as to make a perfect harmony in this chorus. In this diagram we show the globe of Saturn as A, its axis BC; the satellites ED, their orbit FG, and its diameter EAD. (See figure 24.) Comparison of these with what has been said above will give the harmony mentioned.

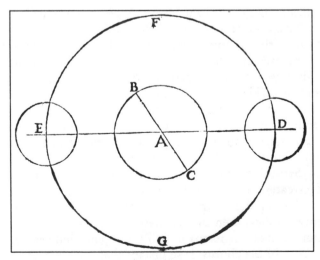

FIGURE 24: *The "Satellites" of Saturn*

COROLLARY

From the said harmonic constitution it necessarily follows that Saturn cannot be illuminated by the Sun on account of its excessive distance, and hence is provided with light by its two satellites. By certain modern calculations it is 20,000,000 *leuci horarii* distant. Certainly our eye would perceive the solar diameter from Saturn as no more than 3'11" of arc; i.e., a little larger than Venus, Jupiter, or a star of the first magnitude; so it would need another light than the Sun. It follows, moreover, that on Saturn we would have a 'day' of 29 days. None of the inner planets would ever be observed to diverge more than 40° from the Sun: Jupiter 40°, Mars 12°, Venus 4°, Mercury 2°, the Moon and Earth barely perceptibly: so all these planets might be called companions of the Sun. And observing the Sun's 29-year course through the Zodiac, on Saturn you might well believe yourself in the center of the universe. All of which I could demonstrate here, only I will reserve it, with the latest discoveries, for the second edition of the *Great Art of Light and Shadow*,[19] to which I refer the reader. This will suffice in the meantime to demonstrate the marvelous harmony of the Saturnine Chorus.

The Choir of Jupiter

It has appeared from frequent observation in very recent times that the Architect of the heavenly harmony has adorned this chorus with a most singular artifice, surpassing all others. Great and glorious Jupiter has now displayed himself to this lower world with a system of four companions, hitherto unheard of and unseen, in that symphony of all things which we have described. This globe is far bigger than the Earth, and so remote from the Sun that, like Saturn, it seems to have found the Sun insufficient for its illumination; and lest its immense mass, imbued with supreme virtues, should lack light and, wasting away in perpetual mourning, should fill the other heavenly bodies with dissonance, it has been decreed by the wondrous council of Nature that four choristers [*Choragi*], perpetually striking with the quills of their rays, should excite it by their harmonic motions into the most exquisite harmony.

Jupiter's first singer [*Phonascus*] has been observed to have a periodic motion of about 42 hours, the second, 3 days and 13 hours, the third 7 days and 4 hours; the fourth and last completes its period in 16 days and nearly as many hours. If we are to believe the calculations of Rheita, the distance from Jupiter of the first one is 20 radii of the Earth, of the second 27, the third 41, and the fourth 69. Rheita's actual calculations of the four satellites of Jupiter are shown in the following table:

Distances from Jupiter of the four Jovial Choristers, or Companions

	1.	2.	3.	4.
Radii of the Earth	20¾	27½	41½	69½
Leuciis horariis	20838	27787	41676	69460
Diameters of Jupiter	3	4	6	10

Whatever is requisite for music certainly lies concealed in these numbers: for the distances of each body correspond precisely to a harmonic quantity: 3 : 4 : 6 : 10.

Insofar as the frequent observations of the Lincei and the remarkable progression of proportions have enabled us to conclude, we have found the first satellite to be approximately equal in size to our Moon, the second to Mercury, the third to Venus, and the fourth almost equal in mass to the whole Earth. Whence it follows again that if we were placed on Jupiter, these satellites would look much larger than the Sun and Moon do from the Earth; the second and fourth would approach the Moon in brightness; the first and third, the splendor of the Sun. We have here, curious reader, the description, position, and sizes of the Jovial Choristers, so that we may see what harmony they make. In the figure the first is marked A, the second B, the third C, the fourth D, and Jupiter in the center V. (See figure 25.)

The Harmost GOD, immense and of incomprehensible wisdom, has decided to stablish the world with an innumerable variety of things and, further, has placed in this cosmic temple a visible chorus equipped with several choristers, namely the planet we call Jupiter. This Jovial Orb is greater by far than the orb of our Earth, and filled with the most eminent virtues which it breathes into our lower world. Yet so far is it from the Earth and the Sun, that it cannot easily absorb any light from the latter: therefore GOD has given it two substitute Suns (1 and 3) and as many substitute Moons (2 and 4), and ordered by his skill and admirable providence that they should continually take turns in illuminating the globe of Jupiter, so that what one Sun and Moon cannot do, the auxiliary ones will supply. The Orb of Jupiter is continuously assailed by changing lights and shades, in harmonic motion, which cause

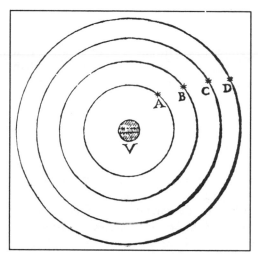

FIGURE 25: *The Moons of Jupiter*

an amazing diversity of qualities in its body. Its days and nights must of necessity vary not only in duration but also in the intensity of their light. Who could doubt that the simultaneous rising of Choristers 2, 3, and 4 would bring more of a day than a single Sun? One can be sure that such a day will be intensely bright. And if all four Choristers were conjunct, they would leave the opposite portion of Jupiter in night: yet a short one, because of the speed of their motion. Lest the harmonious constitution of Jupiter's body be upset by so much light, he is not permitted to rest, but turns ceaselessly around his axis, such that one of these revolutions lasts as many days as there are years in his eccentric orbit: namely 12 years. So if we lived on Jupiter one natural day would last not 24 but nearly 268 hours. As soon as the light of one Chorister wanes that of the second, lunar satellite succeeds it; the third, solar Chorister follows, and as it sets the fourth, lunar one rises, in a perpetual and delightful exchange. Also mitigating the force of the light, very frequent eclipses must attend the motion of the planet. And I reckon that the bands or belts of Jupiter are nothing other than shadows of the lunar bodies cast on the planet's disc. You can find all of this fully and diligently treated under 'Cosmometry,' in the second edition of my *Art of Light and Shadow*.

In comparing all these things, the reader will find an absolutely harmonious arrangement of concords and discords, wondrously tempered so that nothing more satisfying could possibly be imagined. And it is this harmonic constitution which causes the body of Jupiter, supremely beautiful and most wondrously endowed, to give forth the influences which we in this inferior world experience with such admiration.

COROLLARY I

It follows that Jupiter could not have been greater or smaller, nor could it have moved closer to or further from the Sun than its eccentricity permits, without manifest hurt to the harmony of the cosmos.

COROLLARY II

Some say that in places where men are unable to dwell because of the excessive intensity of the light, or because of temperatures incompatible with human nature, there are creatures endowed with a different nature. Since nothing of the kind is known to us, nor can be known, it seems to be fundamentally dangerous to the Faith. Who could regard it otherwise than as a blind and baseless imagination, a novelty and fiction of philosophers? So now having seen the harmony of Jupiter we will proceed to the other choruses.

The Choir of the Sun and Mars

The Solar or Apolline choir includes within itself Venus, Mercury, the Moon, and the Earth, and is somewhat parallel to that of Jupiter, of whose harmony enough has been said. Between the Solar and the Jovial choirs Nature has placed a different arrangement in the harmony of the heavenly choir, namely the globe of Mars. Since this is dissonant with the others, yet necessary to the maintenance of the harmony of the spheres, it is placed in between the consonant Solar and Jovial choirs as a mean, so as to arouse in them the loveliest harmony as if by suspension and syncopation. And in order that this syncopation should achieve a greater effect, it is proper for it to be the most eccentric of all: sometimes it descends below the Sun itself, and at other times it assails the distant empire of Jupiter; now it submits to the Sun's light, now it is seized by Jupiter, lest acting by its own rules it should turn all things topsy-turvy and confuse the harmony of the All. And since it goes its eccentric way alone, associated with no choristers, it is thought to borrow all its light from either the Sun or Jupiter. But this kind of light is not nearly sufficient for exerting its own energy, and so by Nature's secret counsel this globe of Mars is furnished with subterranean fires and becomes the fount of darkness and pestiferous smoke, of excessive and destructive heat, crammed full of evil and venomous vapors. The influence of this poison would not only infect all the other heavenly bodies; it would totally destroy them, had not Nature arranged it in exactly the right place for exerting its force. For as it is the midmost of the celestial choirs, it is as we have said tied and syncopated with its neighbors the Sun, Venus, and Jupiter so as to conspire with them in perfect harmony. And in this ingenious syncopation we see Nature's striving for *euchrasia* in the harmonious symphony of the heavenly bodies and of their influences.

The Harmony of the Celestial with the Angelic Ennead [20]

So as to preserve the lesser world, on whose account all things were made, in perfect harmony, the author and framer of universal Nature has set in motion another ninefold scheme, the system of the celestial worlds, by whose influence the lower world is governed. This comprises the Empyrean heaven, the starry heaven or firmament, the globes of Saturn, Jupiter, Mars, the Sun, Venus, Mercury, and the Moon. Since these bodies are in themselves inanimate by nature and incapable of movement he has assigned to each its Intelligence, whose virtue arouses in them the various motions required to produce various effects in the sublunary world; and the constitution of their bodily nature depends on the influence of higher bodies. The Intelligences placed in the heavens have responsibility for the disposal of souls, while GOD the supreme Choragus brings high, middle, and low into one perfect harmony. The harmony of each will now be clearly revealed.

Observe first that there are nine heavens. The Empyrean heaven, court of the eternal King above the mundane house of the all-mover, is by its order and infinite light not unjustly compared to the Seraphim. From here all illuminations of souls proceed as if gushing forth from a fountain.

Second is the starry heaven or firmament, in which the Artificer has extended himself into numerous instruments for the tasks in hand, and which agrees beautifully with the intelligence of the Cherubic order. For the stars are nothing other than symbols, as it were, of ideas in the human soul, from whose formal concepts knowledge [*scientia*] is born, the property of the Cherubim.

Saturn's, the third globe, corresponds to the Thrones or the intelligence Schabtael. By its influent virtue it generates in us firm, fixed, and unshakable judgment [*prudentia*], as if binding man especially to the throne of his councils. Not without cause do the Hebrews call this *Schabtai*, that is 'my seat.' To this are subject all things in the sublunary world redolent of the saturnine nature: stones, plants, animals, as we have shown elsewhere.

The globe of Jupiter follows fourth, which by nature and properties fits well with the choir of Dominations. His is a beneficent planet, regal and full of majesty, promising domination and rule to those born under him, and his intelligence confers true justice. To his rule submit all things in the sublunary world redolent of the jovial nature.

The globe of Mars follows, which with its fiery and incendiary energy suits best the Virtues: a fire which by its virtue burns up the bile in us, makes us able easily to accomplish hard tasks, and increases our strength and fortitude. To this globe are subject all things in the lower world endowed with a martial nature.

The globe of the Sun, maintaining the central body of the universe, is none other than the King and Leader of all, governing everything by his virtue and power: whence he corresponds with good reason to the middle order of the Powers. His intelligence produces our life, health, and glory, and bodies of a solar nature in the sublunary world concord with him, as we have said elsewhere.

The kindly globe of Venus is the fairest planet, fitting with the Principalities whose intelligence, Haniel instills in us the love of beautiful things. She confers everywhere grace and benevolence, and to her tutelary intelligence are subject all things analogous or corresponding in nature, called venereal, as was shown above.

The globe of Mercury is aptly compared with the Archangels, for just as the Archangels work to mold everything according to the image of the highest principle, and to join them together they are themselves perfectly united, even so Mercury labors to reduce the many formal and rational sounds to consonance, and especially to match rational number with the divine. His Intelligence, Cocabiel, teaches true concord and union with GOD, and strives by this consonance to lead souls to GOD, where joined to him for eternity they will break forth in ceaseless hymns and praises of his eternal majesty. All mercurial things in the sublunary world are subject to him.

The last globe of the celestial Ennead is the Moon, best compared to the Angels, last of the nine orders. Like the Angels it is the closest to us; and as the Moon takes the energies of all the higher planets and then communicates them to the sublunary world, so the Angels receive illumination from the superior intelligences and channel them down to us. Countless things could be said here about the influence of particular virtues, but since we are reserving them for our *Hieroglyphic Theology*[21] we will refer the reader to that when it appears.

Thus we have seen the wonderful symphony and resemblance of the first Ennead with the second. Now we shall see how through the third Ennead, or the cubic power, the Creator of all produces every creature in the sublunary world.

The primeval Monad, the beginning of all things, unfolding itself in time into the Dyad, produced an indefinite duality or matter. From the monad and dyad proceeded numbers, from numbers points, from points lines, from lines surfaces, and finally from surfaces solid bodies, namely the four primary elements, the foundation of things. Since they were fully treated in the first section we shall not expatiate on them here, but merely show how the said four elements, together with five degrees of mixed bodies, constitute a third ennead. The elements are not so much corporeal in nature but rather an intellectual ennead, invented by GOD himself in his own way so that he, the Musician of the Universe, may more clearly reveal it as conceived according to the Archetype.

The four elements plus the five stages of corporeal nature comprise the ennead of nine members, as follows: Earth, Water, Air, Fire; stones and metals, plants, zoophytes, brute beasts, and Man: all of which constitute the lower world. Even though they are corporeal, the wise composer and Harmost has arranged them in such harmony with the other two worlds that the same things occur in each world after their fashion. In Earth is the basis and support of the lower world.

1. There is Earth in the heavens, as we showed in the second Register; even among the Angels it manifests as the unmoving throne of GOD and the footstool at his feet, and in the Archetypal world as the firm and fecund nature of all things, as witness: "Let the Earth open and bring forth Salvation" [Isaiah 45:8], the latter being the fine and clear elemental water, purifying and bathing all things.

2. There is also Water in the heavens, flooding the Earth with its influence, and a compounded virtue such as the Moon, Venus, and Jupiter bestow on man. Among the Angels it is instructive and purifying, as witness: "Who lays the beams of his chambers in the waters" [Psalm 104:3]. In the Creator is truly the water of salvation by which sins are washed away, man is regenerated, purified, and cleansed.

3. The elemental Air is that which we breathe, see, hear, and smell. There is also air of a finer nature in the heavens, displaying to our eyes their splendor and riches; and in the Angelic world there is a space where the concert of blessed souls resounds, a vital breath and a subtle air. In our Maker is the supreme and perfect life and spirit which blows on all creatures, bestowing on them life and breath.

4. What the element of Fire is to us, the Sun is in the heavens, and in the Hierarchy the Seraphic fire and igneous force, as witness: "Thou makest thy Angels spirits, and thy ministers a flaming fire," [Psalm 104:4]; this last being the ideal fire and spiritual light in the Archetypal world by which all things are generated, preserved, sustained, nourished, and increased. So the elements are as it were nurtured in GOD as ideas for the production of things, the first nurslings and originals.

Their powers are distributed among the Angels; in the heavens are the analogous virtues, in Nature the seeds of things, and in the lower world their gross forms.

So this is how the admirable symphony of the world is born: the symphony in which Earth concords with the Moon, the heavenly earth, Water with Mercury and Venus, Air with Jupiter and Venus, Fire with the Sun and Mars, stones with Saturn, metals with Jupiter, zoophytes with Mars, insects with the Sun, birds with Venus, amphibians with Mercury, and quadrupeds with the Moon. All concord with all in a wonderful sympathy, now among themselves, now with the rational soul, whose Ennead is compounded of the five exterior senses and the four interior ones—the common, imaginative, appetitive, and rational senses. These are completed by the Intellect as by a tenth, in the same way as the other, purer Ennead comprising the nine orders of Angels becomes the complete image of the Angels and of GOD by the addition of the tenth source, namely the supreme Father of all. His is the three-fold operation, of such consonance and harmony: the working of the inferior world with Nature; of the heavens with the senses; and of the Angels with the intellect. It appears in the tetrad through the simple power of the soul; in the ennead through the nine senses, five corporeal and four incorporeal; and in the icosiheptad (27) it attains the idea of the whole consummation of things, until by intellection it becomes the All.

COROLLARY

You see, then, how our harmonic scheme shows the perfection of the whole world through unity; through the binary, the variety and multitude of the things contained in it; through the ternary, the analogous unity in the eternal triad, joining all things through fit proportion; through the ennead, the link and binding of the world's three enneads explained a little earlier. The harmony of all, moreover, is compounded of simple, square, and cubic number through the cube, which the Pythagoreans called a harmony; there is no going beyond the cube, since number by its very nature, when it increases in length, breadth, and depth, can extend no further.

Therefore GOD, existing uniquely in himself, extends into multitude just as unity into number, losing thereby nothing of his own property. From this unity he draws together all the 27 strings of the universe as one, reduces them to consonance, tempers, and tunes them, bringing the equal and unequal voices of genera, species, and individuals into a single concord. To the various portions he communicates the proportions of 2:1, 3:1, 4:1, 2:3, 3:4, 8:9, or 9:10, necessarily resulting in the intervals of octave, twelfth, double octave, fifth, fourth, tone, etc. Only he will understand their connections and combinations who can truly say with Ecclesiasticus: "He himself has given of them to me, that I should know the disposition of the globe, and the virtues of the elements, and the end and middle of time; the changing of fortunes, the ends of epochs, the mutations of customs, the dispositions of the stars, the natures of animals, the wrath of beasts, the force of the winds and the thoughts of men, the differences of plants, the virtues of roots, and whatever things are hidden and unforseen I have learned." [Wisdom 7:17–21]. Whether all things are manifest through one's being steeped in the unction of the Holy Spirit, or through the death of the kiss, as the secret theologians say,[22] or, as the Platonists say, through an essential contact with the ideas of the first intellect's own power, one can do thereby all that one undertakes. From this contact as if from intercourse images are conceived, the species of things are learned, the soul is awakened and purged as by ambrosia and nectar from the contagion spreading from the abyss of Lethe. In this harmonious constitution of things you see the divine mind, the Arch-muse, regulating all things by musical numbers. And since the perfection of all things is brought about by their perfect mixture and tempering, nothing in the nature of secret things will be found so abstruse that it may not be penetrated by adapting its consonances to its dissonances in the polychord of this universal system, according to the rules laid down above. I will reveal the secret in a few words. The enneachord of the soul must be aligned with the enneachord of the Hierarchy, and that of the body with the celestial one of the stars, through *isophony*; and you will penetrate, as I said, the secret of secrets, the absolute knowledge of things divine and human, as Saint John says: "You have been anointed by the Holy One, and know all things" [1 John 2:20]. This miraculous music of the supramundane Choragus and the universe is shown in the preceding figure, in which you will easily find in synopsis whatever is said here.[23]

38

✦

Angelo Berardi
c. 1636–1694

Berardi lived his life as a canon and *maestro di cappella* in the towns of
Viterbo, Tivoli, Spoleto, and at Santa Maria in Trastevere, Rome. His
historical importance lies in his theoretical works, which comprise a full
course in Renaissance contrapuntal style (the *prima prattica*) as it was
cultivated by church composers well into the eighteenth century. Covering the techniques of fugue and every kind of canonic ingenuity,
Berardi's theory of counterpoint stands as a bridge between the technique of the Netherlanders (Josquin Desprez, etc.) and that of J. S.
Bach.[1] Interest in and practice of the archaic style often went hand in
hand with a musical antiquarianism and, as in the case of Bach himself,
with a speculative attitude. Berardi was in the Pythagorean tradition as
it was handed on by Kircher, holding a sincerely religious attitude to
music as an image of the heavenly harmony, and of counterpoint as a
symbol of the world process.

The *Miscellanea Musica* is mainly a treatise on counterpoint, on the
modes and their inventors, but in a lengthy introduction Berardi en-
larges the obligatory *laus musicae* (the passage in praise of music's powers) to include many of the speculations of Giorgi, Kepler, Fludd, and
Kircher. Beginning with *musica mundana*, he describes the planet-
intervals, the cosmic monochord, the proportions of the elements, the
zodiacal aspects as consonances. He proceeds to *musica humana*, writing
of the soul's harmony, the symbolism of notation, the inventors of

286

music; then comes this digression on David's Lyre.

Any connections of music with Kabbalism are rare, and in this period Berardi's may well be unique. The two-octave system on which he hangs his chain of correspondences is analogous to those of our Twelfth-century Anonymous and to Giorgi—at least, so one conjectures (see note 6 to Giorgi)—but is not identical to either. Just as Kircher himself would often do, Berardi has here gone to the trouble of assembling and painstakingly explaining a whole scheme only in order to reject it, leaving one wondering whether the whole matter was for him merely an intellectual game. In the end, he says, the lyre of David was not anything mysterious and Kabbalistic, but simply an instrument for the divine and healing power of music.

Source: Angelo Berardi, *Miscellanea Musica* (Bologna, 1689), pp. 31–36. Translated by the Editor.

Composition of David's Lyre, According to the Opinion of the Cabalists

Chapter 10. The Prophet King, with pleasing improvisations [*ricercari*] on harmonious lyre, had power to repress and mitigate the fury of that malignant spirit which tormented Saul, as is recorded in the First Book of Kings: "And it came to pass, when the evil spirit was upon Saul, that David took a lyre, and played with his hand: so Saul was refreshed, and was well, and the evil spirit departed from him" [I Samuel 16:23]. The Demon, who as I have said in my *Documenti Armonici* mocks at all the weapons of the world, was routed and overcome by Music.

The Cabalists do not on any account concede that such was the effect of harmony, but they know well that this proceeds from strings artificially ordered, as one can see in the following system:[2]

1 *Proslambanomenos,* A re
   Being; Essence
2 *Hypate hypaton,* B mi
   Living; Life
3 *Parhypate hypaton,* C fa ut
   Knowing; Sense
4 *Lichanos hypaton,* D sol re
   Thinking; Reason
5 *Hypate meson,* E la mi
   Understanding; Intellect

6 *Parhypate meson*, F fa ut
  Kingdom [Malkuth]   Blessed Souls
  Elements
7 *Lichanos meson*, G sol re ut
  Foundation [Yesod]   Angels
  Moon   Vegetables
8 *Mese*, A la mi re
  Glory, Honor [Hod]   Archangels
  Mercury   Animals
9 *Trite synemmenon*, B fa be mi
  Eternal Victory [Netzah]   Principalities
  Venus   Concupiscent energy
10 *Trite diezeugmenon*, C sol fa ut
  Beauty [Tiphereth]   Virtues
  Sun   Vital faculty
11 *Paranete diezeugmenon*, D la sol re
  Fortitude [Gevurah]   Powers
  Mars   Impulsive virtue
12 *Nete diezeugmenon*, E la mi
  Mercy [Chesed]   Dominations
  Jupiter   Natural energy
13 *Trite hyperbolaion*, Fa ut
  Intelligence [Binah]   Thrones
  Saturn   Body
14 *Paranete hyperbolaion*, G sol re ut
  Wisdom [Chokhmah]   Cherubim
  Firmament   Fire
15 *Nete hyperbolaion*, A la mi re
  Crown [Kether]   Seraphim
  Primum Mobile   Water

So that anyone may understand the foregoing system, I have translated the Hebrew words into Latin, and the Greek into ordinary characters.

The Cabalists, as reported by Pico della Mirandola, are of the opinion that "All the Sephiroth and divine emanations of David's Lyre existed to demonstrate nothing that could refer to music, and they judged that with such a lyre David was capable of expelling not only the fury but all the evil spirits of Saul" (*De Cab.* 71). The strings of this lyre were fifteen in number, the first five corresponding to the five genera of beings: Being, Living, Knowing, Thinking, Understanding, or Essence, Life, Sense, Reason, Intellect; which correspond to all the genera of beings: Insensible, which can only be, like stones; Living, living like plants; Sensitive, being aware like animals; Thinking, like men, Intelligent and purely intellectual, like the angels.

Essence corresponds to A re.

Life corresponds to B mi.

Sense corresponds to C sol fa ut.

Reason corresponds to D sol re.

Intellect corresponds to E la mi.

The ten strings are:

The first, F fa ut, to which corresponds Kingdom, the Divine Name or Sephirot.

The second, G sol re ut, to which corresponds the Divine Name or Sephira Foundation.

The third, A la mi re, to which corresponds Gloria, Honor.

The fourth, B fa be mi, to which corresponds Eternal Victory.

The fifth, C sol fa ut, to which corresponds Beauty.

The sixth, la D sol re, to which corresponds Fortitude.

The seventh, E la mi, to which corresponds Mercy.

The eighth, Fa ut, to which corresponds Intelligence.

The ninth, G sol re ut, to which corresponds Wisdom.

The tenth, A la mi re, to which corresponds Crown.

For further instruction of the curious I will make a little digression, to put in order these ten Sephiroth. The superior one, closest to God, is Kether, Crown: the nearest to us, from whose channel the influences proceed to the Angelic World, thence to the celestial, and from the celestial to the human, is Malkut, Kingdom, and it is the first one united in the triangle, in this manner:[3]

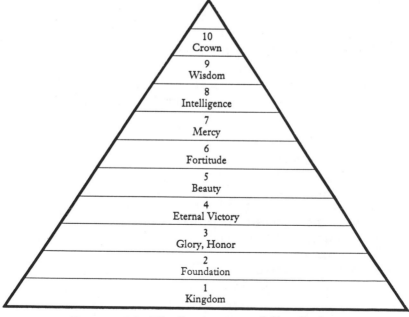

FIGURE 26: *The Instrument of Ten Strings*

The Sephiroth emanate from God, say the Cabalists, and therefore that emanation is called Azilut, whence come:

Briah	Creation
Yezirah	Formation
Assiah	Effect

Returning to the composition of the Lyre, from the first string:

[Briatic World]
Fa ut corresponds to the Blessed Souls
G sol re ut corresponds to the Angels
A la mi re corresponds to the Archangels
B fa be mi corresponds to the Principalities
C sol fa ut corresponds to the Virtues
D la sol re corresponds to the Powers
E la mi corresponds to the Dominations
Fa ut corresponds to the Thrones
G sol re ut corresponds to the Cherubim
A la mi re corresponds to the Seraphim

[Yetziratic World]
F fa ut corresponds to the Elements
G sol re ut corresponds to the Moon
A la mi re corresponds to Mercury
B fa be mi corresponds to Venus
C sol fa ut corresponds to the Sun
D la sol re corresponds to Mars
E la mi corresponds to Jupiter
F fa ut corresponds to Saturn
G sol re ut corresponds to the Starry Heaven
A la mi re corresponds to the Primum Mobile

[Assiatic World]
F fa ut corresponds to the Ever existent
G sol re ut corresponds to the Vegetative
A la mi re corresponds to the Animal
B fa be mi corresponds to the Concupiscible
C sol fa ut corresponds to Vital faculty
D la sol re corresponds to Impulsive virtue
E la mi corresponds to Natural energy
F fa ut corresponds to the Body
G sol re ut corresponds to Fire
A la mi re corresponds to Water

This opinion is false, because David never imagined that his lyre was so full of mysteries, nor half so rich in influences, as the Cabalists have said it was. In order further to confute what they have said, and in order to prove that the Demon fears and abhors harmony, I will quote this wonderful sentence of Saint Thomas of Villanova, Archbishop of Valencia:[4] "Music puts the Devil to flight, and he who (according to Job) spurns arrows as straws, and the hurling of stones as chaff; who mocks even the brandished spear, and weighs the heaviest hammers as nothing, shrinks trembling at the sound of the lyre; and he whom no force can overcome, is overcome by harmony."

39

✦

Andreas Werckmeister
1645–1706

Werckmeister owes his reputation today to his *Orgelprobe* of 1681, a solid work on organ-building and the problems of tuning and temperament. He held posts as organist successively at Hasselfelde, Quedlinburg, and Halberstadt—small East German towns within twenty miles of each other—and his intellectual horizon was accordingly limited. But he had read deeply in Kepler, Kircher, and the older theorists, and was led by his interest in speculative music to translate into German a treatise of Agostino Steffani, *Quanta certezza habbia da suoi principii la musica . . .* (1695).

We here supplement the extracts given in *Music, Mysticism and Magic* with a short treatise, complete in itself, about the musical symbolism of the Days of Creation. This is our first encounter with someone familiar with the phenomenon of the harmonic series—though Werckmeister only seems to have known it as the "trumpet scale." He rightly intuits that musical evolution follows, to a certain extent, the acceptance of successive intervals in the series as consonant: a principle that can obviously be extended far beyond his own day, as whole-tones, semitones, and even microtones are used without the necessity for resolution.

Werckmeister himself attributed the actual changes of musical style less to evolution than to changing planetary influences (see the extract in *MM&M*).

We are soon to move from "baroque" thinkers to "enlightenment" ones; and if the terms make any sense in this context, it is in the change from the literal period of speculative theory, in which it was actually supposed that the World-Soul or the elements had musical proportions *per se*, to a period at once poetic and scientific, in which the Harmony of the Spheres was either treated as a picturesque analogy, or subjected to the findings of quantitative science. Werckmeister was on the brink. The similes in this treatise are remarkable, but, as he realizes, in the last resort only poetic. In this regard his symbolism may be classed with that of Saint-Martin and Peter Singer (nos. 43, 48).

Source: Andreas Werckmeister, *Musicae Mathematicae Hodegus Curiosus, oder Richtiger musikalischer Weg-Weiser* . . . (Frankfurt & Leipzig, 1687, reprinted Hildesheim: Olms, 1972), pp. 141–154. Translated by the Editor.

On Allegorical and Moral Music

All arts and sciences in this mortal state fall far short of comparison with the holy and divine creations, inasmuch as nothing is more incomprehensible and occult than God, who dwells in darkness where no man can enter (I Kings 8:12; Isaiah 45:15). Yet since the Almighty has revealed himself to us not only in the Holy Scriptures but also in Nature and the Arts (Romans 1:19–20; Wisdom 13), we have decided to present a parallel of the holy Creation in music for the admonition and edification of our fellowman.

CHAPTER I

We know that all things have their origin from God as a perfect being, and that everything which comes forth from him strives to return thither, to the most perfect of all. If we consider our music, it is also a mirror and an example through which we may recall the same perfect Being. For all perfect harmony comes from the perfect unison and unity. Now as God makes a good harmony and concord with those closest to him—

we mean the holy angels and blessed Christ—so the unison makes a harmony with its own nearest neighbors, the octave, fifth, etc. And just as the great God makes no harmony with those who depart from him and his holy Word—indeed, if they stray too far and too long, they can scarcely be accepted again, but are cast out—, so it is the case that those dissonances which depart too far from the unison can sometimes hardly be resolved or tolerated without upsetting the harmony.

CHAPTER II

I. If we compare the days of Creation to music,[1] we see that God created heaven and earth in the beginning or on the first day. This beginning is the unity or unison out of which all consonances and dissonances proceed: yea, God himself is the unity, a beginning without beginning or ending. Light is also figured in this unison, which can be grasped by the senses and differentiated from the confused noise. II. On the second day occurred the first division and partition: the water was divided from the dry land. So 1:2 or C to c is the first division in music, between which octave, however, no other consonance is present, but all is formless and void as the earth was. III. After this division, on the third day, God made on the earth grass, herbs, and fruit trees; in the second octave there follow in sequence the proportional numbers 2.3.4, which make the notes c g c'. Just as the herbs and trees were on earth but at the same time between the earth and the clouds, so the g is between the octave c's. IV. But the harmony is yet incomplete: for we hear that on the fourth day God only-wise created light, to be a sign of seasons, etc. The sequence of our numbers of which the third octave consists is 4, 5, 6, 7, 8, i.e., c'e'g'c", showing us a light by which we can glimpse a complete harmony. Likewise we have two thirds, a minor and a major one, similar to a lesser and a greater light, i.e. the Sun and Moon. The different intervals included therein show certain seasons, days, and years; the seventh which occurs here and will not accord shows how times are not always good, but that sometimes a dissonance must be mingled in with this life.[2] This is now the complete structure of heaven and earth, Sun and Moon, stars and elements. V. Now follows the creation of every kind of animal. We see this in the next octave, which consists of the following numbers used in music: 8, 9, 10, 12, 15, 16; c" d" e" g" a" [sic] c'". Among these numbers there are not only good but also bad sounding notes, which cannot always be used in harmony: thus we know that among animals there are some clean, some unclean, which man is ordained to use only at certain times. VI. Finally on the sixth day man himself was created, and appointed as lord over all that lives and moves on earth. This must also be true in music, for a musician should rule over all the above-named numbers or notes, bringing them into good order; and should also live his life so that a good harmony comes from it. Clean and unclean beasts, consonances and dissonances, have their uses at certain times, and must become compatible so that God and man may take pleasure therein.

Chapter III

Other musicians explain the numbers thus: as God took six days for the Creation, so a complete harmony consists of the numbers 1, 2, 3, 4, 5, 6. And as God rested on the seventh day, so the number seven makes no consonance with the others, but must rest and remain silent. It is certain and remarkable that the number seven will not accord with the others, yet if the string of the monochord is divided into seven parts, one part will still make a pure consonance with the other,[3] and have a peculiar completeness in itself, as experience will bear ample witness. On this one can also consult Schwenter and Hartzdorffer in their *Mathematische und Philosophische Erquickstunden.*[4]

Some compare the four octaves with the four elements: the first and lowest with Earth, the second with Water, the third with Air, the fourth with Fire (see Franctrinus ch. 19). This comparison can also be made within one octave: c is Earth, e is Water, g is Air, c' is Fire. We will forego judgment on these comparisons because we wish to be brief. An informed musician will know how to make his own application; presumably since Earth is the heaviest element and has its position beneath, the others must follow after.

Chapter IV

We also know that everything in this world is in an imperfectly tempered and adulterated state. If we consider the elements such as Earth, then the ancient philosophers are of the same opinion. But some moderns posit no elements, at least only ones mixed with water and air: water carries with it air and earth; air contains earth, water, and fire; fire, again, is mixed with air, and so forth. For the sake of brevity we will pass over other things for the present. If we observe the structure of our music, we find it always tempered, and cannot attain the point of perfection on the fretted instruments, such as lutes, violas da gamba, citterns, claviers, and harps.[5] As our senses cannot always detect the temperatures of the elements exactly, it is done by a special device:

similarly our ears cannot always grasp or judge the unity or disunity of voices and harmony. It can be done through numbers, by which we can have all consonances pure and perfect, but not in practice and modulation. This provides us with an excellent moral concerning our mortality and imperfection in this life, of which something has already been mentioned above.

CHAPTER V

We can also make the following allusions from our proportions, whose sequence occurs regularly on the trumpet. If no. 1 is low C or unison, there follows next the second, making an octave with the first. This unity is the image of the unique and eternal being of God from which everything arises. The first octave from C to c indicates God the Father, who dwells in darkness where no man can enter, for no man has seen God (John 1:18; cf. Exodus 33:20). Since no sound can be found within this octave, we can compare to it the time before the creation of the world, when there was no time and nothing was nor had been made.

Next we compare 2 to 3 and 3 to 4. These stations give c g c' on the trumpet, which is the second octave within which only a single note and consonance stands, namely the g. Now in the Old Testament the Holy Trinity had not yet revealed itself, but was only represented concealed in symbols; so the symbol of the fifth contained in this second octave gives a beautiful symbol of the same. For if the fifth is made the mean, we see that it contains the harmonic triad in concealed fashion:

2	5/5	3	or harmonically	3	5/6/2	2
4		5		15	12	10

If we reduce the numbers to 15:10, the major third from C is 12:10, the minor from e to g. To fill out this octave, we have from g to c' a fourth, in the terms 3:4. If these 3 and 4 are taken collectively they give the septenary, which is a remarkable and sacred number. This is a symbol of God's wish to have a holy people (Leviticus 19:2). And as the same fourths cannot stand or sound alone, they rest on their fundamental bass, which in the octave is the fifth below. Even so, the people of God could not persist in their holiness unless they did not depart from the fifth, namely the true service of God in which the Holy Trinity was comprehended (albeit hidden and symbolically), and set their base upon it. We know, too, that heathen music recognized no third, but extended only to the harmonic numbers 1.2.3.4, as the Pythagorean hammers and their successors bear ample witness. Now as they did not know the

Holy Trinity, we see how God also prevented them from discovering any third, hence from being able to acquire any perfect music.

If we proceed to the third octave, which exhibits our natural sequence of numbers as 4.5.6–8., there we have a symbol of the New Testament in which God revealed himself further to us. For 4.5.6 point to the complete triad or threefold unison [*Drey-Ein-stimmigkeit*], in that when we hear it, it is nothing but a unison—yet it is also a three-note chord; it is indeed 'unitrisonus.' Could any clearer likeness be imagined, in which the threefold unity of God's being were better mirrored than in this?[6] Would to God that all good Christians understood music thus: they would find heartfelt joy in this symbol. Just as the Jews did not recognize the true God, three-in-one, so they would not suffer the beautiful triad to be used in their worship—yet in their drinking-bouts it was, alas, very much misused by them.[7]

CHAPTER VI

If we consider this further, we will discover that the middle note in this triad contains two natures in itself: one is glorious, perfect, and majestic; the other humble, imperfect, and servile. The perfect one sounds c e g; the imperfect c eb g. What better comparison could there be with this than the divine and human nature of our savior Jesus Christ? For in his divine nature he was glorious, yet in his human nature humble. If we try to combine the two middle notes in the triad, we find that they can make no harmony, that they refuse to be resolved, and that almost nothing appears more nonsensical to us than to combine them, e.g.:

$$
\begin{array}{lll}
\text{f\# g a f\#} & & \text{f g a f} \\
\text{f} & \text{or} & \text{f\#} \\
\text{d} & & \text{d}
\end{array}
$$

We know from our Christian symbolism that the two natures in Christ could not be entangled or confused with one another, for the divine and the human characters cannot exist together. Now there follows in our natural sequence the seventh number. This has no affinity within this octave, nor indeed with the whole of theoretical music, although on the trumpet the note bb' follows here. After g' follows c', a fourth above; and the terms of their ratio 3:4 add up to seven, thus completing the octave and witnessing to its sanctity, on which account it is called the Sacred Number. In the fourth and last octave we not only have with 8.9.10.12.15.16 a perfect consonance, but can also construct a good melody from it, symbolizing for us the good Christian life

on earth, and the everlasting life and all its joys if we can ever attain
to God in heaven.

CHAPTER VII

Music reminds us of another Christian idea if we let the musician represent God the Father, the instrument or organ God the Son, and the
beautiful chords and sweet harmony that come forth from it the Holy
Spirit. For just as there is no God besides the Trinity, music is incomplete and can have no effect without these three elements. If the music
is vocal, then the larynx must be the musical instrument. For brevity's
sake we will omit the other pleasing applications that could be made
here. All five major points of our catechism could be very clearly symbolized in music: only it must be agreed that the music is not understood to mean anything strange or senseless, of which I do not need to
say anything here. We can also see, meanwhile (and this has been touched
on above), that the great God has revealed certain laws in the arts,
though these are certainly not to be compared with the revelation in
Holy Scripture. Saint Augustine supports this in *De libro arbitrio*, with
the following words: 'Do not hesitate to attribute to God's handiwork
everything in which you see number, measure, and order.' One can also
consult Johannes Manlius,[8] *Collectanea de Deo, Trinitate, et Tribus in Divinitate
personis.* We see, moreover, that the structures of the Old Testament are
often decreed by God to be made harmonically, notably the Mercy Seat
in which is contained the perfect triad 2:3:5, c g e'.[9] This Mercy Seat is
indeed none other than the symbol of Christ, as can tell from the 5 and
the note e' arising from it. If we were to take 4 instead of 5, then we would
again have a representation of the Old Testament's worship, 2:3:4. But
now the 4 is raised to 5 and the triad is completed, so that the proportions
of this Mercy Seat give a perfect harmony.

CHAPTER VIII

From contrary motion we can gather a fine symbol of Christian love:
for as notes come from unity and move from one place to another, and
yet make a lovely sound with other notes, sometimes even meeting each
other pleasantly or returning to unison or unity, so also virtuous Christians are not averse to one another, even though they may sometimes
separate. Harmony and concord remain in faith and love; indeed there
is one will, one belief, one body, one soul, and they are united to God
through faith and love, as Holy Scripture commands us. This harmony
and agreement is found also in good Christian married couples, whom
we could compare to the whole of music.

Others have compared music to worldly authority:[10] for as the proportions and notes stand in association and always have respect to equality, so it is also with a good republic. Whatever strays too far from equality cannot attain unity and can never again be resolved. We see, too, how our dear music symbolizes all kinds of Christian virtues: for as God has openly shown us through his preachers how we should live in love and unity with our neighbors and ever strive toward the one divine Being, so this is done figuratively in music—for which reason, moreover, it belongs in worship.

CHAPTER IX

In conclusion, music is not only a mirror and symbol of the spiritual and divine being, but also of all secular and earthly actions, as Kepler, Bartholi, and others have discussed sufficiently at greater length. And just as the style of music has changed from one period to another over the past century, a similar change has also taken place in the secular state. The older music of a hundred and more years ago was inferior and purer: the political situation was also simpler and more candid, so that not so much ceremony took place, and people could trust one another and understand each other's motives. The music of today is in certain cases so embellished and decorated that one can scarcely find one's way in it: for when one thinks the progression will go one way, it turns out differently, so that few composers, even, can give a rationale for their harmony. Is not our modern everyday life and existence so intricate and confused that sometimes no one knows who is a cook or a waiter, and few consider the reasons for their actions and conduct, but wander after their inclinations? And although a few musicians understand how to adduce some rules of composition, they are neither hot nor cold, and can seldom maintain them. Is it any different with the changes of life today: is not something often turned now this way, now that? Are not the rules about some things merely invented, and sincere naivetés represented as if they could not be otherwise? And if we examine them in the light, they are only words and feathers.

Consequently it is not a good thing, as mentioned above, that modern music is in most cases unfortunately not judged by rules, but that everyone sings, plays, judges, and enjoys it however he pleases or feels inclined. Experience shows all too well what incompetence is thus admitted: and it is to be regretted that so many eminent musicians can produce no reasons or very flimsy ones, when every craftsman, even straw-cutters and broom-binders, know how to acknowledge certain rules in their professions and handiwork. Thus the same musicians abuse their own profession, even sublime music itself, no less than if music were a thing with neither basis nor power! As at every time there have been broken vessels among musicians, and a few errors crept into composition in the past, so sometimes in the worldly affairs of the past a bad deed would occur. But perhaps it was not so excessive as occurs nowadays, both with some musicians and in worldly affairs.[11]

CHAPTER X

We will not dwell longer on this subject. If anything in this treatise should serve to further the glory of God (as we have no doubt) and the benefit of our neighbor, to which all our efforts should be aimed, then I will be all the happier. Just as in music the proportions distant from equality give no harmony and will not concord together, yet are sometimes thrown in thoughtlessly by unschooled musicians, so there are always people who disdain even the best of books because their opinions, understanding, and fantasies are not in equal proportion with those of the author. Hence one can scarcely find a single book in the world which pleases and satisfies everybody: for as their heads are, so is their judgment [quot capita, tot sensus]. Without a doubt this will be the case with the present book, because so many opinions prevail in music that most people judge merely by fantasy, and seldom by its foundation and rules. Whoever is not pleased by this work, whoever stands too far from equality with one or another part of this little treatise and cannot get in tune with it, should seek another foundation on which to build his beliefs and harmony. For as long as the world has stood and will stand, no work of man has ever pleased everyone. Many may also be repelled because the rhetoric of this treatise is very poor, and the logical sequences not always correct. Now and then I have said the same thing more than once, as one does deliberately with simple people. This I willingly confess, and answer thus: I noted down these things as a collection of what I have read and thought about, and because I had little free time, and demanded not a penny of profit thereby, yet out of a desire to please my fellows, I wanted to have it printed, hoping that the dexterity of the gentle reader would look more to the matter and contents than to the words. If perchance a few of these things were dark and not very clearly expressed, in each case I express my opinion for you.

I know all too well from experience that there will be more faultfinders and enviers than true lovers of music. But since I am aware that I can serve some of my fellows thereby, I have not failed to do so, and have no regard for the slanderers and enviers, for they are impelled by Satan, from whom all envy stems, to make one abandon that which redounds to God's glory and the benefit of one's neighbor. Let who will revile and slander: he will spew out and pour his poison not on me but on my goal, which is God's glory and my neighbor's benefit: and let him beware of what may come of that. My comparison of music with other things and of other things with music was done so that simple people can better understand and assimilate it, for sometimes the thing itself is more difficult than music which is compared to it. If there should be a subtler brain and a sharper intelligence which could expand my well-meant work and write and comment more tidily and clearly on this material, it would be very agreeable to me, for it is most necessary that the foundation of music should be maintained, lest barbarism gain the upper hand. Meanwhile the gentle reader, whom I commend to the care of the Highest, will recognize my goodwill, interpret everything for the best, and at least bear in mind this well-intentioned work. And to wish is enough.

V

ENLIGHTENMENT
AND
ROMANTICISM

40

Isaac Newton
1642–1727

It is no secret nowadays that Newton considered his researches into alchemy, biblical chronology, and prophecy fully as important as his work in physics, mathematics, and astronomy. He saw himself less as an innovator than as the rediscoverer of lost knowledge, the wisdom of Antiquity and the *prisci theologi*. In the full daylight of the Scientific Revolution, there hover in the background the familiar shades of Orpheus, Pythagoras, and Macrobius, as Newton seeks to demonstrate that the Ancients already knew the inverse-square law of planetary attraction.

In the 1690s, Newton drafted a number of notes and learned comments (the "scholia") for a proposed second edition of his *Principia Mathematica*; our excerpt is from the classical Scholia to Propositions IV–IX of Book II. In the words of McGuire and Rattansi, who first drew attention to these ideas, Newton here "asserts unequivocally that Pythagoras discovered by experiment an inverse-square relation in the vibrations of strings (unison of two strings when tensions are reciprocally as squares of lengths); that he extended such a relation to the weights and distances of the planets from the sun; and that this true knowledge, expressed esoterically, was lost through the misunderstanding of later generations." (art. cit., p. 115)

Source: Isaac Newton, Classical Scholia on the *Principia Mathematica,* in
Royal Society (London), Gregory Ms. 247, transcribed in J. E. McGuire
and P. M. Rattansi, "Newton and the 'Pipes of Pan'" in *Notes and
Records of the Royal Society* 21 (1966), pp. 108–143.

The Pipes of Pan

By what proportion gravity decreases by receding from the Planets the
ancients have not sufficiently explained. Yet they appear to have adum-
brated it by the harmony of the celestial spheres, designating the Sun
and the remaining six planets, Mercury, Venus, Earth, Mars, Jupiter,
Saturn, by means of Apollo with the Lyre of seven strings, and measuring the intervals of the spheres by the intervals of the tones. Thus they
alleged that seven tones are brought into being, which they called the
harmony diapason, and that Saturn moved by the Dorian phthong, that
is, the heavy one, and the rest of the Planets by sharper ones (as Pliny,
bk. I, ch. 22 relates,[1] by the mind of Pythagoras) and that the Sun
strikes the strings. Hence Macrobius, bk. I, ch. 19, says: "Apollo's Lyre
of seven strings provides understanding of the motions of all the celestial spheres over which nature has set the Sun as moderator."[2]   And
Proclus on Plato's Timaeus, bk. 3, page 200,[3] "The number seven they
have dedicated to Apollo as to him who embraces all symphonies whatsoever, and therefore they used to call him the God of the Hebdomagetes,"
that is the Prince of the number Seven. Likewise in Eusebius' Preparation of the Gospel, bk. 5, ch. 14, the Sun is called by the oracle of
Apollo the King of the seven sounding harmony. But by this symbol
they indicated that the Sun by his own force acts upon the planets in
that harmonic ratio of distances by which the force of tension acts upon
strings of different lengths, that is reciprocally in the duplicate ratio of
the distances. For the force by which the same tension acts on the same
string of different lengths is reciprocally as the square of the length of
the string.

The same tension upon a string half as long acts four times as powerfully, for it generates the Octave, and the Octave is produced by a
force four times as great.[4] For if a string of given length stretched by
a given weight produces a given tone, the same tension upon a string
thrice as short acts nine times as much. For it produces the twelfth, and
a string which stretched by a given weight produces a given tone needs

to be stretched by nine times as much weight so as to produce the twelfth. And, in general terms, if two strings equal in thickness are stretched by weights appended, these strings will be in unison when the weights are reciprocally as the squares of the lengths of the strings. Now this argument is subtle, yet became known to the ancients. For Pythagoras, as Macrobius avows, stretched the intestines of sheep or the sinews of oxen by attaching various weights, and from this learned the ratio of the celestial harmony. Therefore, by means of such experiments he ascertained that the weights by which all tones on equal strings . . . were reciprocally as the squares of the lengths of the string by which the musical instrument emits the same tones. But the proportion discovered by these experiments, on the evidence of Macrobius, he applied to the heavens and consequently by comparing those weights with the weights of the Planets and the lengths of the strings with the distances of the Planets, he understood by means of the harmony of the heavens that the weights of the Planets towards the Sun were reciprocally as the squares of their distances from the Sun.

But the Philosophers loved so to mitigate their mystical discourses that in the presence of the vulgar they foolishly propounded vulgar matters from the sake of ridicule, and hid the truth beneath discourses of this kind. In this sense Pythagoras numbered his musical tones from the Earth, as though from here to the Moon were a tone, and thence to Mercury a semitone, and from thence to the rest of the Planets other musical intervals. But he taught that the sounds were emitted by the motion and attrition of the solid spheres, as though a greater sphere emitted a heavier tone as happens when iron hammers are smitten. And from this, it seems, was born the Ptolemaic system of solid orbs, when meanwhile Pythagoras beneath parables of this sort was hiding his own system and the true harmony of the heavens.[5]

These are passive laws and to affirm that there are no others is to speak against experience. For we find in ourselves a power of moving our bodies by our thought. Life and will are active principles by which we move our bodies and thence arise other laws of motion unknown to us.

And since all matter duly formed is attended with signs of life and all things are framed with perfect art and wisdom and nature does nothing in vain; if there be an universal life and all space by the sensorium of a thinking being who by immediate presence perceives all things in it, as that which thinks in us, perceives their pictures in the brain: those laws of motion arising from life or will may be of universal extent. To some such laws the ancient Philosophers seem to have alluded when they called God Harmony and signified his actuating matter harmonically by the God Pan's playing upon a Pipe[6] and attributing music to the spheres made the distances and motions of the heavenly bodies to be harmonical, and represented the Planets by the seven strings of Apollo's Harp.

So far I have expounded the properties of gravity. Its cause I by no means recount. Yet I shall say what the ancients thought about this subject. Thales regarded all bodies as animate, deducing that from magnetic and electrical attractions. And by the same argument he ought to have referred the attraction of gravity to the soul of matter. Hence he taught that all things are full of Gods, understanding by Gods animate bodies. He held the sun and the Planets for Gods. And in the same sense Pythagoras, on account of its immense force of attraction, said that the sun was the prison of Zeus, that is, a body possessed of the greatest circuits. And to the mystical philosophers Pan was the supreme divinity inspiring this world with harmonic ratio like a musical instrument and handling it with modulation, according to that saying of Orpheus "striking the harmony of the world in playful song."[7] Thence they named harmony God and soul of the world composed of harmonic numbers. But they said that the Planets move in their circuits by force of their own souls, that is, by force of the gravity which takes its origin from the action of the soul. From this, it seems, arose the opinion of the Peripatetics concerning Intelligences moving solid globes. But the souls of the sun and of all the Planets the more ancient philosophers held for one and the same divinity exercising its powers in all bodies whatsoever, according to that of Orpheus in the Bowl.[8]

Cylennius[9] himself is the interpreter of divinity to all:

> The nymphs are water. Ceres corn, Vulcan is fire.
> Neptune is the sea striking the foaming shores.
> Mars is war, kindly Venus is peace, the Bull-born
> Horned Bacchus frequenting gladsome feasts
> Is to mortals and to gods relief of mind from care.
> Golden Themis is guardian of Justice and right
> Next Apollo is the Sun, hurling his darts
> From afar, circling round, the Divines and the Soothsayers
> The Epidaurian God[10] is the expeller of diseases: these things
> All are one thing, though there be many names.[11]

41

Jean-Philippe Rameau
1683–1764

There is no doubt that Rameau was the foremost theoretician, as he would later become the foremost French composer, of his epoch. He was the first to realize fully the dependence of melody and harmony alike on the common chord and on the progressions of the fundamental bass. But while the working-out of his theory was firmly focused on the practices of his day, behind it lay the ancient approach through number. Already in the Preface to his *Traité de l'harmonie* (1722), he had told his contemporaries of his own search: "Notwithstanding all the experience I may have acquired in music from being associated with it for so long, I must confess that only with the aid of mathematics did my ideas become clear. . . ." In the *Nouvelles réflexions*, his Pythagoreanism is more explicit. Encouraged by the theories of Briseux concerning musical proportion in architecture, Rameau has by now come to realize that the same mathematical archetype lies behind all the arts, and to believe that it reaches its clearest expression in the laws of musical harmony.

Rameau did not have the mystical and prophetic character of a Tartini, who in *Scienza Platonica* (see no. 42) abandons all pretense of addressing his audience on its own terms. But he resembles the Italian in his conviction that the "enlightened" eighteenth century could not, in the last resort, dispense with the way to enlightenment that has always served the speculative music theorist: the assumption that number and harmony in some way underlie, and even explain, the phenomena of the universe and the experience of beauty. The exact way in which they are supposed to work changes, of course, as cultural attitudes change; one era may imagine souls in the planetary spheres, another prefer to think of impersonal natural law. But the facts do not alter, and the more acute theorists of every epoch meet here on a certain common ground.

Source: Jean-Philippe Rameau, *Nouvelles réflexions sur sa démonstration du Principe de l'harmonie* (Paris, 1742), pp. 47–54, 58–66. Translated by the Editor.

The Universal Principles of the Arts

It is the same with the sense of hearing as with the judgments we make about objects which strike our other senses: while one easily perceives certain differences, it is not so with the differences between these differences.[1] Often one thinks that one can find none, whereas they do actually exist, and error prevails (unless the contrary can be demonstrated by calculation, in matters susceptible to that). The ear, too, easily appreciates the differences between all the consonances, even the thirds, which unlike the others require special caution. It becomes insensible only to the differences between these differences, but at the same time it finds a means to prevent any error: a means all the more worthy of attention because it must be common to other arts.

Let us not believe that nature has presumed to assign us simple products as guides. And if musicians have at all times fallen into such a trap, at least the most eminent architects have known how to avoid it, in considering the length of the ground plan as the basis for all the parts of a building: a length from which all the beauties of the elevation are derived through division, made in the same proportions as those of music.

I derive this latter observation from Monsieur Briseux, architect,[2] who is to publish forthwith a learned treatise on this subject, wherein he reckons to demonstrate among other things that the beautiful buildings of the ancient Greeks and Romans, whose precious relics are still admired in every nation, are founded upon proportions all drawn from music. This justifies the notion I have had for a long time, namely that the principle of all the fine arts resides most certainly and most sensibly in music. In what other art, in fact, could this basis of architecture be better established, since it is Nature alone who has performed the first operations? I refer to the division of the string into regular parts, from which are born the proportions, each in its order of pre-eminence or subordination, and consequently the progressions which one only has to follow. It is noteworthy that division here precedes multiplication: which might lead one with more certitude than one had before to reasonable conjectures on some of the points of a higher and more sublime philosophy.[3]

Examining the matter rather more closely, one will see that it is not for nothing that Nature has chosen the sense of hearing above all others to judge her most general rules—*superbissimum auris judicium* [the most excellent judgment is the ear's]. She does not summon vision or touch as witnesses here; and are not all our senses modifications of touch? We do not generally judge phenomena *[effets]* except by the phenomena themselves, whereas in music one judges them from evident and sensible causes: one sees in them a basis in geometrical proportion, in which each term itself contains the entire harmonic edifice which is to accompany it (such is my "fundamental bass"). It is only thanks to this bass that the feeling of its harmonic edifice arises in us; from its succession are formed the modes and all possible genera, which affect us only by means of this same succession. The correct relationship of consonances is determined therein, so that the ear does not bother with their differences except as they serve to lead it from one consonance to another: it has no need to bother with the differences between these differences, which are thus of no use to it.[4] What does it matter to the ear, after all, whether the tones or semitones are major or minor, when it is not from this difference that it receives the feeling of relationship which exists between successive consonances, but from the fundamental progression which determines their correct relationship both with itself and with each other? And this is what should be observed above all: then one will see that the necessary temperament cannot be an obstacle to the feeling of correct relationship among these consonances. Experience proves this, moreover, when different instruments play together, each having its own particular temperament, while the voice they accompany is in no way distracted from the correctness it observes between these consonances.[5]

The necessity for temperament does not concern music alone: how many other arts use it in compromises, borrowings, repayments, approximations? The architect whom I cited told me, on this subject, that in his art approximations are only possible by means of the numbers 8, 9, 10, 11, and 12. They use 11 only if really necessary, preferring wherever possible the arithmetical mean between 11 and 12, because it approaches closer to the harmonic number 12, the octave of 6 and 3. This relates to the fourth, which horns and trumpets cannot play except as an eleventh, and which is very out of tune:[6] but by forcing the air they can form a sharp from it which gives them practically the arithmetical mean in question, and which leads to the twelfth by a pleasant enough semitone.

If I have spoken of measure, where one sees the resolution forming between every two tones, as in the diatonic tetrachord of the Greeks, B C D E, whose fundamental bass G-27 C-9 G-27 C-9 gives the absolute resolution from G to C,[7] and at the same time sets the limits of the mode, it is not only to draw from it the fullness of harmonic phrases, where more or less definite resolutions are felt every two measures, or from four to four, the first of these two successive resolutions almost forming what is called in verse a hemistich; but even more to infer from it that the very natural feeling for even numbers in Poetry, especially for hemistiches, as well as for symmetry in Architecture and other arts of that kind, may well find its source in Music, where additionally the octave, the most perfect interval, is in double progression, and where the *genera*, as well as the resolutions, owe their origin solely to two fundamental tones.

What fruitfulness there is in this phenomenon! What consequences arise from it of their own accord! Could one refuse to regard so unique a phenomenon, so abundant, so reasoned (if I may use the term), as a common principle of all the arts in general, at least all the arts of taste?

In fact, is it not reasonable to think that Nature, simple as it is in its general laws, might have but a single principle for so many things that seem to be so connected, exciting in us virtually the same sensations, as the arts destined to give us the sentiment of beauty?

If one considers that all our senses are in truth only modifications of touch, the same order should subsist in all these modifications. In exercising the sense of sight, as in architecture and other objects apt to give us pleasure through that sense alone, should we not be affected with the sentiment of beauty by the same general order of organization as causes us to experience the same sentiment by means of harmony when we exercise the sense of hearing?

From another point of view, if one considers the infinite relationships between fine arts destined, as I have said, to excite in us the sentiment of beauty; and if almost everywhere it is the same proportions which give them their rules, is one not persuaded that they all have but one and the same principle? And is not this principle now revealed and demonstrated in the harmony which derives directly from it, as clear to our perceptions as we could ever wish, to convince our reason and leave it no room for doubt?

If Monsieur Newton, for example, had known this principle, would he have chosen a diatonic system—a system of simple products, and moreover full of errors—to compare to colors?[8] Would he not have examined beforehand whether these colors should be considered as each forming a basis, a generator, and (forming groups with each other) an agreeable assembly? Would he not have chosen first those which could be compared to octaves and to fifths? And after recognizing the superiority of these fifths in harmony and in succession, doubtless he would have followed the consequences.

Let no one be mistaken. The arts which are known as "arts of taste" have less about them that is arbitrary than this title has given to suppose up to now. One cannot dispense nowadays with recognizing that they are founded on principles: principles the more certain and immutable in that they are given us by Nature. Knowledge of them lights up talent and rules the imagination, while ignorance, on the contrary, is a source of absurdity in mediocre artists, and of misleading in men of genius.

I leave it to those persons more generally versed than I in all the different arts and sciences to pursue this parallel. I am content if, in offering them the fruits of sixty years of practice and meditation on my art in particular, the discoveries which I have made may put them on the way to generalizing its application with certitude and utility for the other sciences and arts. I do not believe that anyone can oppose to the principle which I have discovered and recognized as the basis of my art any other one comparable to its in its manifestness, its richness, and in the superiority which it derives from Nature herself, as I flatter myself to have demonstrated.

42

◆

Giuseppe Tartini
1692–1770

Tartini's fame has rested since his death, as it did in his lifetime, on his unrivalled achievement as player, composer, and teacher of the violin. But whereas his compositions were altogether of his own time, breaking new ground both formally and technically, his theoretical works show that he was a Hermetic Christian Platonist, more akin to Giorgi or to Kepler than to anyone of his epoch in his mystical arithmology and his conviction of a personal revelation of overwhelming importance.

Tartini's musico-mathematical theories are so complex, and in some respects so confused, that it is impossible to give any details of them here.[1] His theory arises from two different things: (1) his practical discoveries in acoustics, namely of some of the harmonics of a vibrating string, and of the difference-tones resulting from chords; (2) his adaptation of a geometrical theorem (Euclid, VI, 13) concerning the geometric mean as found in the circle to the equation that relates the three means, Arithmetic, Harmonic, and Geometric.[2] He refers to these in this extract from the *Trattato* as the "practical" and the "demonstrative" principles.

The *Scienza Platonica* was written shortly before his death, and remained in manuscript until 1977. Here Tartini repeats much of what he had said in the *Trattato*, but now it has become a pretext for a journey into mathematical futility as he tries to pin down the incommensurablity of the circle and the square by using ever larger numbers. Tartini's mathematical discoveries seemed to him all the more remarkable since they were worked out laboriously in whole numbers without the help of algebra, much less of calculus. This caused him to recapitulate some of the efforts of earlier civilizations to calculate tuning-systems in the same manner, and aptly enough Tartini believed himself to have redis-covered the lost key to the universal science of Antiquity: it had been lost, he thought, since the Greek philosophers, but this had not pre-vented composers from following it intuitively. In the opening pages of *Scienza Platonica* he defines this lost science as the "Harmonics" recom-mended by Plato (in Bk. VII of the *Republic*) as essential to philosophers but insufficiently cultivated in Greece. This harmonic science, says Tartini, "was so named by the Egyptians, from whom it was transmit-ted to Greece by Pythagoras and Plato," and its purpose is to under-stand the whole cosmos: "The Harmony of the universe is the whole tree; music is one of its branches, and necessarily of the same nature and root. . . . One can see the possibility of discovering the tree from the branch, the whole from the part, as in fact the author has suceeded in doing." (ed. cit., p. 4)

Working in virtual isolation from the learned world of his own time and (like Rameau) acknowledging no theorist later than Zarlino (see no. 32), Tartini regarded himself, in all Christian humility, as the sole modern recipient of an ancient, secret knowledge that could shake the world if only it would listen to him. Unfortunately he seems to have been unable to make clear to the reader any connection between his dry calculations and the living application of his insights, which were un-doubtedly a reality to him.

Source: Giuseppe Tartini, *Trattato di Musica secondo la vera scienze dell'armonia* (Padua, 1754), pp. 20–22; *Scienza Platonica fondata nel cerchio,* edited by Anna Todeschini Cavalla (Padua: Cedam, 1977), pp. 80–84. Translated by the Editor. Also consulted: German translation of the *Trattato* by Alfred Rubeli (Düsseldorf: Verlag der Gesellschaft zur Forderung der systematischen Musikwissenschaft, 1966).

On the Nature and Significance of the Circle

Chapter 2. When a system is to be established, it is necessary to unite the two realms, the physical and the mathematical *[dimostrativo]*, in such a way that they are inseparable, and to form from them a single principle. Only thus will any system endure; and we will be convinced of this when we understand clearly what we mean by "a single principle." It means that the calculation used for the mathematics must be intrinsically deduced from the physical nature of the thing demonstrated. Thus and in no other way will the two realms, physical and mathematical, form a single principle between them. The law is strict, but just; and in virtue of this law, which is the touchstone of any physico-mathematical system, very few systems will be found to which no exception can be taken.

I will explain it better. Up to now, the quantitative sciences have all been based on quantity, but in different ways. There is the science of arithmetic: it is based on quantity made from equal rational parts, and consequently uses the common arithmetical numbers 1, 2, 3, 4, etc. There is the science of harmonics: it is based on quantity made from unequal rational parts, hence of fractions: 1, ½, ⅓, ¼, etc. There is the science of geometry: it is based on continuous quantity, whence arises the irrational which cannot be expressed in whole numbers or in fractions.[3] There are many other sciences, based on other aspects: algebra, differential and infinitesimal calculus, etc. All are adaptable to physics as such, but the converse does not apply, because it is not true that physics as such, according to its own intrinsic nature, is adaptable to all these sciences.

The plainest example is that of the string with three tones.[4] All the known sciences of quantity are adaptable to this string in their respective ways. But if the string as it sounds reveals itself independently of human will as harmonic, since it divides itself in 1, ⅓, ⅕, then physically it excludes any other sort of quantity.[5] Thus the adaptation of any other sort of quantity to the string, which declares itself physically harmonic, will be a pure and simple fallacy. Certainly no one would deny the possibility in general of adapting to it any other sort of quantity: considering the string in the abstract, as a straight line, it could doubtless be both the object and the quantity in any science. But one must see whether or not it refers concretely to some particular object and subject, as is precisely the case in the above example. In such a case nature is stronger than human will; and calculation will surely be wrongly applied if it is not precisely the sort demanded by nature. Consequently, the mathematical deductions may be true, but none of the physical ones. Thus the above law will always be true: that for the establishment of any physico-mathematical system, the two realms of physics and mathematics must be united in such a way that they are inseparable, and form from them a single principle.

I will follow the law. Consequently, I have to find the same unity in the mathematical realm as in the physical. For our demonstrations we will need the geometrical figures, and above all possible figures there is and can be no other than the circle, which is one in itself. It is one because the infinite radii drawn from the center to the circumference are equal, and these are nothing other than the very unity which automatically forms the circle according to the opening of the compasses, such as does not, and cannot, exist in any other figure. Thus the circle is one from its first principle, and is intrinsically one in comparison with all possible figures.

It is not sufficient that the circle should be one in general. I must now demonstrate it as one in its harmonic unity, because harmonic unity is the physical principle.[6] This is an easy matter, though, as far as I know, unnoticed by geometricians. Proceeding with all strictness with the mathematical figures which represent faithfully the physical data, one will need straight lines and curved ones. Therefore our figure will show the square and the circle, the latter being the most curved of all. The physico-harmonic phenomena show the necessity of such a figure. A hanging string is in itself a straight line, like the diameter of a circle or the side of an exscribed square. The vibrations of the hanging string will be curved in nature.[7] We do not know to which of the two figures the straight line belongs, since it is common to both; but the two figures are necessary because they have the straight line in common. The same can be said in general about the string stretched on the monochord: although up to now the circularity of its vibrations have not yet been proved. The same can be said of the third tone resulting from two strings sounded simultaneously.[8] The two given strings are jointly two straight lines, physical and acoustical; the two volumes of air moved by the two strings are jointly two spheres. For in plane [geometry we need] the straight and the circular line, and consequently the figures of square and circle, having the straight line in common. Why the square should be exscribed and not inscribed will appear for many reasons as we proceed: suffice it to say, for now, that the straight line has natural priority over the circular: something physically and mathematically true: physically, in the whole universe and in all its parts; mathematically, in the construction of the circle which is impossible without presupposing a straight line. So it follows: exscribed square, inscribed circle.[9] This is therefore the necessary position of the two figures; and now we must compare them to one another in order to deduce their nature.[10]

Announcement of a New Science

The author repeats many things here which have already been given to the public in his *Trattato di Musica*, and one of these is the present matter of the four odd numbers 5:7:9:11. But since he was then considered by the mathematicians as a musician who did not acknowledge their lofty science, but only the simple arithmetic of common numbers, the learned world passed on his *Trattato* a judgment somewhere between indifference and contempt. But now that the author has changed his role from that of musician to that of an interpreter and also discoverer of the science of Plato which he calls the Arithmetic of the Philosophers, he believes that, thanks to this role, he is entitled to the full attention of the mathematicians for his discovery of the falsity of their science. It has come to the point at which either their indifference or their contempt condemns them, because he has discussed so much, it being a matter of the greatest weight and the gravest consequence that has ever been, or could be, proposed. Moreover, considering the impotence of common mathematical science to persuade the present day of even the smallest fallacy, the author is confident with a confidence that far surpasses common mathematical science.

It is clear that the present science is in agreement with the common science as to consequence and effect, and is therein equal: it is not so in the antecedent, and for a reason always known in my science, always unknown in any other. Therefore the conviction produced by this one is so far superior to the common ones that someone who possesses it, even though he may utterly lack any notion of the common ones, experiences its own strength in so eminent a degree that he considers it as nothing to expose himself on trial against the whole mathematical world, even if it were composed of Archimedes' and Euclids.

The author is obliged to make this declaration here no longer as a violin player but as a possessor of this science, of which, considering the present need, he is explaining only that part which can in very truth be called its simple grammar. This serves as a sure notice to the learned world of today to conduct itself as it should in an altogether unfamiliar situation, in which he is employed in bringing to life, after thousands of years, the greatest of all the sciences,[11] with the sole exception of revealed theology. In order to conduct itself properly, the following precautions are necessary:

I. Do not be concerned with the means which Providence chooses to use for such a discovery, and remember that "God has chosen humble things, and confounded the strong."

II. On no account believe that such an argument is the author's fanaticism and overenthusiasm. Apart from the fact that anyone may ascertain for himself everything explained and demonstrated here, one can readily go to the substance with the assurance that the author's character naturally abhors vanity and glory; which, if he were fond of them, he would have gained better by his own art. If he had been master of himself, he would have given up his own life sooner than taking on such an enterprise, from which he has gained, and will gain, nothing but "labor et dolor" [labor and sorrow].

But there is something worse: that after his death people should unfortunately proceed from speaking of subjects grand beyond dispute, to speaking of facts that actually happened, incomparably more important than what it is permissible to give the public, but essentially dependent on this science. Would that the author could impose perpetual silence on those concerned, as he declares himself repentant of his exuberance of heart; but it is too late, and he can only hope that through the honesty of those in the know he may be spared the trouble of being discovered. He is no longer surprised by the necessity of Pythagorean silence: he has learned it at his own cost. Superior powers do not lead him, they drag him to this public appearance; and for himself he would sooner stand apart from human commerce, and thus lose neither his name nor his face. It is thus impossible that a man in such a situation should be capable of overenthusiasm.

III. The author is old, beyond his seventieth year. If Providence grants that the learned world should believe the resurgence of this science to be useful and necessary, there is no time to lose. The author must be quickly aroused to finish the undertaking by those who hold authority in the most respectable classes of mathematical physics. This is the only way to urge the author before his death to the last and greatest labor, which on his own, owing to his fullness of years and weariness, he does not have the courage to face.

IV. A doubt may easily arise in the more deeply learned, so that they draw the conclusion that all that is explained of this science here consists of affections and peculiarities of common numerical arithmetic. If they conclude thus, they may agree completely with the author's conclusion, but they should not stop at this, but should continue by considering carefully three things:

First, that more important affections and peculiarities than those explained here, and ones undoubtedly new, will occur to them; and that as many more new ones, in fact unknown to them, will occur to them which remain to be disclosed.

Second, that these serve as a simple material foundation to the present science, whose formal foundation (which they see and touch with their hands here) consists of ratios and proportions between them, joined in a perpetual bond of relationship, so that in this science no disjunct ratio or proportion whatsoever signifies anything, nor forms a system. This is the substantial difference between the common science of ratio and proportion and the present science. In the former, every ratio or proportion can be divided by another, and form a class by itself; in the latter it is quite impossible, because not only must they be joined in such a way that one proves another, and all together form the same proof, but in addition all together must be resolved into their first principle of all: which is the greatest universal proof, without which no particular proof is worth anything.

The author knows very well (and knows to his cost) that if a science is complete it will contain such a complex of things as to overwhelm the human mind, so that those gifted ones who can attain its total comprehension are few. This truth neither can nor should be denied. But one should not then blame the whole difficulty of this science on the author, as followed so unfairly on his *Trattato di Musica:* it would be far more just for them to thank heartily the Divine Author who has deigned to make the human race worthy of such a gift. If then he has not made everyone's intellect equal to such a gift, this increases all the more the debt to God of him who is so furnished at need, and also increases the merit of this science, founded by God, through the impossibility of its being made common. But one should not on this account spurn (as too often occurs) what one does not understand. No one is to blame for the lack of gifts necessary to understand it: and certainly it is an unbearable fault to presume that one is more capable than one in fact is, and to be unwilling to measure one's powers with any other gauge than that of self-approbation and contempt of others.

Third, that Plato calls number a divine gift; and therefore one should see first of all whether his authority is founded on truth or falsehood. It is easy enough to see this from the great amount the author himself has written in his works, in which the authority of his text is verified by positive and actual facts and by authentic proofs, since it is only through number that one can succeed here in demonstrating that which one could never have achieved through any other demonstrative science. In the second place, one must investigate this presumed foundation of truth in order to find what precisely constitutes the divinity of this gift, as something superior to any idea of purely human science. The text of Plato indicates not so much that the gift itself is one of these possessions in general, which are all divine gifts, as that the character of this gift is distinct from others by so singular an excellence that it should be separated from the class of those human things which are acknowledged as divine possessions and gifts; and this is the true sense of Plato's text.

But here one should make no other response than that on this point Plato spoke admirably of metaphysical philosophy, and divinely of moral philosophy, when he said that the acquisition of this science requires a profoundly able intellect and a soul religiously inclined, because it is more a wisdom than a science. To what a heathen has said, a Christian can add nothing more or better. From which the most profoundly learned conclude that beyond the affections and properties of the material basis which is number; beyond the ratios and proportions of the formal basis which is in the relationships and their connections; beyond what has not yet been said, but will be said in its place on numbers as the real signs of specific natures, which belongs to number as regards its nature, and not to those affections and properties which belong to number as regards ratio; beyond all this there is in this science something else fundamental of greater excellence and significance. Believe the author to be all too sincere, and free from the narrow prejudice of possessing these sciences which distinguishes them in their most respectable classes, but believes them to be most sublime when in reality they are not, let them deepen themselves by the study of this science, sparing no toil to understand it to its depths, having the intention of consecrating their pains and the faith they have in the author.

Up to now the Christian author has spoken to Christian philosophers with such truth and sincerity of heart, that he has no fear if he is answerable before God. But next one will see him declare, with equal truth and with as much courage against the class of those pretended philosophers who affect the name of strong spirits: "procul este profani."[12] It is one of their foolishnesses if they presume to attain possession of this science by the sole power of their cleverness, because if the measure of their cleverness is not sufficient to make them know the One, the True and the Good in the spectacle of the universe, it suffices far less to make them understand the doctrine of the universe. It is, moreover, one of their blasphemies if they presume that they are aided in the comprehension of this doctrine by that light to which they are directly opposed. Here, in all truth, one may say "evanuerunt in cogitationibus suis."[13] They remain in their blindness because they love it, either through pride in their intelligence or through depravity of will, or through both together.

This work, which essentially destroys their principles, will be the prime target of their slander, oppositions and literary cabals, nor will they spare anything in reducing it to worse than nothing, which is what abuse does. But their plan will fail, because there is no plotting against God. The author has for many years resisted God in another sense by not making the present public presentation to the world on this subject, and by remaining silent and secretive—and he has not succeeded.

43

✦

Louis-Claude
de Saint-Martin
1743–1803

The life and work of "The Unknown Philosopher," as Saint-Martin signed his works, opens a window on the subculture of the Age of Reason, and on the degree to which esotericism penetrated the highest ranks of French society both before and after the Revolution of 1789. Saint-Martin entered the world of occultism through his initiation by Martinez de Pasqually, a Kabbalist, theurgist, and Freemasonic reformer. Freemasonry had its origins in the tangled skein of the seventeenth-century Rosicrucian movement, and besides its political and fraternal aspects, continued to be a vehicle for hidden knowledge in the tradition of Cornelius Agrippa, John Dee, and Robert Fludd. In *Des erreurs et de la vérité*, Saint-Martin's first book, some of these doctrines were again set before the public. It is a long and difficult explanation of the true nature of the human being, much of it devoted to symbolic numbers and to the idea of a primordial language.

Aristocratic by birth, Saint-Martin entered first on a successful career in the salons and the peacetime army. These he subsequently abandoned in favor of the solitary life, spent as an exile in Switzerland and provincial France after the Revolution. His second initiatory meeting came with the discovery of Jakob Boehme's works in the 1780s, which he immediately set himself to translate into French. Saint-Martin's inner experiences led him away from the more outward manifestations of occultism, such as evocations and alchemy, toward a Christian theosophy. He believed that we possess an inner eye that, once opened, can reveal the truth of all things, since it is nothing less than an organ of God himself. Thereupon he dispensed with all ritual and sectarian belief, accepting the Christ who said: "The Kingdom of God is within you." Saint-Martin's spiritual lineage is that of the theosophers and introverted, systematizing mystics: Dionysius the Areopagite, Hildegard of Bingen, Boehme, Swedenborg, etc.—illuminated individuals always at the limits of orthodoxy (or beyond it) who explore through their inner visions the intermediate realms of the divine Imagination.

Saint-Martin was an amateur violinist and a lover of music and drama, but he was not technically educated, as can be seen from the first part of this extract: a rather labored, though entirely original, symbolization of the common chord. In the second, more fluent part, where he turns to meter, several currents can be detected. There is the Kabbalism that he learned from Martinez de Pasqually, according to which every thing has a primordial name given it by Adam, and every name is a number. There is Platonic aesthetics, in which the artist strives to imitate the superior realities of the intellectual world in the intractable world of matter. There is the potential of human perfection and the restoration of man to his birthright, proposed by Jean-Jacques Rousseau (whom Saint-Martin admired) but here given a more spiritual meaning. The ideas on rhythm are particularly interesting in the light of subsequent musical developments.

Source: Louis-Claude de Saint-Martin, *Des erreurs et de la vérité*, par un Ph . . . Inc . . . (Edinburgh [actually Lyon], 1775); reprinted Paris: Le Lis, 1979, pp. 507–531. Translated by the Editor.

On the Common Chord

We come now to examine one of the productions of that true language[1] whose conception I am trying to recall to mankind: it is that which joins with our verbal expression, which regulates its strength and to an extent its pronunciation—it is in fact that art which we call Music, but which among mankind is as yet but an image of true harmony. Verbal expression cannot use words without making audible sounds; so it is the intimate relationship of both that forms the fundamental laws of true music. It is this that we imitate, as far as lies within us, in our artificial music, by the care we take to depict in sound the sense of our conventional words. But before displaying the principal defects of this artificial music, we shall survey some of the true principles that it offers us. From these we will discover surprising connections with all that we have proved up to now, sufficient to convince us that everything derives from the same source, and that from then onward it is within the province of man. In this examination we will also see that, however admirable our talents in musical imitation, we always fall infinitely short of our model. From this it will be seen whether this powerful tool was given to man for the sake of his childish amusements, or whether it was originally destined for a nobler use.

In the first place, that which we know in music under the name of the common chord [accord parfait] is, for us, the image of that first unity that embraces everything and from which everything comes forth. This chord is single and unique, entirely self-contained without need of any note other than its own; in a word, it is unalterable in its intrinsic value, like unity.[2]

Secondly, this common chord is the most harmonious of all: it is the only one that satisfies the human ear and leaves nothing else to be desired. The first three notes which comprise it are separated by two intervals of a third, distinct but linked to one another. Here we have the repetition of everything that happens among sensible things, where no corporeal being can receive or continue its existence without the help and support of another being, corporeal like itself, which restores its strength and maintains it.

Lastly, these two thirds are surmounted by an interval of a fourth, whose terminal note is called the octave. Although this octave is only the repetition of the fundamental note, it is nevertheless the one which completely defines the common chord. It belongs essentially to it, in that it is included in the primordial notes that a sounding body makes audible above its own fundamental.[3]

This interval of a fourth is thus the principal agent in the chord; it is placed above the two thirds, to preside over them and direct all their action, like that active and intelligent cause which we have seen dominating and presiding with a double law over all corporeal beings. Like that cause, it can suffer no adulteration, and when it acts alone, like this universal cause of time, it is certain that all its results will be correct.

I am, however, aware that this octave, being in fact only a repetition of the fundamental note, can be suppressed if necessary, and need not be counted among the notes of the common chord.

I know, however, that this octave, being in truth but a repetition of the fundamental note, can under necessity be withdrawn, and not enter the enumeration of notes which constitute the common chord. But, first, it is the one which essentially terminates the scale; beyond that, it is indispensable to admit this octave if we would know what is the alpha and omega, and have evident proof of the unity of our chord. All this is justified mathematically in a way which I cannot express otherwise than by saying that the octave is the first agent, or the first organ, through which Ten has been able to enter our consciousness.[4]

One should not insist, in the sense-picture which I am presenting, on an entire uniformity with the Principle of which it is merely the image, for then the copy would be equal to the model. All the same, although this sense-picture is inferior, and can moreover be subject to variation, it nevertheless exists in no less complete a manner, it nonetheless represents the Principle, because the instinct of the senses supplies the remainder.

It is for this reason that having presented the two thirds as bound one to another, we by no means claim that it is indispensable for them both to be heard; we know that each of them can be sounded separately, without the ear suffering. But the Law is no less true for that, for the interval thus sounded always conserves its secret correspondence with the other notes of the chord to which it belongs; so it is still the same picture, but one does not see more than a part of it.

We can say the same, moreover, if one chooses to omit the octave, or even all the other notes of the chord, and only to keep any one of them, because a note heard alone is no burden to the ear, and furthermore it can in itself be considered as the generating note of a new common chord.

We have seen that the fourth dominates over the two lower thirds, and that these lower thirds are the image of the double Law which directs elementary Beings. Does not Nature herself show us thus the difference which exists between a body and its Principle, in making us see one in subjection and dependence, while the other is its chief and support?

These two thirds, in fact, symbolize for us by their difference the state of perishable things in corporeal Nature, which subsists only through the union of diverse actions; and the last note, formed by a single quaternary interval, is a new image of the first Principle; for it recalls to us its simplicity, its grandeur and immutability, as much by its rank as by its number.[5]

It is not that this harmonic fourth is any more permanent than all other created things: so long as it is sensible, it must pass away; but this does not prevent the fact that even in its ephemeral action it depicts for the intelligence the essence and stability of its source.

One finds, then, in the assembled intervals of the common chord, all that is passive and all that is active: that is to say, all that exists and everything of which man can conceive.

But it is not enough for us to have seen in the common chord the representation of all things in general and in particular: we can also see there, through further observations, the very source of these things and the origin of this distinction which was made before time began between the two Principles,[6] and which ever manifests itself within time.

To this end, let us not lose sight of the beauty and perfection of this common chord, which draws all its virtue from itself alone. We will readily judge that if it had remained forever in its natural state, order and just harmony would have lasted perpetually, and evil would have been unknown because it would never have been born; that is to say, none but the faculties of the good Principle would ever have been manifested, since it is the only real and the only true one.

How then was it possible for the second Principle to become evil? How could evil have taken birth and appeared? Was it not because the superior and dominant note of the common chord, namely the octave, was suppressed, and another note introduced in its place? And what is this note which was introduced in place of the octave? It is the one which immediately precedes it, and we know that the new chord which results from this change is called the chord of the seventh. We know, too, that this chord of the seventh tires the ear, holds it in suspense, and demands (in aesthetic terms) to be saved.[7]

It is therefore through the opposition between this dissonant chord and all those derived from it, and the common chord, that all musical works are born: for they are nothing other than a continuous play—not to say a combat—between the consonant common chord and the seventh chord, or all dissonant chords in general.

Why should not this law, thus shown us by nature, be for us the image of the univeral production of things? Why should we not find therein the Principle, as we have found above the assembly and the constitution in the order of intervals of the common chord? Why, I say, should we not touch with finger and eye the cause, the birth, and the consequences of the universal temporal confusion, since we know that in this corporeal nature there are two Principles which are ceaselessly opposed, and since nature could not survive without the help of the two contrary actions from which proceed the combat and the violence that we see: a mixture of regularity and disorder which harmony represents to us faithfully by the assembly of consonances and dissonances of which all musical works consist?

For all that, I flatter myself that my readers should be sufficiently intelligent to see here only images of the lofty facts to which I direct them. Doubtless they will appreciate the allegory when I tell them that if the common chord had remained in its true nature, evil would not yet have been born; for, according to the established principle, it is impossible that the musical order in its particular law should be equal to the superior order which it represents.

The musical order being, moreover, founded on the sensible order, and the sensible being only the product of manifold actions, if the ear were offered nothing but a series of common chords it would not be shocked, it is true; but aside from the monotonous boredom that would ensue, we would not find therein any expression, any idea. It would not, in fact, be music for us, because music, and in general everything that is sensible, is as incompatible with unity of action as with the unity of agencies.

In thus acknowledging all the laws necessary for the constitution of musical works, we can still apply these same laws to verities of another level. It is on this account that I shall continue my observations on the seventh-chord.

In putting this seventh in place of the octave, we have seen that it would be placing one principle beside another principle, from which, according to the light of the sanest reason, only disorder can result. We have seen this even more clearly in observing that this seventh which produces the dissonance is at the same time the note which immediately precedes the octave.

But this note which is a seventh in relation to the fundamental note can also be regarded as a second in relation to the octave which is a repetition of that note. So we see that the seventh is by no means the only dissonance, but that the second also has this property; and that thus every diatonic step is condemned by the nature of our ear, and that whenever it hears two adjacent notes sounding together, it will suffer.

Since therefore there is in the whole scale absolutely nothing but the second and the seventh which can be found in this relationship with the lowest note or its octave, it becomes clear to us that every result and every product in music is founded on two dissonances, whence arises every musical reaction.[8]

Transferring this observation to sensible things, we will see with equal clarity that they could never and can never be born except through two dissonances; and whatever effort we make, we will never find another source of disorder than the number attached to these two sorts of dissonances.

Even more, if one observes that what is commonly called a seventh is in effect a ninth, given that this is the combination of three quite distinct thirds, we will see if I have misled my readers in telling them previously that the number nine is the true number of extension and of matter.[9]

If, on the other hand, we look over the number of consonances, or notes which accord with the fundamental, we will see that they are four in number, namely the third, fourth, perfect fifth, and sixth[10] (for here we must not speak of the octave as an octave, since it is a matter of the particular divisions of the scale, in which the octave is no different in character from the fundamental of which it is the image, unless one views it as the fourth of the second tetrachord—which does nothing to alter the number of four consonances that we have established).

I cannot continue, much as I would wish, on the infinite properties of these four consonances, and I truly regret this because it would be easy for me to show with striking clarity their direct relationship with Unity; to show how universal harmony is connected to this quaternary consonance, and why it is impossible for any Being to subsist in good order without it.

But at every step prudence and duty prevent me, for in these matters one point leads to all the others; and I would not even have undertaken to treat a single one, if the errors by which the human sciences poison my kind had not forced me to take up its defense.

I am nevertheless committed not to end this treatise without giving some more detailed explanations on the universal properties of the quaternary. I have not forgotten my promise, and I propose to make it good as soon as I am permitted to do so; but for the present let us return again to the septenary, remarking that if it is this which creates a diversion with the common chord, it is also through this that crisis and revolution occur out of which order must come again and the peace of the ear be restored, because after this seventh one is indispensably obliged to return to the common chord. (I do not regard as contrary to this principle what is called in music a sequence of sevenths; for that is nothing but a continuum of dissonances, and one can never avoid terminating with the common chord or one of its derivatives.)

It is this very dissonance that again repeats for us what takes place in corporeal nature, whose course is nothing but a sequence of derangements and rehabilitations. Now, if this observation has indicated to us precisely the true origin of corporeal things, and made us see today that all the beings of nature are subject to this violent law which presides over their origin, their existence, and their end, why cannot we apply the same law to the universe in its entirety, and recognize that if it is violence which has caused it to be born and which sustains it, then violence too must work its destruction?

It is thus that we see that at the moment of termination of a piece of Music there is ordinarily a confused beating, a trill, between one of the notes of the common chord and the second or seventh of the dissonant chord, which latter is indicated by the bass which usually holds its fundamental note in order then to restore the whole to the common chord or to unity.[11]

One can see, moreover, that just as after this musical cadence one necessarily returns to the common chord which restores all to peace and order, it is certain that after the crisis of the elements, the Principles which have fought over them will also regain their tranquility. And applying the same to man, one must see how the true knowledge of music might preserve him from fear of death: for this death is only the trill which ends his state of confusion, and restores him to his four consonances.

I have said enough for the intelligence of my Readers, and it is for them to extend the bounds which I have set myself. I can presume in consequence that they will not consider the dissonances as vices in regard to music, since it is from them that it draws its greatest beauties, but only as the sign of the opposition which reigns in all things.

They will also realize that within the harmony of which sensible Music is only the image there must be the same opposition between dissonances and consonances; but that, far from causing the least fault in it, they are its nourishment and its life; and intelligence will see there only the action of several different faculties which sustain one another even though they fight together, and which by their reunion give birth to a multitude of results, ever novel and striking.

This has been only a very much abbreviated account of all the observations of this sort I might make on music, and on the relationships which exist between it and important verities; but what I have said will suffice to give a glimpse of the reason of things, and to teach men not to isolate their different branches of knowledge: for we show that they all come from the same tree, and that the same imprint is everywhere.

Need one now speak of the obscurity in which the science of music still remains? We might begin by asking musicians what their rule is for taking their pitch: that is to say, what is their *A-mi-la* or their diapason;[12] and if they have none, and are obliged to make up one, how can they believe they have anything of this sort fixed? So, if they have no fixed diapason, it follows that the numerical ratios which one can derive from their spurious diapason, together with the notes which should be correlative to them, are no longer true ones, and that the principles which musicians claim as true for the numbers which they have accepted can equally well be true for other numbers, according to whether *A-mi-la* is pitched higher or lower. This renders absolutely uncertain the greater part of their opinions on the numerical values they attribute to various notes.

However, I speak here only of those who have tried to evaluate these different notes by the number of vibrations of strings or other sounding bodies; it is here that a fixed diapason is necessary for the experiment to be accurate, and one accordingly needs sounding bodies which are essentially the same, if one is to base anything on their results. But since man is denied these two means, seeing that matter is only relative, it is evident that everything which may be established on such a basis will be susceptible to many an error.

It is not in matter that one should seek for the principles of harmony, for, after all we have seen, matter, being never fixed, can never supply the principle of anything. But it is in the true nature of things that everything is stable and forever the same, and one only needs eyes to read the truth therein. For, in the end, man should have seen that he has no other rule to follow than that which is found in the double ratio of the octave, or in this famous double ratio which is inscribed on all beings, from which the triple ratio has descended; it would have recalled to him once more the double action of nature, and this third, temporal cause universally established on the other two.[13]

I will here conclude my observations on the deficiency of the laws which human imagination has managed to introduce into music; for all that I could add would only point to this first error,[14] and it is now obvious enough that I can concern myself with it no longer. I will simply counsel the inventors to reflect well on the nature of our senses, and to observe that the sense of hearing, like all the others, is susceptible to habit; that thus they may have been deceived in good faith, and have made up rules from things guessed, and from assumptions which time alone can have made to seem to them true and regular.

It now remains to examine the use which man has made of this music with which he is almost universally engaged, and to see whether he has ever suspected its true application.

Quite apart from the innumerable beauties of which music is capable, we know that it has one strict law which is this rigorous meter from which it absolutely cannot escape. Does not that alone proclaim that it has a true principle, and that the hand which directs it is above the power of the senses, for these have nothing fixed?

But if it holds to principles of this nature, it is certain that it will never need any other guide, and that it was made to be forever united to its sources. Now, as we have seen, this source is the primal and universal language which indicates and represents things in their natural state: so one cannot doubt that music was once the true measure of things, just as writing and the word once expressed their meaning.

It is therefore only by cleaving to this fertile and invariable principle that music can preserve the laws of its origin and fulfill its true duty; it is thus that it could have painted lifelike pictures which would have satisfied completely all the faculties of those by whom it was heard.[15]

In a word, it is thus that music could perform the prodigies of which it is capable, and which have been attributed to it in every age.

Consequently, in separating it from its source, in seeking its subject-matter only in artificial sentiments or vague ideas, it has been deprived of its primary support and of the means by which it could be shown forth in all its splendor.

And what impressions, what effects does it produce in the hands of men? What ideas, what sense does it offer us? Excepting those who compose it, can many ears understand that which they hear expressed in the music which enters them? And even the composer himself, after giving himself up to his imagination, does he not always lose the sense of what he has painted, and of what he wished to express?

Nothing is more deformed and defective than the use which men have made of this art, and this solely because being so little occupied with its principle they have not sought to support the one with the other, and because they have thought they could make copies without having the model before their eyes.

It is not that I censure my fellows for seeking in the infinite resources of artificial music the charm and ease which it can offer, nor do I wish to deprive them of the succor which this art, even when defective, can always give them. It can, I know, sometimes help to revive in them a few of these clouded ideas, which being further refined may be their only nourishment, and which alone may help them find a basis. But for this to happen I would urge them always to lift their intelligence above what their senses experience, for the first principle of man is certainly not in the senses. I would urge them to believe that, however perfect their musical compositions may be, others exist *of another order and more perfect*; and that it is only by reason of its greater or lesser conformity to these that artificial music touches us and causes us more or less emotion.

When I touched on the precision of meter to which music is subject, I did not lose sight of the universality of this law; I intended, on the contrary, to return to it in order to show that while it embraces everything, it still has a distinct character in each place. And there is nothing here which does not conform to all that we have set out: one can see meter take its place among the intellectual faculties of man, and enter into the number of Laws which rule him; one can judge hereby that since these intellectual faculties are themselves semblances of the faculties of the higher Principle from which man derives everything, that Principle must also have its meter and its particular Laws.

Hence, if superior things have their meter, we should not find it surprising that the inferior and sensible things which they have created are also subject to it; and consequently we should find in this meter a strict guide for Music.

But if we reflect a little on the nature of this sensible meter, we will soon see its difference from the meter which regulates things of another order.

In Music we see that meter is always constant; once the tempo is given, it perpetuates and repeats itself in the same form and with the same number of beats; in short, it appears to us so regular and so exact that it is impossible not to sense its law, nor to deny its necessity. This regular meter is, moreover, so well suited to sensible things that we see men applying it to all of their productions which involve a continuous action; we see that this law is for them a basis on which they rest satisfied; we even see them using it in their roughest work, and it is thus that we can judge the advantage and usefulness of this powerful aid; it helps to ease labors which without it would seem unbearable.

But here again there is something that can help to instruct us on the nature of sensible things, for in offering us such equality in action, and I might say such servitude, it tells us clearly that the Principle which is in these things is not the master of this action, but that everything in it is constrained and forced: which goes back to what we have seen in different parts of this work on the inferiority of matter. It consequently offers us only a marked dependence, and all the signs of a life which we can recognize as passive, which is to say that, having no action of its own, it is obliged to await and receive it from a superior law which provides its action and commands it.

Secondly, we can observe that this law which rules the progress of music manifests itself in two ways, through two sorts of meter known by the names of duple and triple meter. (We do not count quadruple meter, nor any of the other subdivisions which can be made, being only multiples of these two primary meters. Even less can we allow a meter of one beat, because sensible things are not the result nor the effect of a single action: they are born and subsist only by means of several united actions.)

Now it is the number and quality of these actions that we find revealed in the two different sorts of meter allocated to music, according to the number of beats which these two sorts of meter contain. And indeed, nothing could be more instructive than to observe this combination of two or three beats in relation to everything that exists corporeally; we would see there afresh and clearly the double and triple ratios directing the universal course of things.

But these points have become too detailed: I should only invite men to appraise what surrounds them, and should never communicate to them knowledge which can only be the reward of their own desires and efforts. In view of this I will conclude forthwith what I have to say on the two sensible meters of music.

In order to know which of these two meters is used in any piece of music, one has to wait until the first measure is complete, or, which is the same thing, until the second measure has begun; only thus can the ear decide and feel certain of which number it can rely on. For if a measure were not completed in this way, one could never know what its number might be, since it is always possible to add a beat to the preceding ones.

Does this not show us in Nature herself that hackneyed truth, that the properties of sensible things are not fixed but only relative, and that each is sustained by another? For if that were not so, a single one of their actions when manifested would carry their true character with it, and one would not have to wait for a comparison in order to know it.

But such is the inferiority of artificial music and of all sensible things, that they only comprise passive actions, and their meter, although determined in itself, cannot be known to us except in relation to the other measures with which we compare it.

Among things of a higher order, absolutely beyond the sensible, this meter makes itself known in nobler form: each being, having its own action, also possesses in its laws a meter proportional to this action; but at the same time, as each of its actions is ever new and different from those preceding and following, it is easy to see that the adjacent measure can never be the same; and therefore one should not look in this class for the uniformity of meter which reigns in Music and among sensible things.

In perishable nature all is in dependency, and shows only blind action which is nothing but the forced connection of divers agents under the same law. All converging on the same end, in the same manner, they can only produce a uniform result when they suffer no disturbance or obstacle in the accomplishment of their action.

In imperishable nature, on the other hand, all is living, all is simple, and hence every action carries all its laws with it. That is to say that superior action determines its meter itself, whereas it is the meter that determines inferior action, or that of matter and all passive nature.

No more is needed to understand the infinite difference that there must be between artificial music and the living expression of that true language which we declare to men as the most powerful of the means destined to restore them to their rights.

Let them learn here to distinguish this unique and invariant law from all the spurious productions which they continually put in its place; the one, bearing its laws within itself, always has only those which are just and conformable to the Principle which uses them; the others are spawned by man while in the darkness, where he does not know whether what he is doing agrees or disagrees with the superior

Principle from which he is separated and which he knows no more.

So when he sees the works of his hands vary and multiply to infinity the abuses which he makes of language, as much in his use of the word as in that of writing and music; when he sees all human languages born and die, one after another; when he sees that down here we know only the number of things, and almost all of us die without having ever known their names, he will still not believe that the Principle after whom he gives birth to his works is subject to the same vicissitude and the same obscurity.

On the contrary, he will avow that, being capable at present of nothing except imitation, his works will never have the same solidity as genuine works. Then, observing that not everyone can possibly view the model from the same place, he will understand why its copies are all different; but he will nonetheless know that this model, being in the center, remains always the same, like the Principle whose laws and will it expresses; and if men were courageous enough to approach it closer, they would see all these differences vanish, which only existed on account of their distances.[16]

◆

44

Johann Friedrich Hugo
Freiherr von Dalberg
1760–1812

Dalberg is one of the great unknowns of speculative music, responsible for a body of work without parallel in its period. In a half-dozen essays he covers the whole field in exemplary Platonic fashion, mixing myth and dream with allegory, theory with experiment, and criticism with a vision of artistic and moral renovation for his own time. Rather like his contemporary Fabre d'Olivet, Dalberg combined speculation with a gift for composition in modern style, and with an interest in the ancient religions of East and West. He translated into German the pioneering treatise on Indian music of Sir William Jones, and contributed original scholarly work on the Parsees, the Persian myth of the Phoenix, and the Meteor cult of Antiquity. In outer life he was a canon of Worms and Trier cathedrals; a scion of an ancient family, slightly handicapped physically and eclipsed by the fame of his two brothers (the archbishop-politician Karl Theodor, and the theater director Wolfgang Heribert). The biographical sources are equally divided as to his birthdate, some giving 1752, others 1760.

When in 1806 Dalberg reissued his fantasy-essay of 1787, "A Composer's Glimpses of the Music of Spirits" (see *MM&M*), he prefaced it with this short "allegorical dream" of heavenly and earthly music.

How far is it to be taken as fact? Probably further than meets the eye, as Dalberg throws out a challenge to the myth of music's origin as told by Jean-Jacques Rousseau (in *Essai sur l'origine des langues*, Ch. 9). For Rousseau, as for certain ancients, music had its beginnings from water: from meetings at the wells, to be precise, where the "noble savages" greeted each other with speech that turned naturally into song. Dalberg's primitives, on the contrary, are wretched creatures on the brink of extinction, only rescued from the animal state by the heavenly dispensation of music. Intervention from above, rather than growth from below, is the rule for Platonists and Christians alike as they try to imagine the beginnings of the arts and sciences; while, as we have heard from Isaac ben Haim (no. 25) and others, earthly music is still only a means to the end of hearing the celestial song. With his contemporaries Tieck, Wackenroder, and Novalis, Dalberg is a pioneer in the Romantic revisioning of the arts, their origin, and their purpose for mankind.

Source: "Die himmlische und irdische Musik. Ein allegorischer Traum." In Friedrich Hugo von Dalberg, *Fantasien aus dem Reiche der Töne* (Erfurt: Beyer & Maring, 1806), pp. v–xiv. Translated by the Editor.

Heavenly and Earthly Music: An Allegorical Dream

In the immeasurable ether above the stars Urania holds sway, and rules with golden scepter the circling of the spheres: the silent night and the young day rejoice in the magic of her melodious voice. At her summons the sisterhood of the Horae begin their lightly-winged dance, the solemn melody of the creation sounds forth in antiphonal choirs, the strings of the heavenly lyre ring out, and a thousand harmonious voices echo the sublime hymn. Not only the heavenly scale, but also the realm of morality is set under her law; all beings are tones of the universal symphony, single notes which a primordial, all-uniting Spirit creates, orders, and tunes, that by their pleasant song they might enhance the harmony of the whole.

Long lay our earth bereft of the joy of song. In the first days of her birth, scarcely escaped from chaos, the firstborn children of men wandered, battling with the elements, in a state of untamed savagery. Without morals or laws they scarcely sensed the God who had made them; unknowing, artless, the speech of these wild ones was naught but a childish stammering; feelings such as pity, goodwill, and love lay undeveloped in their hearts, as brutal battle and the sating of animal needs alone occupied the first inhabitants of the earth.

Then Urania[1] looked down pityingly upon the sad vales of earth; she saw the distress of the new-made race, heard their sighs and their cries of woe; and, moved, she spoke to the Father of Gods:

"Divine Zeus, will you send no comfort or relief to the new race? For believe me, it will sink through brutality to the level of beasts, and men will destroy themselves; only through the magic of song can their souls be made mild, their habits more gentle."

"Daughter!" answered Jupiter, "I too am shocked by the condition of men whom Deucalion raised at my behest from the barren earth; I would send you yourself to them as tutelary goddess, only your tones are too lofty for their unschooled souls. Mnemosyne, the divine mother, shall send them as teacher one of your sisters, decked with all the charms of beauty, all the magic of song, and friendly of mien."

And Polyhymnia the friendly singer was elected for this mission, accompanied by three young sisters of the Muses: Poetry, Mime, and Dance. The heavenly choir sank down to earth on a light cloud and was greeted as a wonder from heaven.

Through music, in fair companionship with poetry, theater, and dance, the motherly spirit of Mnemosyne first taught raw humanity to know itself and its world, its origin and the goal of its being; by arousing in men's hearts the feelings of love, goodwill, and decorum, she changed raw and beastly creatures into moral beings, working to a goal. Their passions were purified, their taste for virtue increased. Through music and poetry mankind was first taught to honor the gods, their ancestors, and the fatherland that bore them in solemn and moving songs, and to impart like valor and strength to the generations to come.

So Polyhymnia was the benefactor of mankind—and she still is. Whereas Urania leads the music of spiritual beings and higher spheres, the younger sister teaches the inhabitants of earth with soothing song. She purifies their morals, impresses tender feelings on their hearts, and while with heavenly tones she transports their souls out of the earthly sphere into higher regions, she ripens the spirit of mankind that knows its divine origin to begin to perceive Urania's songs: to follow into higher regions the rhythm of a nobler music than is ever heard on earth.

45

◆

Arthur Schopenhauer
1788–1860

Some of the most significant passages in Schopenhauer's *Die Welt als Wille und Vorstellung* ("The World as Will and Representation," published 1818) are devoted to music—significant, not only for the authenticity of their insight into the nature of the art, but for the inspiration that they gave to Richard Wagner. These were discussed in *Music, Mysticism and Magic*, to whose sampling of Schopenhauer's text we here add a complementary excerpt.

What is remarkable about this is the way in which the philosopher, after discoursing in somewhat abstract terms, "gives his mind entirely up to the impression of music in all its forms," then tries to put into words what he has experienced there. He moves straightway into a far different mode of discourse, which one might say was borrowed from the Romantic poets, did one not recognize in it the perennial doctrine of correspondences which those poets may have embraced, but did not invent. The analogy of the musical range with the stages of creation, or objectification of the cosmic Will, has its roots in all those chains of being, from the Greeks onward, which set out a hierarchical universe alongside a musical scale. For an audible illustration of this passage, one need only listen to the Prelude of Wagner's *Das Rheingold.*

Later in the work (Bk. III, Ch. 39, "On the Metaphysics of Music"), Schopenhauer takes his parallel still further into the concrete. He says there that "The four voices, or parts, of all harmony, the bass, the tenor, the alto, and the soprano, or the fundamental note, the third, the fifth, and the octave, correspond to the four grades in the series of existences, the mineral kingdom, the vegetable kingdom, the brute kingdom, and man. This receives an additional and striking confirmation in the fundamental rule of music, that the bass must be at a much greater distance below the three upper parts than they have between themselves . . . " (ed. cit., vol. III, p. 231) A comparison of this analogy with the musical allegories of the naive and pious Andreas Werckmeister (no. 39) and Peter Singer (no. 48) shows that Schopenhauer was inhabiting the same imaginative realm, and places him as a link in the chain of German speculative theorists.

Source: Arthur Schopenhauer, *The World as Will and Idea,* translated by R. B. Haldane and J. Kemp (Boston: Osgood, 1883), vol. I, pp. 335–339.

Music and the Objectification of the Will

Chapter 52. Now that we have considered all the fine arts in the general way that is suitable to our point of view, beginning with architecture, the peculiar end of which is to elucidate the objectification of will at the lowest grades of its visibility, in which it shows itself as the dumb unconscious tendency of the mass in accordance with laws, and yet already reveals a breach of the unity of will with itself in a conflict between gravity and rigidity—and ending with the consideration of tragedy, which presents to us at the highest grades of the objectification of will this very conflict with itself in terrible magnitude and distinctness; we find that there is still another fine art which has been excluded from our consideration, and had to be excluded, for in the systematic connection of our exposition there was no fitting place for it—I mean *music.*

It stands alone, quite cut off from all the other arts. In it we do not recognize the copy or repetition of any Idea of existence in the world. Yet it is such a great and exceedingly noble art, its effect on the inmost nature of man is so powerful, and it is so entirely and deeply understood by him in his inmost consciousness as a perfectly universal language, the distinctness of which surpasses even that of the perceptible world itself, that we certainly have more to look for in it than an *exercitum arithmeticae occultum nescientis se numerare animi*, which Leibniz called it.[1] Yet he was perfectly right, as he considered only its immediate external significance, its form. But if it were nothing more, the satisfaction which it affords would be like that which we feel when a sum in arithmetic comes out right, and could not be that intense pleasure with which we see the deepest recesses of our nature find utterance. From our standpoint, therefore, at which the aesthetic effect is the criterion, we must attribute to music a far more serious and deep significance, connected with the inmost nature of the world and our own self, and in reference to which the arithmetical proportions, to which it may be reduced, are related, not as the thing signified, but merely as the sign. That in some sense music must be related to the world as the representation to the thing represented, as the copy to the original, we may conclude from the analogy of the other arts, all of which possess this character, and affect us on the whole in the same way as it does, only that the effect of music is stronger, quicker, more necessary and infallible. Further, its representative relation to the world must be very deep, absolutely true, and strikingly accurate, because it is instantly understood by everyone, and has the appearance of a certain infallibility, because its form may be reduced to perfectly definite rules expressed in numbers, from which it cannot free itself without entirely ceasing to be music. Yet the point of comparison between music and the world, the respect in which it stands to the world in the relation of a copy or repetition, is very obscure. Men have practiced music in all ages without being able to account for this; content to understand it directly, they renounce all claim to an abstract conception of this direct understanding itself.

I gave my mind entirely up to the impression of music in all its forms, and then returned to reflection and the system of thought expressed in the present work, and thus I arrived at an explanation of the inner nature of music and of the nature of its imitative relation to the world—which from analogy had necessarily to be presupposed—an explanation which is quite sufficient for myself, and satisfactory to my investigation, and which will doubtless be equally evident to any one who has followed me thus far and has agreed with my view of the world. Yet I recognize the fact that it is essentially impossible to prove this explanation, for it assumes and establishes a relation of music, as idea, to that which from its nature can never be idea, and music will have to be regarded as the copy of an original which can never itself be directly presented as idea. I can therefore do no more than state here, at the conclusion of this third book, which has been principally devoted to the consideration of the arts, the explanation of the marvelous art of music which satisfies myself, and I must leave the acceptance or denial of my view to the effect produced upon each of my readers both by music itself and by the whole system of thought communicated in this work. Moreover, I regard it as necessary, in order to be able to assent with full conviction to the exposition of the significance of music I am about to give, that one should often listen to music with constant reflection upon my theory concerning it, and for this again it is necessary to be very familiar with the whole of my system of thought.

The (Platonic) Ideas are the adequate objectification of will. To excite or suggest the knowledge of these by means of the representation of particular things (for works of art themselves are always representations of particular things) is the end of all the other arts, which can only be attained by a corresponding change in the knowing subject. Thus all these arts objectify the will indirectly only by means of the Ideas; and since our world is nothing but the manifestation of the Ideas in multiplicity, though their entrance into the *principium individuationis* (the form of the knowledge possible for the individual as such), music also, since it passes over the Ideas, is entirely independent of the phenomenal world, ignores it altogether, could to a certain extent exist if there was no world at all, which cannot be said of the other arts. Music is as *direct* an objectification and copy of the whole *will* as the world itself, nay, even as the Ideas, whose multiplied manifestation constitutes the world of individual things. Music is thus by no means like the other arts, the copy of the Ideas, but the *copy of the will itself*, whose objectivity the Ideas are. This is why the effect of music is so much more powerful and penetrating than that of the other arts, for they speak only of shadows, but it speaks of the thing itself. Since, however, it is the same will which objectifies itself both in the Ideas and in music, though in quite different ways, there must be, not indeed a direct likeness, but yet a parallel, an analogy, between music and the Ideas whose manifestation in multiplicity and incompleteness is the visible world. The establishing of this analogy will facilitate, as an illustration, the understanding of this exposition, which is so difficult on account of the obscurity of the subject.

I recognize in the deepest tones of harmony, in the bass, the lowest grades of the objectification of will, unorganized nature, the mass of the planet. It is well known that all the high notes which are easily sounded, and die away more quickly, are produced by the vibration in their vicinity of the deep bass-notes. When, also, the low notes sound, the high notes always sound faintly, and it is a law of harmony that only those high notes may accompany a bass-note which actually already sound along with it of themselves (its *sons harmoniques*) on account of its vibration. This is analogous to the fact that the whole of the bodies and organizations of nature must be regarded as having come into existence through gradual development out of the mass of the planet; this is both their supporter and their source, and the same relation subsists between the high notes and the bass. There is a limit of depth, below which no sound is audible. This corresponds to the fact that no matter can be perceived without form and quality, i.e., without the manifestation of a force which cannot be further explained, in which an Idea expresses itself, and, more generally, that no matter can be entirely without will. Thus, as a certain pitch is inseparable from the note as such, so a certain grade of the manifestation of will is inseparable from matter. Bass is thus, for us, in harmony what unorganized nature, the crudest mass, upon which all rests, and from which everything originates and develops, is in the world.

Now, further, in the whole of the complemental parts which make up the harmony between the bass and the leading voice singing the melody, I recognize the whole gradation of the Ideas in which the will objectifies itself. Those nearer to the bass are the lower of these grades, the still unorganized, but yet manifold phenomenal things; the higher represent to me the world of plants and beasts. The definite intervals of the scale are parallel to the definite grades of the objectification of will, the definite species in nature. The departure from the arithmetical correctness of the intervals, through some temperament, or produced by the key selected, is analogous to the departure of the individual from the type of the species. Indeed, even the impure discords, which give no definite interval, may be compared to the monstrous abortions produced by beasts of two species, or by man and beast. But to all these bass and complemental parts which make up the *harmony* there is wanting that connected progress which belongs only to the high voice singing the melody, and it alone moves quickly and lightly in modulations and runs, while all these others have only a slower movement without a connection in each part for itself. The deep bass moves most slowly, the representative of the crudest mass. Its rising and falling occurs only by large intervals, in thirds, fourths, fifths, never by *one* tone, unless it is a bass inverted by double counterpoint. This slow movement is also physically essential to it; a quick run or shake in the low notes cannot even be imagined. The higher complemental parts, which are parallel to animal life, move more quickly, but yet without melodious connection and significant progress. The disconnected course of all the complemental parts, and their regulation by definite laws, is analogous to the fact that in the whole irrational world, from the crystal to the most perfect animal, no being has a connected consciousness of its own which would make its life into a significant whole, and none experiences a succession of mental developments, none perfects itself by culture, but everything exists always in the same way according to its kind, determined by fixed law.

Lastly, in the *melody*, in the high, singing, principal voice leading the whole and progressing with unrestrained freedom, in the unbroken significant connection of *one* thought from beginning to end representing a whole, I recognize the highest grade of the objectification of will, the intellectual life and effort of man. As he alone, because endowed with reason, constantly looks before and after on the path of his actual life and its innumerable possibilities, and so achieves a course of life which is intellectual, and therefore connected as a whole; corresponding to this, I say, the *melody* has significant intentional connection from beginning to end. It records, therefore, the history of the intellectually enlightened will. This will expresses itself in the actual world as the series of its deeds; but melody says more, it records the most secret history of this intellectually-enlightened will, pictures every excitement, every effort, every movement of it, all that which the reason collects under the wide and negative concept of feeling, and which it cannot apprehend further through its abstract concepts. Therefore it has always been said that music is the language of feeling and of passion, as words are the language of reason. Plato explains it as "a movement of melody, imitating the feeling of the soul" (*Laws*, Bk. VII), and also Aristotle says: "Why do the rhythms and melodies of music, which are voices, cause similar behavior?" (*Problemata*, ch. 19)

46

◆

Fabre d'Olivet
1767–1825

The author of perhaps the most important work of speculative music of
his time here sets out some of the foundations of his musical and cosmological thought. Our first two extracts are taken from Fabre d'Olivet's
book *La Musique expliquée comme science et comme art* ("Music explained
as science and art," probably written in 1813–15), for which he failed
to find a publisher in his lifetime. In the 1840s, the manuscript came
into the hands of the Escudier brothers, Parisian music publishers, who
printed the greater part of it in their periodical *La France musicale*. The
third excerpt was discovered after World War II in the unpublished
papers relating to a cult that Fabre d'Olivet founded in the last years
of his life, called "La Vraie maçonnerie et la céleste culture" ("True
Masonry and Celestial Culture"). It was apparently a lecture or instruction for the members, probably prepared in 1824. Taken together, these
three chapters give the essentials of Fabre d'Olivet's musical theory.

Several themes underlie the approach to music of this author, as of
several of his contemporaries in France. They are: (1) the inferiority of
modern music; (2) the superiority of the Ancients and their music,
especially the Egyptians, to whom the Greeks are thought of as having
been largely indebted; (3) the rediscovery of the archaic wisdom of the

Chinese, including their musical system, and the proof it gives of a primordial universal civilization; (4) the correspondence of the seven Chaldaean planets with the notes of the scale, to which the key is furnished by the days of the week; (5) Pythagorean intonation, founded on the powers of three, as the correct tuning of the scale; (6) the evils of equal temperament, responsible for our loss of music's powerful therapeutic and ethical effects; (7) the priority of melody over harmony, and of the voice over artificial instruments. To these we would add the Hermetic principle alluded to at the end of the third extract: that the universe is created by the interplay of the expanding and the contracting force, here called Love and Chaos, which are no doubt the same as Robert Fludd (no. 35) symbolized with his light and dark pyramids. In applying these forces to the imagined monochord string, Fabre d'Olivet is doing the same as Fludd, who stretched his string between heaven and earth. In both cases, the musical scale becomes a way of understanding the cosmic hierarchy, though the details of interpretation are quite different.

Source: Fabre d'Olivet, *Music Explained as Science and Art*, translated by Joscelyn Godwin (Rochester, Vt.: Inner Traditions International, 1987), pp. 94–109; 167–170. Originals in *La France musicale*, 9 and 16 June 1850, and *La Vraie maçonnerie et la céleste culture*, edited by Léon Cellier (Grenoble: PUF, 1953), pp. 69–71.

Survey of Sacred Music

The number 12, formed from the ternary and the quaternary, is the symbol of the Universe and the measure of tone. In expressing myself thus, I simply speak as the interpreter of the ancient philosophers and the modern theosophers, and say openly what the hierophant of Eleusis and of Thebes confided only to initiates in the secrecy of the sanctuary. What is more, it is by no means merely an opinion maintained by a single people, at a certain time, in a particular country of the earth; it is a scientific and sacred dogma accepted in all ages and among all nations from the north of Europe to the most eastern parts of Asia. Pythagoras, Timaeus of Locris, Plato, in giving the dodecahedron as symbol for the Universe,[1] were expounding the ideas of the Egyptians, the Chaldaeans, and the Greeks. These peoples had long since attributed the government of Nature to twelve principal gods. The Persians followed in this regard the doctrine of the Chaldaeans, and the Romans adopted that of the Greeks. Even at the extremities of Europe, the Scandinavians, in admitting the duodecimal division, also counted twelve rulers of the Universe whom they named the Ases. When Mani[2] wished to take over the Christian religion in order to allegorize it and to call a halt to its still uncertain forms, he did not fail to apply the dodecahedron to the Universe, recalling the supreme Governors of the Ancients which it represented, filling the immensity with a celestial harmony and strewing flowers and eternal perfumes before the Father. It is not long since a German theosopher, a shoemaker named Boehme, a man of extraordinary genius but lacking in erudition and intellectual culture, examining on its basis elemental Nature and the system of the Universe, was compelled as by irresistible instinct to take the zodiacal number as constituting the regimen of the world.[3] He did more: he saw in this number what I do not think anyone had seen since the extinction of the Mysteries of Antiquity: a double rulership, celestial and terrestrial; one spiritual, intelligible, and ascending, the other creaturely, sensible, and descending.

The institution of the Zodiac is due to the application of the number 12 to the highest sphere. This institution, according to a learned modern astronomer,[4] was not unknown to any of the world's peoples. The ancient temples, considered as images of the Universe in which ruled the immutable Being to whom they were dedicated, all bore the same number and the same division. The Peruvian architects had ideas in this regard no different from those of the Egyptians, the Persians, the Romans, and even the Hebrews. The number 12, thus applied to the Universe and to all that represented it, was always the harmonic manifestation of the natural principles 1 and 2, and the mode under which their elements were coordinated. It was at the same time the symbol of the coordination of tones, and as such applied to the Lyre of Hermes. Boethius speaks of it in clear enough terms,[5] and Roussier has interpreted his opinions very well.

After the number 12, product of the multiplication of 3 and 4, the most generally revered number was the number 7, formed from the sum of 3 and 4. It was considered in the sanctuaries of Thebes and Eleusis as the symbol of the Soul of the World unfolding itself in the bosom of the Universe and giving life to it. Macrobius, who has transmitted many ancient mysteries to us, tells that this soul, distributed among the seven spheres of the world which it moves and animates and from which it produces the harmonic tones, was designated emblematically by the number 7, or figuratively by the seven-holed flute placed in the hands of Pan, the God of the Universe.[6] This number, revered by all peoples, was specially consecrated to the God of Light. The emperor Julian speaks enigmatically of the god with seven rays, knowledge of whom is not given to everyone.[7] The Brahmins taught, again, that the Sun is composed of seven rays; their sacred books represent its genius, Surya, riding a chariot yoked to seven horses. The ancient Egyptians, in place of a chariot, imagined a boat steered by seven genii; and Martianus Capella, who acts as their interpreter, places the Sun god in the middle of this boat, holding in his hands seven spheres, which like so many concave mirrors reflect the light which he pours out in great waves.[8] The Chinese scholars meditated much on the number Seven. Like the Pythagoreans, they attributed profound ideas to it. One of their sacred books, the *Liu-Tzu*,[9] says that it is a number of overwhelming wonder. Finally, even the first Christians, although in everything they distanced themselves from ancient ideas, nonetheless divided into seven gifts the influence of the Holy Spirit which is hymned in the Catholic churches.[10] Quite recently a Christian theosopher,[11] examining the properties of the number 7, taught with great conceptual force, though otherwise he was unlearned, that there can be no spiritual movement that is not septenary, because this is the number of the provinces of the Spirit; and because force and resistance, which are the universal pivot of every action, are themselves the two constitutive bases to which the septenary number owes its existence.

It is, I believe, unnecessary to multiply citations to prove the unanimous agreement of peoples on the recognized influence of the numbers 7 and 12, productions of the numbers 3 and 4 by simple addition or by multiplication. Now I will continue my dogmatic synthesis.

The fundamental principles B and F, developing in inverse directions either by fourths or fifths, that is, proceeding from 4 to 3 or from 3 to 2, produce two identical sets of tones.[12] It is this identity that constitutes the musical septenary, and which causes these notes to be called *diatonic* to distinguish them from all the other tones that can still be born from the two fundamental principles, but which no longer resemble each other, going outside the diatonic order to enter the chromatic and enharmonic ones. The diatonic septenary of music, born from the union of the two principles, is applied in celestial harmony to the planetary septenary (though they did not infer from this in the sanctuaries that there are only seven primitive planets, identical and really influential in our zodiacal system, the others being only secondary like the chromatic and enharmonic tones in our system). The fundamental tone B represents Saturn, the furthest from the Sun of the primordial planets. The fundamental tone F represents Venus, the closest of these to the Sun.[13] The first has a rising motion by fourths, the second a descending motion by fifths as follows:

Saturn	Sun	Moon	Mars	Mercury	Jupiter	Venus
B	E	A	D	G	C	F

This planetary septenary, moving in the universal dodecahedron represented by the radical number 12, is its perfect measure, and constitutes the diatonic order of tones and of the musical modes that follow from it. I will represent the image of this motion after having made some preliminary observations.

The first is that a string measured off in quarters to give the fourths B, E, A, D, G, C, F, cannot at the same time be measured in thirds to give the fifths F, C, G, D, A, E, B; hence two strings are needed to represent the two principles B and F.

The second observation is that these two strings, supposing them otherwise to be equal, will be unequal in length, since the F proceeding by fifths needs a greater distance to reach the B than the B needs in order to reach F by means of fourths.[14]

Consequently, and this is the third and most important observation, supposing that these two strings are bent in an arc to represent the universal sphere, and applying to them the zodiacal measure 12, the two hemispheres will be far from equal, although they give respectively identical tones, because the two strings, incommensurable with one another, enclose areas or spaces which, though one cannot measure one by the other nor ever express them in physical numbers, will nevertheless be in the relationship of the musical fourth to the fifth. This will serve to prove that the Universe is by no means contained, as the vulgar seem to think, in a perfect circle, but in a sort of oval, which the Orphics rightly depicted in the form of an egg, and that the individual spheres of the planets, conforming to those of the Universe, are not exactly circular but describe a more or less elongated ellipse, according to the portion of the harmonic string that serves them as measure.[15]

Survey of Celestial Music

Let us pause a moment on the celestial sphere, and, penetrating as far as prudence permits into the secrecy of the ancient sanctuaries, let us suppose that we are listening to one of the wise Eumolpids[16] speaking: "Seeing the seven primordial planets forming a sort of circle around a common hearth," he says, "the vulgar imagine that the earth is placed at the center of this hearth, and that it sees not only the planets turning around itself, but even the supreme sphere that encloses it; but this is mere appearance, a gross illusion of their senses that they take for a truth. It is prudent to leave them to their error until they can rid themselves of it; for as they cannot grasp the truth if it is presented to them before their mind is properly prepared, by relieving them of their error one would only be throwing them into chaos and making them incapable of guiding themselves through the darkness in which they would suddenly be enveloped. The Earth is no more at the center of the Universe than Jupiter or Mercury; it is only a planet, like them. The Moon has its place in the planetary order, and when the initiates speak of the Moon, they always mean the Earth, because they know that the Moon, the Earth, and Tartarus, or the Earth of the Earth, are but one and the same thing under three different names.[17] For them it is the triple Hecate: Proserpine in the underworld, Diana on the Earth, and Phoebe in the heavens. If the Earth is central, it is only when one considers it as constituting a particular system within the universal system, and takes it as the tonic of a musical mode. On the other hand, the philosophers, having seen that the Earth cannot occupy the center of the Universe, place the Sun there, and explain by mathematical abstractions the phenomena of the celestial motions. But," the Eumolpid continues, "that is still only the system of the Lesser Mysteries of which one now and then permits a part to be divulged to the people, so as to attack unconsciously the multitude of their errors. Although it is certain that the Sun is infinitely better placed at the center of the Universe than at any point on the circumference, it is none the less true that this star, seen from the Earth, should never be considered as a planet. Listen carefully to the reason, and do not reject without a lengthy examination what I am going to tell you. It is that in its central place, it is invisible to us. If it manifests itself to our eyes, it is by the reflection of its light. The Sun that we see is only a sensible image of the intelligible Sun,[18] which from the center imparts movement to the Universe and fills it with light. Those of its rays that reach us illuminate us only thanks to a sort of circumferential mirror that corporifies them and adapts them to the feebleness of our organs.

"It is not necessary to know any more of this for the understanding of the musical figure with which we are concerned, and it would be beside the point for me to go further into this matter. Suffice it for you to know that the calculations of our astronomers relating to movements, to mass, to the respective distances of the celestial bodies, to their inner nature, are excellent as far as physical relationships and civil usage are concerned, and deduced, for the most part, with rare talent; yet they are vain when one comes to apply them to knowledge of the truth. Calculations based on terrestrial illusions are never accurate except on that basis, and vanish as soon as one tries to detach them from it. The movements of the stars are a consequence of those attributed to the Earth, and have no other certitude. Thus if the Earth were not to have the motions the astronomers believe it to have, or if it has other motions, everything in their universal system would change instantly; they calculate distances by solar parallax, which is entirely unknown to them because they seek the center of that star where it is not, and they weigh masses by means of relations they establish between the Moon and the Earth, without knowing that since the Moon is in no way different from the Earth, these relations are identities; instead of two terms, as they believe, they never give them more than one.

"For the rest, these calculations, although there is nothing true about them, are still very useful, as I have said, when one applies them solely to the necessities of life; they become vain or dangerous only when one tries to transfer them from the sensible to the intelligible, and to give them a universal existence which they lack. It would be the same if after having established, like our initiate sages, an intellectual system founded on celestial music, one tried to submit the results to the calculation of physical numbers. For knowing from the first principle that there is the ratio of a fourth between Saturn and the Sun, and between the Sun and the Moon—so that the Sun is the central and tonic point of the other two planets—does not enable one to express in physical numbers the respective distances of these luminaries, their size and movement, because the musical ratio of a fourth can be given by strings infinitely varying in length, thickness, and vibrations, according to their inner constitution and the more or less homogeneous nature of their parts.

"One must therefore avoid unwisely substituting one system for the other. The physical system serves to calculate by approximations that seem exact the apparent courses of the celestial bodies, and to predict the return of phenomena; the intellectual system, to make known by constant ratios the cause of these movements, and to evaluate the phenomenal illusions that they produce. The first is knowledge of the external and visible effects, the second of the internal and hidden principles. Science consists of uniting these systems and of using each for its own object. This is where true philosophy lies. In contemplating them both, this science teaches that the first of these systems, unvarying like the Cause of which it reveals the principle, disappers as the intellect is dimmed; while the other, bound to the variation of forms, changes with the times, peoples, and climates, so as to serve at least to enlighten people again in the moral darkness wherein their own will and the vicissitudes of Nature often plunge them."

After having meditated for a moment on this discourse of the Eumolpid, let us pass on to the diatonic development in music. This development works by opposing the fundamental strings that give the two primordial tones B and F.

B	E	A	D	G	C	F
4096	3072	2304	1728	1296	972	729

F	C	G	D	A	E	B
5832	3888	2592	1728	1152	768	512

We find in the opposition of these two strings the ratios existing between all the diatonic intervals, and the identity of the tones is irresistibly proven by the union established on the D, which is the median tone of the two strings. In the planetary spheres, this unison on D corresponds to the planet Mars.

If we now transpose the strings B and F to their higher octaves, always moving them by the appropriate contrary progressions, we will forthwith obtain the series of diatonic tones following the rank given them by Nature.

Saturnian Diatonic System

Saturn	Jupiter	Mars	Sun	Venus	Mercury	Moon
B	C	D	E	F	G	A

Cyprian Diatonic System

Venus	Mercury	Moon	Saturn	Jupiter	Mars	Sun
F	G	A	B	C	D	E

From all that I have said it follows that the diatonic tones as we have received them from the Latins and Greeks are in no way arbitrary, either in their ratios or in their rank, and that the Egyptians, who equated the number to that of the planets and who ranked them in the same order, followed in this regard a respectable tradition founded on truth, or else were themselves inspired by a profound wisdom. These tones, as we are convinced, owe their identity to the contrary unfoldings of two principles, and their ranking order to the reconciliation of these same principles. Their ratios are established by mathematical proportions of rigorous exactitude, of which one can alter nothing without throwing everything into confusion. We can thus accept them in all certitude and make them the unshakeable basis of our system.[19]

Saturnian Diatonic System—Fundamental string B

	B		C		D		E		F		G		A
from	2048	to	1944	to	1728	to	1536	to	1458	to	1296	to	1152

Cyprian Diatonic System—Fundamental string F

	F		G		A		B		C		D		E
from	2926	to	2592	to	2304	to	2048	to	1944	to	1728	to	1536

The only thing still left to mention about this diatonic order concerns celestial music. One should recall that the Egyptians, having represented the planetary septenary by the fundamental string B and conceived its ascending development according to the progression by fourths, considered this progression as divine and spiritual, and gave to the progression by fifths the name of terrestrial and corporeal; they also preferred the diatonic order given by this string, all the more since it assigns to the planets the same order as they have in ethereal space,[20] as follows:

Saturn	Jupiter	Mars	Sun	Venus	Mercury	Moon
B	C	D	E	F	G	A

It is because of the idea that the Egyptians had of the superiority of the Saturnian principle B over the Cyprian F, that they made its progression by fourths govern the seven days of the week, and its diatonic course the 24 hours of the day, as Dion Cassius says expressly in his *Roman History* [XXXVII, 18].

Here is this order for the days of the week:

Saturday	Sunday	Monday	Tuesday	Wednesday	Thursday	Friday
Saturn	Sun	Moon	Mars	Mercury	Jupiter	Venus
B	E	A	D	G	C	F

For the hours of morning and afternoon:

	1	2	3	4	5	6	7	8	9	10	11	12	
Saturday Day of Saturn	B	C	D	E	F	G	A	B	C	D	E	F	a.m.
	G	A	B	C	D	E	F	G	A	B	C	D	p.m.
Sunday Day of the Sun	E	F	G	A	B	C	D	E	F	G	A	B	a.m.
	C	D	E	F	G	A	B	C	D	E	F	G	p.m.
Monday Day of the Moon	A	B	C	D	E	F	G	A	B	C	D	E	a.m.
	F	G	A	B	C	D	E	F	G	A	B	C	p.m.
Tuesday Day of Mars	D	E	F	G	A	B	C	D	E	F	G	A	a.m.
	B	C	D	E	F	G	A	B	C	E	F	G	p.m.
Wednesday Day of Mercury	G	A	B	C	D	E	F	G	A	B	C	D	a.m.
	E	F	G	A	B	C	D	E	F	G	A	B	p.m.

Thursday	C	D	E	F	G	A	B	C	D	E	F	G	a.m.
Day of													
Jupiter	A	B	C	D	E	F	G	A	B	C	D	E	p.m.
Friday	F	G	A	B	C	D	E	F	G	A	B	C	a.m.
Day of													
Venus	D	E	F	G	A	B	C	D	E	F	G	A	p.m.

Thus, by making the musical diatonic septenary operate within the harmonic septenary, applied to the seven days of the week after dividing each of these days into twice twelve hours, the Egyptians found a way to distinguish the different and respective relationships of the two principles B and F, which had combined their actions in the zodiacal number 12, and demonstrated the identity of their products by forming a series of similar diatonic tones; then they distinguished these tones within the horary number 24 by coordinating them in different ways among themselves, and opposing them to each other according to whether they were taken alternatively as the principle of a series, or, musically speaking, as the tonic of a mode. The result of this new movement is that one can recognize seven diatonic modes, which form fourteen since one can consider them as primordial or secondary; but, as I shall explain later, these seven primordial modes are reduced to five, because the principles B and F, acting separately, can never constitute true modes in the meaning I give to this term.

Brief Exposition of the Musical System

Every tone that sounds can be conceived under the form of unity. Every tone includes all tones. But the string that gives it can be divided into parts, and from the moment it is divided, it produces other tones that are analogous to the generating tone, but whose analogy is harmonic or inharmonic. For the notes produced to be harmonic, the division of the string must be made according to geometrical proportions. The quaternary of Pythagoras 1, 2, 3, 4, supplies the only proportion admissible in music. These proportions are remarkable in that they proceed according to an arithmetical and geometrical progression. Every other progression than that contained or produced by the quaternary 1, 2, 3, 4, gives only tones that are inharmonic, false, and heterogeneous.

A string conceived in its unity gives a certain tone which acquires its properties and a name only from the relation it has with other tones. A tone must necessarily be considered as producer or product. But a tone can only produce other tones through the division one makes in the string which gives it; and it can be produced only by means of the division one has made of a generating string to which it belongs.

Let us then take a string as producer, and begin by submitting it to the quaternary progression 1, 2, 3, 4. We will call this string B. Divided from 1 into 2, it will give its own octave and not depart from its diapason: thus we will have done nothing for the musical system, for B is no different from B. And however much one may raise or lower this B from octave to octave, it will never produce any melody. This proves that the two principal principles 1 and 2 cannot act in their essence. They can only act in their faculty. Now the faculty of 1 is imparity (oddness), and the faculty of 2 is parity (evenness). The number 3, being the first number, thus displays the faculty of 1; and consequently 4 displays the faculty of 2, whose power it is. But if a string be divided in 3, it cannot be divided in 4; for evenness and oddness are incompatible. So we need two strings, the one to represent the principle 1, which we shall divide into 3; the other to represent the principle 2, which we shall divide into 4. But what are these strings which we are to divide thus? They must necessarily be those strings which in producing themselves reciprocally, produce reciprocally all the other tones, without exceeding the musical septenary given by Nature. Now the two strings that fulfil these conditions are F and B. These strings form between them an irrational and incommensurable interval. They are opposed to one another as even is to odd. So let the F string be divided into 3, representing the principial principle 1. This string thus divided produces its fifth C by proceeding from 3 to 2. Next let the B string be divided by 4, representing the principial principle 2. This string thus divided produces its fifth E by proceeding from 4 to 3. By continuing the progression from 3 to 2 for the string F, it develops from fifth to fifth: F, C, G, D, A, E, B. In continuing the progression of 4 to 3 for the B string, it develops from fourth to fourth: B, E, A, D, G, C, F. Thus these two strings mutually produce each other by opposite paths; and in so doing they give birth to all the notes of the scale. One should observe that the note on which the two strings meet is D. The note D is thus the archetype of unison. It represents Mars in the planetary system. This system is conceived starting from the B string as follows:

B	E	A	D	G	C	F
Saturn	Sun	Moon	Mars	Mercury	Jupiter	Venus
Saturday	Sunday	Monday	Tuesday	Wednesday	Thursday	Friday

In conceiving this system in the diatonic order, one obtains:

B	C	D	E	F	G	A
Saturn	Jupiter	Mars	Sun	Venus	Mercury	Moon

Thus the Sun is at the center of the Universe, a fourth distant from Saturn and a fourth from the Moon. But the string that gives the fourth from Saturn to the Moon is far shorter than that which gives the fourth from Saturn to the Sun.

If one continues the progression of the F string from 3 to 2 it produces the sharp and destroys itself by the action of the B string of which the sharp is the direct production and latent principle.[21] B produced by F represents Love or the expansive Force; F produced by B represents Chaos or the compressive Force: the primordial principles of the Universe. These musical notations are sufficient for theory.

47

◆

Alphonse Toussenel
1803–1885

Born in Montreuil-Bellay (Maine et Loire), Toussenel was in turn a journalist, newspaper editor, civil commissioner, and socialist politician, until he decided to retire to the country to farm and observe animals. He wrote three books on politics, and three on natural history, of which this one, on "passional zoology," ran to four editions and was translated into English.

The mainspring of all Toussenel's activities was his admiration for the ideas and ideals of Charles Fourier (1772–1837), which were so influential in the politics of the Left before 1848. Fourier himself was not a well-organized writer, and it would take many more of his pages to explain his musical cosmology.[1] Toussenel's is one of the best and briefest accounts of it.

Basing his philosophy, as he would have us believe, purely on observation, Fourier saw the whole universe as a play of passions or emotions, and the goal of human existence as the living of these passions to their fullest extent. In this he set himself explicitly against the Christian contempt for the body and its desires—though by no means against Jesus Christ, whom he saw as having come to further the evolution of mankind. At the present time, says Fourier, we have come out of the state of barbarism into that of civilization: that is to say, from one disgusting condition into another; and thanks to the arrogance of philosophers who refuse to study the laws of Association and Attraction, we have been stuck in this civilized state for 2,300 years longer than necessary. We have far to go before we attain our birthright as "harmonian" beings, but once attained we will enjoy it not for a millennium but for 70,000 years.[2]

There is a note of contemporary relevance in Fourier's conviction that sickness assails both mankind and the earth, the one involving the other. But what is one to say of his cosmology, which promises to restore the earth's original five moons and to correct its polar tilt, as soon as the harmonian era commences? Perhaps that it is no further removed from current science than the Ptolemaic spheres. In any case, Fourier's cosmic music is as traditional as could be, with his assumption that the universe works through the law of correspondences and that the ordering of one realm of existence (the scale) explains, or is explained by, another (the planets and their moons).

Source: Alphonse Toussenel, *L'Esprit des bêtes. Vénerie française et zoologie passionnelle* (Paris: Librarie Sociétaire, 1847), pp. 11–16. Translated by the Editor.

A Summary of Charles Fourier's Cosmology

Man is a passional[3] keyboard with 32 keys, a series of the third degree[4] like the planetary series. But this keyboard, unfortunately, most often functions only as a keyboard of 12 keys, like a series of the second degree. Twelve is the number of *simple* harmony, and thirty-two the number of *composite* harmony. The passional keyboard of man is a keyboard momentarily in eclipse. Of all man's senses, in fact, the sense of hearing alone possesses its integral keyboard of 32 notes. Every series of 32 notes comprises 24 notes of the scale or octave (12 in major and 12 in minor), 4 notes of transition or "ambiguous," and 4 subpivotal ones.[5]

Why is the human series so often reduced to working as a series of the second degree (12 terms), or, lower still, as one of the first (7 terms)? Ask the Earth which has engendered man, and which has necessarily made him in its own image, as Hippocrates says.

Man is the product of a *gelded* creation, that is, one interrupted at its best moment. Man is the lastborn child of a fallen globe, all of whose kingdoms are marked with the seal of omission and miscarriage. Man is the king of a planet banished, as it were, from its vortex because of contagious disease, dragging behind it a filthy corpse of a satellite: whereas it should stride forward escorted by a glorious procession of five living moons.[6] One can see why, under such conditions, God would have looked twice before allowing earthly man, like those of Saturn or Jupiter, the free and full scope of the harmonial keyboard; and why he thought it proper, in the great majority of cases, to reduce the passional keyboard of the king of the earth to a lesser number of notes. In so doing, the supreme director of things has wisely made man's attractions proportional to his temporary fate. Let us not blame the Eternal for his parsimony; earthly humanity has quite enough with its twelve passions, because it already has more than it can feed. The important point is that the power of the passional lever which remains to us is sufficient to goad us into reaction against our present misery, and to prepare us for a pleasanter future.

Before passing on to the sad story of the miseries of the Earth and of the fall, and before explaining the reason for the horrors of the latest creation, I will finish my account of the passional scale of man.

The passional scale or series of man is made essentially from 12 radical notes in double rank. These 12 notes are divided into three groups, like every series of 32 terms.

1. The group of *cardinal* or *affective* passions corresponding to the moving principle, which one could call the mainsprings of the human heart. These cardinal passions are four in number: Friendship, Love, Familialism, and Ambition.

2. The group of *sensitive* passions corresponding to matter, whose number is inevitably fixed by that of the five senses.

3. Last, the group of *distributive* passions, corresponding to the neutral or regulating principle. The distributive passions, three in number, are charged with directing the ranks of the general keyboard, controlling the chords and discords of the other passions. They bear the following names, drawn from their uses: *Cabalist*, a *deliberate* ardor, the passion of emulation and intrigue; *Composite*, enthusiasm, blind ardor, the passion of concords; *Butterfly* or *Alternating*, the passion of changeability, the source of charm and preventive of the boredom caused by uniformity.

The cardinal or affective passions tend towards the *Group*; the sensitive or material ones, to *Luxury*; the distributive passions to the *Series*, which distributes harmonies.

The cardinal passions have two keyboards[7] or two modes, like the series: the *major* and the *minor* mode. The major mode comprises two passions in which the spiritual impulse exceeds the material, Ambition and Friendship; the minor mode, the other two in which the material impulse dominates the spiritual, Love and Familialism. Ambition is called the cardinal *hypermajor*,[8] Love, the cardinal *hyperminor*; Friendship, *hypomajor*, and Familialism, *hypominor*. The major mode modulates by odd numbers, 7 and 5; the minor mode, by even numbers, 8 and 4. These details, which may seem trivial, are of immense importance in the study of passional movement.[9]

All the passions of man are combined and summarized in a single pivotal or focal passion, called *Unitism* or the passion for unity, a religious sentiment. In just the same way, all the colors of the prism unite to form white, the color of unitism.

The planetary passional keyboard corresponds exactly to the human passional keyboard, and the series of planets is in perfect accordance of types and names with the series of human passions. The principal parts of the sidereal framework bear the same names as those of the human framework; only their ranks are always double, or rather composite, whereas the rank of human passions is only simple, half the time.

The planetary keyboard is also composed of twelve radical keys in major and minor, divided likewise into three groups as are the keys of the human keyboard.

The sidereal keyboard at full strength displays all the pieces in its double rank. Major scale, 12 satellites: 7 to Saturn, cardinal of Ambition; 5 to Earth, cardinal of Friendship.[10] Minor scale, the same number of satellites: 8 to Herschel,[11] cardinal of Love; 4 to Jupiter, cardinal of Familialism.

Twenty-four satellites, plus the four cardinal or sub-pivots, making 28; plus the four ambigous ones, making 32 planets in all: the series of the third power with the Sun as pivot or general focus of aromas.[12]

I am well aware that the astronomers of the Institute are not in agreement with me about the normal number of 32 planets, and that they contest the eight satellites of Herschel and the five of the Earth; but I am not concerned in the least with these somewhat finicky objections. In the first place, an Institute which had the least notion of passional astronomy would understand straightway that a cardinal planet of Love could not accommodate a retinue of four or six moons, seeing that Love can have nothing to do with these two numbers. Next, it is sufficient that the telescope of analogy has discovered the eight satellites, for me to regard as immaterial the protests of a few poor myopic observatory telescopes,[13] and I refute those protests with this simple reply. Analogy had proclaimed the telescopic planets before the invention of the telescope, and the planet Leverrier[14] (Sappho) before the birth of that astronomer. As for the five satellites of the Earth, I do not deny the strength of the objection. The Earth does not have a retinue of five satellites, it is true, but it might have it: the proof is, that it did have it once.

I would remark in passing on the unsuitability and impropriety of the names given by civilized science to the planets of the solar vortex. This unsuitability of terms seems actually to have been pushed to the verge of insult in the naming of the cardinal planet of Love, which they have saddled with the name of a man. A man's name, for a cardinal planet of Love! That is one of those blunders which suffice to give one an idea of the disorder that troubles the minds of our poor humanity. How much better the Greeks understood the fine points of language and the regard due to the weaker sex, they who called every noble abstraction, every virtue, by a woman's name, and who feminized even the divinity of war!

A strange anomaly, and one which shows man's imperfect and unfinished state and the shortcomings of his organism, is that while one of his senses, that of hearing, possesses its entire keyboard of 32 keys, the sense of sight only enjoys a keyboard of 7 to 12 keys, while the other senses scarcely show the rudiments of the serial hierarchy.

The musical keyboard, the keyboard of the ear, is in fact complete in man; it has its 32 notes with 12 passional types, its two major and minor scales or 12 notes each, supported by their 4 subpivotal notes and their notes of transition or ambiguity. Man's musical keyboard corresponds perfectly in types and numbers with the planetary keyboard: it is easy to see there the 4 cardinal notes of C (friendship), E (love), G (familialism), and B (ambition); the 3 ambiguous or distributive ones of D (cabalist), F (butterfly), and A (composite); the 5 sensitive or inferior passions, being the 5 semitones. However, we should note as a sign of the weakness of the human ear that this organ is incapable of perceiving the five semitones in the first vibration of the scale.

Alas, our eye does not even perceive twelve colors in the solar spectrum, although they are all there, and more. We only see seven rays, corresponding to the four cardinal and three distributive passions: violet (friendship), blue (love), yellow (familialism), red (ambition), indigo (cabalist), green (butterfly), and orange (composite). Our polluted atmosphere not only intercepts the five semi-colors analogous to the five semitones of the musical scale and to the five leading-tones; it completely deprives us of perceiving the minor keyboard.

A single observation will suffice to show the imperfection of the sense of sight in man. Man is the only animate being on his planet who cannot stare at the sun. Owls, who are blinded by excessive daylight, have a compensation that man lacks, the faculty of seeing at night; if they lack *cosolar* vision, they at least have *conocturnal* vision. Ovid was right when he said that God has given man a sublime face, *os sublime*, and that he has directed him to look at the sun. Only his definition applies to normal man, to the man of harmony, and not to the man of civilization, who cannot look the sun in the face. Animals, being aware of this shortcoming in their master, like to tease him about it. A cock who sees that you are paying attention to him never fails to turn aside his head to flash a glance at the sun, as if to say in his ironic language: "King of the Earth, try to do *that!*"

The scales of taste and smell, and the scale of the pivotal sense of touch, are barely sketched out in man. It is painful to think how different things are for men of other planets. On all the globes that have reached the state of harmony, man perceives distinctly 32 colors in the solar spectrum; he has in addition a keyboard set up for the senses of taste, smell, and touch. The varied play on these new instruments multiplies infinitely and incredibly the enjoyment and pleasure of the harmonians, such that the imagination of the civilized man cannot begin to calculate it without being immediately overcome by a profound feeling of envy. Poor Earthlings, whenever will we inspire in our turn the sentiment of envy, after having so long inspired pity?!

48

✦

Peter Singer
1810–1882

Peter Alkantara Singer was a Franciscan monk and musician who spent his life in Austria, working mainly in Salzburg. He composed much church music in a simple, sentimental vein, but his contemporary fame rested much more on his organ improvisation and on his ingenuity in building instruments. Wagner, Liszt, and Spohr were among those who came to see his two-manual "Pansymphonikon" of 1845, which could reputedly imitate every instrument in the orchestra.

"Pater Peter" seems to fall into a line of musical churchmen who combined an interest in speculative music with a delight in practical construction: one thinks of Athanasius Kircher, designer of mechanical instruments; of Père Castel, color theorist and inventor of the "ocular harpsichord;" of the Abbé Roussier, a Pythagorean theorist who encouraged the building of a quarter-tone harpsichord; and, nearer Singer's own time and métier, of the organ virtuoso Abbé Vogler, whom our author himself acknowledges here. But Singer's own theory and learning did not go much beyond Rameau, the evident source for his basing of all music on the triads of tonic, dominant, and subdominant.

What is original, and intriguing, about Singer's one theoretical work is his absolutely naive musical symbology. He believes that one can explain the mathematical and acoustical mysteries of the musical system by pointing out correspondences with Christian doctrine. The last music theorist to do so had probably been Andreas Werckmeister (see no. 39), though Louis Claude de Saint-Martin came close in his early work, excerpted here (no. 43). Given Singer's own circumscribed universe, what he says is perfectly valid; it evidently gave music its deepest meaning to him, and he never intended to force it upon the world. (The treatise was published on the initiative of his friend Georg Phillips.) The most illuminating parallel I can offer is with another pious Austrian organist, though a better-known one: Anton Bruckner.

Source: Peter Singer, O.S.F., *Metaphysische Blicke in die Tonwelt, nebst einem dadurch veranlassten neuen System der Tonwissenschaft*, edited by Georg Phillips (Munich: Literar.-artist. Anstalt, 1847), pp. 1–12. Translated by the Editor.

On the Analogy Found in the Tone-World with the Godhead and the Whole Created Universe

1. When the intelligence is brought to bear on the tone-world, it finds therein a singular image of the primordial Unity that rules in the universe and orders all things from first to last. Not only that: there opens to it a whole immeasurable canvas, reflecting like countless mirrors the uncreated harmony, reposing eternally in itself, of God the Three-in-One. It sees also the harmony created in time, destroyed by the sin of our first fathers, and restored through the incarnation of the eternal Word—the truly realized harmony of the moral and physical creation, concentrating itself in the most manifold ways to glorify the Trinity. The intelligence is moved thereby to admiration, while in the heart there awakens a foretaste of the Beyond, arousing a tender longing for its true homeland, the land of eternal harmony.

2. The harmonic triad has already been called by that deep thinker the Abbé Vogler the "trinitarian harmony," because in it three different tones sound like a single one. In its whole musical existence, this triad seems to us an unmistakable and most impressive picture of the harmony of God the Trinity, resting eternally in itself, which is at the same time the ground of all creation and its goal and end. This triad, this original trinitarian harmony, consists of three tones, of which the first appears as the root, grounding the whole harmony; the second as manifesting the first as root and as characterizing the whole harmony; and the third as the tone that unites the first two in a complete

harmonic whole. Indeed, on perfect instruments one can perceive that every first tone, played alone, brings forth the other two out of itself. The second becomes audible like its reflection, and the third, in the middle, appears as the heart of the trinitarian tone; and in this elementary form it takes its tones from three different octaves, so that the reflection shows itself as the outermost, and the uniting tone in the middle:

FIGURE 27: *The Trinitarian Harmony*

These two tones, coming forth from the first, have the same duration as it, sounding no sooner and no later: so long as it exists and sounds, the other two go forth from it and represent its perfection.[1]

These three tones, although quite different from one another, sound like a single and perfect tone, completely at rest in itself.[2]

3. This trinitarian harmony is also the one and only harmony that is independent and reposes only in itself: all the others find no rest in themselves, are not self-sufficient, consonant and dissonant alike. They all strive toward this primordial harmony in which alone they find rest: the dissonances advance turbulently, the consonances with calm recognition of their dependency and incapacity to satisfy themselves, toward their resolution in the triad.

4. The triad is also the source of all other harmonies, as will later be shown; whereby the practical study of music will be unspeakably facilitated, since an entirely new and extremely simple system will unfold, in which all harmonies are shown to derive step by step from this primordial harmony, in coordination and subordination dictated by nature herself.

5. Finally, this primordial harmony is also the goal of all harmonies and melodies: all of them, consonances as well as dissonances, must contribute their part to the glorification of the primordial harmony. Thus in the created universe everything must eventually contribute its own for the glorification of God Three-in-One: the worm in the dust as well as the seraph at the throne of God; the damned no less than the citizens of heaven, the latter through everlasting rejoicing, the former through an everlasting howl. Both proclaim with a voice that pervades the eternal halls: "How great, how good, how holy is God Three-in-One!" "What eternal bliss I have won!" cries the one; "What eternal bliss I have lost!!!" howls the other. Even the rests in music must contribute their part to the glorification of the triad, just as sin itself

contributes in the end to the glorification of God. For the great Saint Augustine said: "That which in music is rests, in the universe is sin," and vice versa.

6. Moreover, this primordial harmony has such an overwhelming force in the human ear—as in the organ of the tone-world—that even the savage, having at his disposal only a few imperfect tones, seeks out this harmony above all else, and as soon as he finds it, takes his only pleasure therein; and the first of composers likewise takes this harmony as the point of departure in all his wondrous creations, realizes its beauty in the most varied ways, and after all his efforts brings his art to rest in it, as the only goal and end.

7. Here, too, it is very remarkable that this harmony shows forth a most significant image of the godhead, in that it manifests his endless mercies in the work of redemption, by which the second person of the Trinity humbled himself and chose to take on the weak vulnerable form of a man. Now this triad also appears in suffering form, not such that all three tones are suffering equally (for the first and third tone remain in their original form), but the second tone alone is lowered, takes on a suffering form, and thus founds as it were a new world in the great realm of minor keys (also called soft keys on account of the lowered and suffering third); indeed, the whole tone-world suffers important changes thereby, which have already stirred many a profound composer and metaphysician to such creativity that new laws have entered into the tone-world: laws that to outward appearance seem like mistunings and defects, but on closer examination contain a mighty consecration, for into them is poured an image of the highest of all God's works. The death of the God-man on the cross is also, regarded from one point of view, the most outrageous fact in the universe and the culminating point of sin, in which men murdered their god who died for them. But from the other side and from a higher standpoint this is actually the most sublime fact in the universe, in which its new harmony was realized; it is the true culminating point of divine holiness and justice, as well as of endless love and compassion. It is precisely in this worst mistuning that the harmony of the outer glory of the threefold God attains its greatest splendor, and the same fact calls with a voice that fills heaven, earth, and all eternity—calls with the voice of the eternal Word: "What a sacred being is this, who in his own person suffers such a death out of love for his offenders, in order to earn for them—his murderers—eternal life!"

The apparent faults or unpleasantnesses in the tone-world should be judged analogously, since in the whole of physical creation we meet with no clearer trace or image of the work of redemption than in the toneworld; and it is most fitting that the supreme achievement of love should inspire it with compassion: for there is scarcely any creation in the sensible world to which the human heart is more accessible than music, and scarcely anything stands in such intimate relations with it.[3]

8. This also clarifies the choice of the twelve elementary tones, and the circle of the twelve keys around which the whole of today's music turns. Each one is governed by a certain mistuning, so that even the great triad no longer appears in its primordial form, but its second tone is a little displaced. The realization of this circle, without which the whole of today's musical art would collapse, takes place only at the cost of the second tone of the triad, with cooperation from the third.[4] For example, if one tunes perfectly the second tone, E, in the triad C E G, that E will be unusable as a fifth of A, being too low to make the triad A C# E bearable. Therefore, to make the interval with A as good as possible, the E has to be tuned a little higher, displacing it to the exact degree that makes it a pivot to the other keys. One proceeds likewise all the way around the circle of fifths.

What we have said here about the note E applies to all the other thirds, or second tones, of the triads of the circle. And although in this way the fifth, or the third tone, of the triad will sound a vibration lower, in actual practice on an instrument which is equally tempered in all keys it happens that only the second tone (the third) is mistuned, while the third tone [the fifth] seems more concurrent. Moreover, every fifth is at the same time the third of another triad.

9. This noble analogy conceals, and reveals, further secrets that hold sway in the tone-world, as we shall show, which no hypothesis has hitherto explained. Hence they have been deplored as defects, either in the tone-world itself, or in the human ear, since the researcher finds nothing analogous to them in the whole of physical creation. For while in all the visible universe, everything occurs according to very precise, harmonic, and unalterable laws, in important and incidental respects alike, in the tone-world one finds not only incidental discrepancies but quite essential ones. In fact, the substratum of the whole art of music, the twelve basic tones with the key-areas built upon them, only exist at the price of pure hearing and of the mathematical vibration-ratios (which are altogether lost), being mistuned as we have explained in the previous section.

The musical metaphysician feels compelled by this to assume two distinct creations in the tone-world, since everywhere he meets intervals that point to, or at least symbolize, such a twofold creation. Of the original creation, he encounters only a few traces: in this primordial period, all harmonies are perfectly euphonious, without the slightest mistuning (the numbers of vibrations being also in the correct ratio); however, each interval belongs, not to a circle of twelve keys and to the relationships that follow from that, as governs today's music, but reaches, so to speak, into the infinite. For example, if one were to sound the C triad perfectly tuned, according to the finest ear and the exact math-ematical relationships of the vibration-numbers of the tones, then the G triad, and after that the D triad, and so forth, one would never come back to C, but would travel on and on and lose oneself in a world of innumerable triads.

In this primordial period there must have been as many keys as there were tones, and in each key an independent triad with its two adjunct harmonies in perfectly pure tuning. These ratios would not have served for the further modifications of these harmonies as found in today's music, since the latter have only arisen through the realization of the musical circle. If we had at our disposal an instrument that could play all possible tones, we could even now put these primordial harmonies into practice. But every key would be limited to the three harmonies we have mentioned, and to melodies arising from them in their perfect and pure form. One could not include even so close a relationship as that of the A-minor chord to C major, since the E as second tone of the C triad differs by a perceptible fraction from the E that must serve as uniting-tone (or fifth) to the A. For in these absolutely pure primordial intervals no divergence could occur, but following the law of unity each tone must be one and the same; thus the minor chord itself would have no pure existence.

These intervals could be employed only by strings, certain wind instruments, and the voice, in pieces that used a single key and only those three harmonies. Also those triads that are used in larger organs to constitute single perfect tones contain the pure primordial tuning, and would alone be usable as such.

It seems that in these primordial intervals, still existing in the tone-world, the Creator has given us an eloquent picture of the original creation; for according to the most learned theologians, the original harmonies continued so long as sin had not entered into the world. Angelic and human nature were to have lived in recognition and love of the threefold primordial being without compulsion, in joyful recog-nition of their dependence and in free-willing realization of their final goal; they were to add their part to his outward glorification as they passed through their period of trial; then after their course was run, they would be taken up into eternal bliss—into the unmediated con-templation and the eternal joy of their highest and only good.

Thus the perfect triad manifests here as image of the all-perfect godhead, and the two adjunct chords (the seventh-chord on the upper dominant or fifth, and the $^6_5$ chord on the lower dominant or fourth), created for its direct service and carrying its own image within them, as representing the angelic and human nature in their original condi-tion. The melodies and transitory harmonies that arise from it would be an image of the other, transitory creations of the universe.

With regard to the second creation, or the establishment of a new realm in the tone-world, as many images come forward as there are tones and relationships in present-day music; but only the tones formed under this new law are included in this new realm as subjects and contributing members; the others are excluded, because they do not have the constitution of the new intervals. The totality of these selected tones is contained in the twelve elementary tones, which are chosen out of the innumerable ones for the realization of the new realm, just as the totality of these chosen for the new kingdom of God are expressed in the Old Testament by the twelve tribes of Israel, and in the New Testament by the twelve Apostles, as the twelve founding pillars on which the new realm of grace is built.

Each of these elementary tones forms its own realm, and in each of these twelve tone-realms there appears the autonomous triad—in its majestic as well as its compassionate form—as the only lord, who gathers to his direct or indirect service and to his glorification all the specific harmonies and melodies that make up his court, as it were, and in their beautiful ordering constitute his realm. Although each is an independent, autonomous, and complete kingdom, they stand in very definite relationships, closer or further removed from one another; and while principally they all realize the same image of the one great Kingdom of God, they ultimately make up a single, great, perfectly organized empire.

They all owe this organization to the rather prominent second tone of the triad, through which in the whole, as well as in every part, a new complete unity is manifested. Through the entry of this tone, "qui fecit utraque unum," [which makes one of both], the same E, for example, is the third of C and the fifth of A; the G# is truly the same tone as Ab, C# the same as Db, etc., distinguished henceforth not in fact but only in the way they are written, whereby their derivation, origin, etc., are evident. And it is only with this unity that modern music can exist: without it, the loveliest combinations of tones and harmonies remain unlawful and absolutely impermissible; for perfect unity is, and must be, the first unalterable law of nature in music, and actually its soul.

10. After the king of the tone-world—the all-ruling triad—, deeper research encounters a second harmony, which carries the whole image of the triad in itself, yet contains a characteristic auxiliary tone which makes the whole harmony unsatisfied in itself, yet which furthers its resolution in the triad, in whose service it exists. This is the so-called seventh chord, which is formed on the third tone of the triad (the fifth or upper dominant) and bears in its first three tones (fundamental, major third, and fifth) the pure image of the three-in-one primordial harmony, but is characterized through its fourth tone (the minor seventh,

which also gives it its name) as a harmony not at rest in itself, but existing in the service of the triad. Even at first glance—and still more so after understanding the nature of this harmony and its relationship to the primordial harmony—the seventh chord reveals itself as a living image of the angelic nature—the pure, created world of spirits—which as a pure spiritual being, among all beings in the created universe most carries in itself the image of the Godhead. The only thing which characterizes it as merely a creation is that it receives its whole being not from itself, but solely from the Creator, and that it cannot rest in itself, but only in the Creator, for whose glorification and unmediated service it was made.

11. This harmony, just as it most resembles the primordial harmony, is also its nearest direct handmaid and habitual partner, and after the triad most clearly engraves itself upon the organ of hearing. Every natural musician who has at his disposal an instrument containing all twelve elementary tones, even without artificial aid or guidance beyond that of experience, discovers this first adjunct harmony after the primordial one: the first three tones to begin with, then also their fourth tone. In his little pieces, however inadequate they may otherwise be, he obeys faithfully this basic law: that this harmony can never be content with itself, can never be regarded or used as a final chord, but always leads back to the primordial harmony. So much the more has every educated musician obeyed this law since time immemorial, and will always obey it, such that one cannot point to a single example of deviation from it: a proof that this law is stamped definitively and precisely on the human organ of music, and which makes this harmony valid only as an auxiliary, never as a chord of resolution.

12. It is especially striking that while this harmony is not pathetic (since it always contains the major third), it can occur as handmaid to the major or to the minor triad: a phenomenon that offers music extraordinary patterns, quite divergent from normal rules, since in all minor keys a note completely foreign to the scale must be used to form the first significant auxiliary chord. In the whole physical creation, we meet with no analogous example. Only in the idea advanced above does this secret of nature reveal itself: then the matter appears so simple and clear that to demonstrate the contrary would involve greater difficulties. For the angelic nature, too, which is depicted in this harmony, did not suffer through the Incarnation and the raising of a new realm of grace, and it also serves and honors the God-man without changing its situation, as it honored the most holy Trinity before the incarnation of the eternal Word.[5]

49

◆

Albert Freiherr von Thimus
1806–1878

Albert von Thimus represents that almost extinct type of scholar, the obsessionally dedicated amateur. Like his contemporary Edouard Coussemaker (one of the first revivers of medieval music and theory), von Thimus was a lawyer; he also worked as a judge, and sat with the center party in the Prussian Reichstag. In his student days, he had come under the influence of Georg Friedrich Creuzer, whose *Symbolik und Mythologie der alten Völker, besonders der Griechen* claimed that all mythologies have a common source. Von Thimus had the insight that the one unifying symbolic factor behind all ancient lore and learning might have been the discovery of the harmonic series and its association with mathematics. He pursued this idea his life long, across a field of learning that few of his readers, and none of his critics, have mastered: it included the *I Ching* and *Tao Te Ching* in Chinese, Egyptian hieroglyphs, all of Greek philosophy, the church Fathers and Doctors, and music theory since the Middle Ages. Like Tartini (see no. 42), von Thimus was conscious of being the first to disinter the esoteric doctrine of Antiquity from the oblivion into which it had fallen, not only through centuries of ignorance but through deliberate concealment by initiates. But the sophistication and erudition of von Thimus made him a more persuasive advocate for his cause than the elderly violinist.

The two volumes of his great work, published in 1868 and 1876, were intended to be followed by a third one, on the rebirth and completion of the ancient doctrines in the Christian era, but this was never completed. Von Thimus was almost totally ignored until Hans Kayser began to study his book in the 1920s, and to develop his own holistic worldview on the basis of the "Pythagorean Table" or *Lambdoma* diagram in which von Thimus had expressed the dual sequence of harmonics and subharmonics.[1]

Die harmonikale Symbolik des Alterthums vies with Proclus' Commentary on the *Timaeus* (no. 12) and Giorgi's *Harmonia Mundi* (no. 30) as the most difficult text represented in this collection. In this extract von Thimus goes to the core of his subject, and ours: the harmonic foundation of the Music of the Spheres. For him, it is an example of how the esoteric harmonic doctrine came first, then the cosmic picture—in this case, that of the heliocentric system as known in Antiquity. Von Thimus' contentions have yet to be disproved, or even answered; if accepted, they would indeed shake the learned world, as every discoverer of ancient esoteric doctrine hopes to do. But the learned world, like the oyster, well knows how to protect itself against such a threat.

Source: Albert Freiherr von Thimus, *Die harmonikale Symbolik des Alterthums*, vol. I (Cologne: DuMont Schauberg, 1868, reprinted Hildesheim: Olms, 1972), pp. 177–188. Author's footnotes omitted. Translated by the Editor.

The Pythagorean Table

The harmonic structure of the upward and downward rows of notes grouped into a single tone-system comes, as the development of the lambdoma-table[2] from the abacus has shown, out of a sequence of major and minor chords together, whose construction according to the numbers of the senary[3] derives its proportions from the multiples of the increasing series and the fractions of the decreasing ones. The number of instances of the arithmetic and harmonic progressions[4] (resembling the links in a chain, joining them together in larger series) out of which these major and minor chords arise never exceeds six. There are only three prime numbers,[5] which, in their multiplications and divisions with themselves and with each other, provide the number-forms to whose musical apprehension the intuitive power of our soul—very narrowly circumscribed, here below—is receptive. For beings of a higher sort

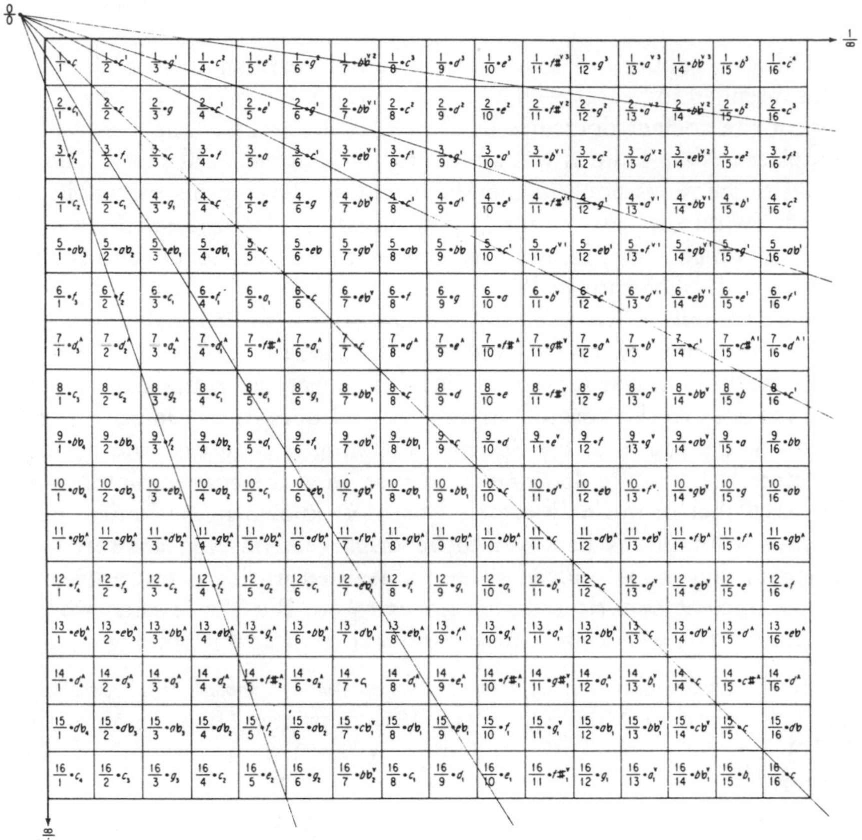

FIGURE 28: *The Pythagorean Table*

(and doubtless also for our souls in their erstwhile community with the choirs of holy Spirits called to the contemplation of God) the direct intuitive perception of the laws of symmetrical beauty which find their expression in numbers is certainly not limited by such narrow boundaries. As the sensible expression of the endless proportions of the harmonies concealed within the depths of the very Godhead, the two number-series of increasing and diminishing quantities, and of accelerating and decelerating regular movement,[6] find in the concepts of the infinitely large and the infinitely small not so much their limits as their unitary starting-point. Before the Lord God the infinitely large is Nothing and the infinitely small is All: both forms of this infinite duality are One. To Man, created in God's own image, is given, even during his existence in bondage to the sense-world here on earth, the capability at least of mounting upward toward the vision of that Truth through the thinking that releases him from the fetters of corporeality.

In order for the spiritual contemplation of harmonic numbers (not the physical perception of tonal phenomena) to bring to consciousness the formative norms of harmonic laws, the series of the *perissos*-ratios of the major chord and those of the *artios*-ratios of the minor chord,[7] must ideally begin not with the measuring of one actual number as always large or small, but with the concept of the One in itself and with the forms of the Unbounded Duality of infinite quantity itself, $^\infty\!/_1$ and $^1\!/_\infty$. The arithmetic *perissos*-series presents, with its real first member $^1\!/_1$, the latter of these two forms as a picture of the beginning of all increasing quantities. Transcendent thought can express the extreme point of its increase only through the other one of the two forms of the Infinite, $^\infty\!/_1$. The harmonic *artios*-series, on the contrary, puts $^\infty\!/_1$ at its beginning (because only the sequence of terms $1/0$ [$= ^\infty\!/_1$], $^1\!/_1$, $^1\!/_2$ corresponds to the law that finds its expression in the archetype of harmonic proportionality $^1\!/_1$, $^1\!/_2$, $^1\!/_3$), and as the equally spaced denominators of the aliquot parts of this progression increase to infinity, the latter ends ideally with the term $^1\!/_\infty$. As the beginning of all harmonic measurements, the ideal beginning-tone A♭ or the Earth-number $^1\!/_{720}$[8], serves for the intellectual view as the infinitely small $^1\!/_\infty$. Beyond the ideal end-tone G#^[9] of the harmonic Heaven-number $^{720}\!/_1$ stands the infinitely large $^\infty\!/_1$.

In the sequence of twelfths[10] of the rising major and falling minor chords, there arise from the A♭ chord at the bottom the harmonies E♭ major, B♭ major, F^ (F) major, C major, G major; then rising to the following twelfth through the central tone D, they cross the descending minor harmonies of the other side, so as to meet there the harmonies D minor, A^ minor, E^ minor, B (B^)[11] minor, F#^ minor, and C#^ minor, and be affiliated to the major chords of the same name from these fundamental tones. From the C#^ minor chord emerging from the parent-tone of the high G#^, the minor harmonies just mentioned proceed downward so as to cross at D with the major harmonies of the other side, in the lower half of the structure; the aforesaid first six major chords emerging from the A♭ affiliate with the minor harmonies of their fundamental tones.

If the frequency of the parent-tone, low A♭, is taken as the independent measure of an arithmetical progression and as such expressed by 1, the ratios of the said major harmonies proceeding up from the A♭ major chord will give a sequence of arithmetical progressions whose members all take the form of multiples of the chief series by the positive powers of three, $^3\!/_1$, $^9\!/_1$, and $^{27}\!/_1$, etc. And if one takes the frequency of the parent-tone of high G#^ as the new measure $^1\!/_1$, of a descending harmonic series, then the first six fractions of the negative ternary powers $^1\!/_3$, $^1\!/_9$, $^1\!/_{27}$, etc., of this falling chief series will furnish the fractional expressions for the ratios of the minor chords proceeding from the latter's sequence of descending twelfths. But each of the derived arithmetical *perissakis-perissos*[12] progressions 1 x 3, 2 x 3, 3 x 3, 4 x 3, etc., 1 x 9, 2 x 9, 3 x 9, 4 x 9, etc., 1 x 27, 2 x 27, 3 x 27, 4 x 27, etc. begins from the ideal standpoint, with 1/∞, like the simple *perissos* sequence 1, 2, 3, 4, etc. For the zero point or first term of these derived series is always an imaginary term: never the first number (e.g., 0 x 3, 1 x 3, 2 x 3, 3 x 3, etc., 0 x 9, 1 x 9, 2 x 9, 3 x 9, etc.).

The *artiakis-artios*[13] progressions, on the other hand, all begin like the primary series with ∞/1, because the true beginning term of any of them is an imaginary term of the form 1/0 x n. The six chords on the left, belonging to the tonally "male" major mode, A♭ major, E♭ major, B♭ major, F^ (F) major, C major, G major (and naturally also the major chords proceeding upwards in the upper half of the diagram) and the tonally "female" minor harmonies C#^ minor, F#^ minor, B minor, E^ minor, A minor, D minor, proceeding from the parent tones G#^, C#^, F#^, B (B^), E^, A^ (and naturally also the minor harmonies proceeding by lower twelfths from these minor chords in the lower half of the diagram) appear, like the chief series itself, as so many images of ideal, boundless types of harmonic number, albeit presenting themselves through sense-perception to the human intuitive power only within the narrowest bounds (namely the bounds of the senary). They are the finite embodiments of infinite Aeon-series.

Even the crazy fantasies of heretical Gnosis[14] concerning the six male-female syzygies of the Dodekas of Aeons, generated in second place by Anthropos and Ecclesia, rest on the content of speculative theorems of harmonic number-doctrine; but like the other senseless allegories of the adherents of Gnosis that relate to this, they are grounded only on one of the traditional meanings of the mystical as well as technical content of this symbol, and are equally distant from it, being a mindless and arbitrary perversion of the same.

From the Pythagorean Table to the Planetary Spheres

If one writes out the notes of the tone-diagram developed above over the range of two octaves and a whole-tone,[15] placing the reciprocal terms of the left and right-hand sides at equally spaced intervals along a straight line coming from the crossing-point at the center, and then draws a system of concentric circles around the center with their radii as the distances of first the note G1, then the note A^1, and successively the notes B1, CC^, D, EE^, F^, and G, then the outermost circle will connect the note a^ at the other end with the note G1. The second circle, going through A^1 on the left, will meet the note g on the right. The next circle will connect in turn the notes B1 and f^, CC^ and ee^, D and d, EE^ and cc^, F^ and B, G and A^.

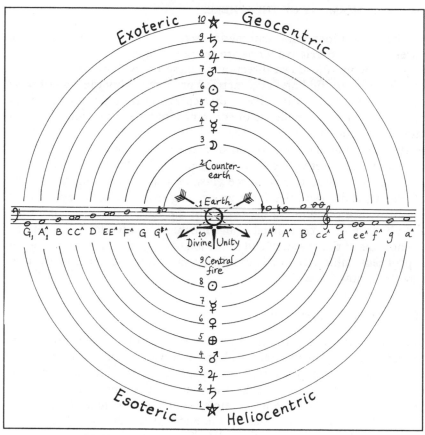

FIGURE 29: *Two Interpretations of the Cosmos*

If one begins numbering the circles thus drawn from the center outward, and places in the middle the sign of the Oth-Aleph[16] —$\underset{'}{\overset{O}{-}}$—, then the ring of the *coronula* of this Taw-letter can be regarded as the innermost and first of the circles. A circle drawn through the half-steps of the middle G#^ and A♭ would represent the second of the rings; the third would then be the circle connecting the diazeuktic[17] notes G and A^. The outermost one, drawn through the proslambanomenoi G₁ and a^, would be the tenth, counted in this manner.

One can also regard the ten concentric circles of the diagram as the plane projection of as many concentric spheres, as in the world-system conceived by Philolaus, to be treated shortly; also as the ten spheres of the Sephiroth nesting within one another, as conceived in the theosophical-cosmogonic allegories of rabbinic Kabbalah.

The notes of the scales constructed out of two octaves plus the diazeuktic tone which served for the above development of the diagram are, in their number ratios, the representatives of two rival series of numbers[18]—an increasing arithmetical series and a decreasing harmonic series—whose correlative terms are related to one another as reciprocal values. The up-and-down direction of these notes, both forms of progressive quantity finding in geometrical proportionality the laws of their balanced adjustment, can serve as the representative types for the interplay and the interdependence of contrary categories of various forms of movement related to one another, and also as a picture of the contrary relationship and the cooperation of archetypal cosmic forces, fighting with each other yet acting as complementaries.[19] One might think here of the effect of an attractive force grasping matter according to the law of gravity, and a repulsive force proceeding from the contrary effect of the resistance of colliding masses; or of the school theorems of a time already past of a centripetal and centrifugal motive-force; or of the phenomena of the radiation and reflection of light, depending on whether it proceeds unhindered or is caught by an opaque obstacle; or of heat, now free, now latent.[20]

Ancient Heliocentricity

Several sayings of the ancient Pythagorean school, handed down to us in fragments by later writers, and the observations on the cosmic system of Antiquity appearing in several places in the Platonic writings, leave us in no doubt that the diagram we have described, or at least one very similar to it, was the harmonic symbol which formed the basis for the astronomical speculations of the ancients on the so-called *musica mundana*. The reporters of the following age, in their exoteric efforts at explanation, understood the incomplete apothegms of the school on this subject, which apparently came to them without the appropriate figures, in a childish way: they took them in all seriousness as an application of concrete musical numbers to the distances of the individual planetary spheres from one another, and as sounds actually produced by the planets and the heaven of fixed stars rotating in universal space.

Curiously enough, they could not get it clear whether according to the doctrine of the ancient Pythagoreans the outermost of the planet-spheres, namely that of Saturn, or the innermost, that of the Moon, gave forth the deepest pitch, the *Hypate* of the bottom tetrachord of the Greek tone-system, and whether therefore the Moon, or on the contrary Saturn, should be assigned the highest pitch corresponding to the *Nete* of the tone-system. If they had had a correct notion of the technical methods of development of the diagram, they would have understood that if the outermost ring, joining the two *proslambanomenoi* (G_1 and $a^\wedge$) with each other, were assigned to the heaven of fixed stars, and the next ring, joining the lowest tone of the tetrachord $A_1^\wedge$ B_1 $CC^\wedge$ D with the highest one of the tetrachord d ee^ f^ g, to Saturn, but that the Moon would be placed on one of the inner rings, perhaps on the ring of the diazeuktic tone GA^, then in the lower half of the diagram Saturn, regarded tonally, would assume the lowest place and the Moon the highest, but in the upper half, on the contrary, Saturn would be highest and the Moon lowest. Hence they busied themselves all the more thoroughly with the question of the remaining conditions of the spheres of this cosmic sphere-system, and Aristotle himself seems to have professed the opinion that they are of a firm impenetrable nature and formed from a glasslike material.

It is self-evident that the esoteric doctrine of Antiquity made use of the harmonic cosmic diagram which concerns us only as a symbolic indication of the numerical laws of regular movement which also appear in the laws of musical sound-relationships, at the same time providing it as an *ideographic hieroglyph* for the logical connection between this musical law and the analytical laws of the cosmic structure in general. If the contemplation of the students were to be directed to the phenomena of the *apparent* movements in the starry sky as they strike the senses—what is up and what is down, the risings and settings, the fixed points of the Pole, the great circles of the heavenly sphere, the axes of the hemisphere of the sky defined by these circles, the hemispheres and zones of this sphere and the intersection of the circles and their axes—then they would think of the Earth as being placed in the center. The center of the Earth would then be the center of the Universe and the crossing of the Arrows of the *Teli*[21] then depict the crossing of the two world-axes in the center of the Earth. In this sense, and in this only, Job could say of the Earth, with regard to the crossing of the arrows, "He hung the Earth up over the void and *crossed* it on nothing."[22] The theory expressed by this sentence is not a placement of the Earth in the middle as the center of our solar system or as the support of the universe, any more than our astronomers today are giving up the Copernican system and returning to the Ptolemaic when they speak of risings and settings, the right ascension, declination, and culmination of the fixed stars, the sidereal periods of revolution of the planets, the Sun's path in the sky, etc. In the middle of the world, so long as the term "world" is understood as meaning our planetary system, is placed, *esoterically*, the Sun: but in a higher speculative meaning connected with the word "world" עוֹלָם comprising the whole Creation, not just the Earth אֶרֶץ and our own heaven consisting of the visible stars, there is a Central Sun invisible to us around which our Sun circles, together with other suns, counter-suns to our own.

In regard to such a fire in the center and to our visible Sun as the image of this invisible Central Sun, Psalm 18:6 says of the Creator:

"He has placed his tabernacle in the Sun"—but we will see in the *Sepher Yezirah* that in the theosophical-cosmic symbolism of Hebrew wisdom-doctrine the midpoint of our diagram is called "the Abode מעון of the Maker," "the Holy Dwelling" מעון קדש and "the Palace of the Sanctuary in the Center" היכל הקדש באמצע. The consequence to be drawn from this parallel for the esoteric view of the world-structure is self-evident.

The Pythagoreans, as will soon be shown, had also told the esoterics something of a central fire in the middle and a hearth of the world-fire. But in order not to reveal the secret teaching of the school they said teasingly that around this central fire there turned not the Sun and a Counter-Sun, but our Earth, perpetually (and very rudely!) turning its backside to the central fire, and a Counter-Earth perpetually on the opposite side of the central fire and hence never visible to the Earth's inhabitants. But the central fire is still supposed to be the actual source of light and heat for our half of the Earth, for its beams are captured by the porous, mirror-like discs of the Sun and Moon, and by the reflection of these astral light-sifters is thrown back to our half of the Earth, always turned away from the central fire.

It is obvious, in our opinion, that these idiocies do not contain the true cosmic doctrine of the Pythagoreans or of Plato. There is in the *Eclogues* of Stobaeus a report of Philolaus' dissertation regarding the cosmos which, though originating unmistakeably from a merely exoteric source, nevertheless contains so much that is correct that it furnishes a basis from which fairly certain conclusions on the actual situation of this branch of Pythagorean doctrine can be drawn. It reads: "Philolaus placed Fire in the middle around the center, which he called the 'Hearth of the Universe,' and the 'House of Zeus,' and the 'Mother of the Gods' [i.e., the heavenly forces and powers], also the 'Sacrificial Altar' *(bomon)* and the 'Bond' *(synochen)* and the 'Measure of Nature.'" Note here that Philolaus also assigned to fire another place (to be described shortly) "above the Enclosure." Then the report continues: "He designated as first-born the middle one, circled by ten bodies of a divine quality: by the fixed-star heaven of Ouranos, by the planets, after these by the Sun beneath which the Moon follows, but beneath these by the Earth and by the Counter-Earth situated under the Earth. Lastly comes fire, which also assumes at the midpoint the position of hearth for the universe." The uppermost (outermost) part of the enclosure, in which the archetypal elements are found in their perfect purity, he called *Olymp*, but the space this side of the path of Olympos, to which the five planets as well as the Sun and Moon are assigned, he called *Kosmos*. That part next below the latter, which extends beneath the Moon and around the Earth, and comprises the realm of the changeable and unstable existence, he is said to have called *Ouranos*.

According to this presentation of the Philolaic teaching, which is doubtless not at all to be taken as that of the Pythagorean School, the Pythagoreans distinguished three separate parts of the cosmic system, of which they called *Kosmos* only the portions assigned to the planetary system and, as the exoteric reporter believes, to the Sun and Moon, likewise the planetary courses of the nesting spheres; but to the middle region, which the reporter, from his standpoint of misplacing the Earth and Counter-Earth in the direct vicinity of the central fire, calls exoterically "hyposelenic" (situated beneath the Moon) and "perigeic" (extending around the Earth), he gives the name *Ouranos*, i.e., Heaven; and to the universal space beyond the outermost sphere of the fixed stars the name *Olympos*, i.e., again virtually Heaven.

If we reinterpret the exoteric content of the report in Stobaeus in the esoteric sense, by placing the Earth at the Sun's position on one of the concentric spherical layers of the planetary spheres and substituting our Sun for the Earth, Moon, and so-called Counter-Earth in the vicinity of the central fire (the Moon, being merely a satellite of the Earth, is not assigned any of the planetary spheres), then in our diagram, if the circle joining the two Proslambanomenoi G_1 and $a^\wedge$ represents the sphere of the fixed stars, the space lying outside this circle is to be taken as Olympos, and the three first rings $A^\wedge_1$-g, B_1-$f^\wedge$, $CC^\wedge$-$ee^\wedge$, as the so-called upper planets Saturn, Jupiter, and Mars. The fourth of these six ring-spheres, D–d, will belong to Earth. The fifth and sixth can be assigned to the two so-called lower planets, namely $EE^\wedge$-$cc^\wedge$ (the sixth, counting from the fixed star-sphere G_1-$a^\wedge$) to the planet Venus, and $F^\wedge$–B (the seventh, counting inclusively from the fixed stars) to Mercury.

We have seen that, just like the inexpressible number belonging to the *harmonia aphanes* of the ekmelic middle tone of the diagram, the real inexpressible numbers of the ratios of the primary notes D and d at the ends of the Dorian-Hypomixolydian octave-scale become the center of the mean proportions of endless involutions of continual geometric proportions.[23] For if we substitute here, for brevity, the musical letter-names for the numbers of the notes concerned, as $D^\vee$ is to D, so D is to $D^\wedge$; as $C\#^\wedge$:D, so D:E♭; as $CC^\wedge$:D, so D:$EE^\wedge$; as B_1:D, so D:$F^\wedge$; as $B♭_1$:D, so D:$F\#^\wedge$; as $A^\wedge_1$:D, so D:G; as $A♭_1$:D, so D:G#^\wedge$; as $G\#_1^\wedge$:D, so D:A♭; as G_1:D, so D:$A^\wedge$, etc.; and the selfsame proportions will appear around d, only an octave higher owing to the doubling of the previous numbers, as we see completely expressed in the tone-system of the Ancients by this order of notes and half-steps.

Since the popular view, stemming from what strikes the eye but approved even by an Aristotle, regards the Earth as the center of the world, the ring D-d serves, for the purpose of this play of proportions, to be shown to the exoterics as the position of the Earth. One can therefore explain the hieroglyphs of the diagram as if D represented the Earth and d the so-called Counter-Earth, but G#^-Ab the central fire. And if, replacing the crossed arrows of the center at G#^-Ab for form's sake with D-d, one were to draw from this last point new systems of concentric circles determined by geometrical proportion, one could also obtain two rings G#^₁-Ab and G#^-ab on which then the Sun, instead of standing still in the center, might move in orbit around D or d. And in the remaining rings of one or other of the two new ring-systems the exoterics could in any case be shown the symbolic representations of the spheres of the Moon, planets, and fixed stars rotating around the Earth and Counter-Earth: if for example in the following tone-scale one connected, with circles concentric around D, the first member G#^₁ with Ab, the second member Ab₁ with the penultimate G#^, and so on in pairs: A^₁-G, Bb₁-F#^, B₁-F^, CC^-EE^, C#^-Eb, and D°^-D^:

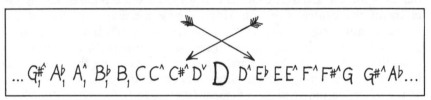

FIGURE 30: *The Scale Centered on the Earth*

If the hypothesis put forward here on the esoteric content of the cosmological doctrine of the Pythagoreans and on the relevance to this doctrine of the symbol that can be developed out of the Dekas-scale is a valid one, the ring[24] drawn through the notes G-A^ of the diazeuktic tone in the middle will represent the division between the planetary region of the cosmic system, called by Philolaus by the name Kosmos, and the solar region to be understood under Ouranos: the inner one in the esoteric sense of the word. The diazeuktic ring G-A^ then appears as the place of the Sun. Because it contains the rings of the mysterious G#^-Ab of the chroma in the center and the crown of the Oth-Aleph (we will later notice that the latter symbol, belonging to the Hebraic-Semitic wisdom-doctrine, also represents the midpoint in the secret Pythagorean teaching), it would serve very well as a symbol of the immeasurable orbit, present only to speculative thought and evading all perception and observation, that our visible Sun (and with it probably other suns, too) describes around the Leader in the fire of the center, whose position the ring G#^-Ab signifies. The coronula of the Oth-Aleph, as the innermost of the ten rings, then appears as the transcendent symbol of the "One in the Fire of the Center," i.e., as the position (if this expression is permissible) and image of the divine, creative Unity itself. The words of the *Sepher Yezirah* refer to this symbol:

> Ten numbers besides the inexpressible Nothing; ten and not nine, ten and not eleven; understand with wisdom, be wise with understanding; prove them and search them out; and place the signified (דבר literally translated, "the Word" [*ton logon*]) in its isolation, and bring the Creator back to his place.

50

✦

Isaac Rice
1850–1915

Born in Bavaria, Isaac L. Rice came to Philadelphia when he was six years old and remained to become one of the most remarkable of American music theorists.[1] His education included three years in Paris (1866–1869), where music was one of his main subjects, and a degree from Columbia Law School (1880), where he taught for six years before becoming an independent counselor and lawyer. Amassing a fortune, he devoted himself to electrical inventions, magazine publishing, and chess, leaving a lasting reputation in the latter field as inventor of the Rice Gambit.

It is extraordinary that Rice, at the age of twenty-five, should have developed an approach to music quite at variance with the attitudes of his time. *What Is Music?* is a Pythagorean work, in the sense of regarding music as a universal and cosmic phenomenon as well as an emotional and acoustical one. One can only suppose that he came into contact with such ideas in Paris. But far from being a high-flown mystical theorizer, Rice supports his views with impressive erudition, both in the artistic and scientific fields, showing in particular a knowledge of the music of the Oriental nations. He has a chessplayer's combination of analytic intellect with intuitive flight.

In the first of these chapters, Rice presents to his audience the full range of *musica mundana* just as Boethius defines it: the music of the stars, of planets, of times, and of the seasons of the earth. His following chapter, on vibration, explains the foundation of the universe as a continuum of which we perceive two very different portions by means of the eye and the ear. In the twentieth century, this observation has become a commonplace of speculative writing on music. In "Colors and forms" he continues the eye-ear analogy, which is also that of Space to Time. We have seen the reciprocal relationship of these two framers of human experience illustrated in the Pythagorean Table of von Thimus (see no. 49); the succession of tone-numbers can be interpreted either as increasing lengths of string (hence progressively lower pitches), or increasing frequencies of vibration (rising pitches). The two series are perfect mirrors of one another. Rice would surely have enjoyed this tabulation if it had been available to him. Finally, with "Internal government," Rice touches on the two favorite themes of our subject, taken in its larger dimension: Pythagorean cosmology, and the dual forces of nature. He makes a laudable effort—all too rare in our collection—to relate the latter directly to "real" music, choosing two moments in Beethoven. Thus speculative music becomes a kind of lens through which to appreciate the subtleties of compositions—and perhaps that is, in the end, its purpose for our time.

Source: Isaac L. Rice: *What Is Music?* (New York: D. Appleton and Co., 1875), pp. 58–85.

Space and Time (Rest and Motion)

Space and Time are the prime elements of the cosmos. The genesis of Nature may be attributed to Time acting on Space. What they are, what may be their true essence, whether they be real or ideal, whether they be things or merely names, whether they exist or be nothings—all these questions concerning them in the absolute do not come within the scope of the present little work. I shall merely consider them from the relative point of view, and only so far as is necessary to define the position of music in the cosmos. For this purpose let us cast a glance at their *fundamental characteristics.*

Space is rest, Time is motion. Space is lifeless, Time is life. Space is rest, lifeless, yet in consequence knows no death. Time is motion, life,

yet what is life but change, and what is change but death? And still—
eternal antithesis!—though in all things the contrary of each other,
they are yet the counterparts of one another; though by their very
definitions the opposites, they are yet wonderfully similar; though,
subjectively, essentially antagonistic, they are, objectively, but different
manifestations of an identical idea! For what is Time? It is the Space
of motion, the Space of existence!

Both exist forever. In Space, this forever is called infinity; in time,
eternity. Yet each particle of Space is infinitesimal but eternal, while
each particle of Time is eternitesimal but infinite. Each particle of
Space is infinitely small, but it remains the same, unchanged through
all ages; each particle of Time is infinitely short, yet it extends through
the whole universe, immeasurable even in the imagination—it is infinite! Space is the limitation of matter; all things material must occupy
some space. Time, however, is the limitation of the spiritual, our very
thoughts are bounded by it. Space may therefore be considered as the
essential limitation of all things material—Time the essential limitation
of all things whatsoever. Ideas consequently do not exist in space—
ideas are life, they exist in time. The inert matter only exists in space.
Matter is lifeless, inactive, put in motion by forces. Forces are ideal,
they exist in Time. Consequently time and Space constantly act upon
each other. But Time is but another name for Space—it is the Space of
motion.

Now, *what is music?* The *beautifier* of *Time*, is the simple and categorial
answer—an answer, too, from which further answers to all questions
springing from the original question may be deduced; an answer that
serves as the corner-stone of the fundamental theory of music itself. It
is to adorn the ever-moving Space of existence that music was generated and the germs of its development were placed within it. In the
Space of Rest, in visible Nature, Nature itself has undertaken the task
of beautifying. And there she has lavished beauties untold and unnum-
bered. Beauty reigns on the mountain and in the valley, on the hill and
in the dale. It is present in the gentle grove as well as in the mighty
forest. It is in the little brook and in the magnificent ocean. It is in man
and woman, in the birds, in the plants—anywhere, everywhere, it meets
our eyes, if we will but see. There are beauties of all kinds and degrees,
from the sublime to the graceful, from the magnificent to the picturesque. All this has Nature done for Space—and to do something similar
for Time is the grand and holy object of music.

The materials of which music is composed exist only in Time, and
here we have the explanation of many of the characteristics of music.
Time is motion, is life, yet the sure bringer of change, of death. As it
is motion, its influence upon us is emotional, agitating; as it constantly

tells us of change and death, it awakens the feelings of melancholy within us. Music, as it beautifies the passing moments, yet tells us that they are passing, and consequently it is so prone to cause sadness.

We may divide the pleasures arising from the contemplation of the beautiful into two classes—pleasures productive of joy, and those productive of sadness. There is nothing paradoxical in this division. A thing that is beautiful will give pleasure at all times, though it may at the same time cause sadness. A great tragedy will give pleasure, though it may not put us in a joyful mood. So it is with a beautiful poem on a tragic subject. In fact, intense and exquisite delight is perfectly compatible with a frame of mind strongly tinged with melancholy. Nay, even more, joyful sensations, when they become ecstatic, generally have a background of deep sadness.

Now, the characteristic state of mind accompanying the contemplation of things in Space is that of serene joy. Space being rest, does not excite the more powerful emotions, it does not agitate. It has, on the contrary, the effect of calming and quieting the mind. I am, of course, speaking of the beautiful, purely as such, without admitting of associations of ideas. These, of course, often exercise a powerful influence, and cause emotion by their own force. But they exist as much for music as they do for things in visible Nature, and their consideration at present would only cause useless complications. That beauty in Space has the tendency—and very strongly—to create a serene frame of mind, any one desirous of doing so can easily test. A landscape must be entirely covered with clouds, be exceedingly gloomy, before it causes us to be sad. Let but the sun appear and shine upon the clouds, and they will be tinged with bright colors; the scene will appear more cheerful even than were there no clouds. The opposite is the case in music. Often a *single* minor or diminished chord, introduced into a gay melody, will change its entire expression, rendering it melancholy. Space knows naught of death, its particles exist forever; its beauties are therefore prone to create joy. Time speaks constantly of change and death; its particles are infinitely short, its beauties create sadness.

Then, again, as Aristotle has already said, what are the emotions but motions? And as music is motion, its effect on them must be great for motions exercise an enormous influence on like motions, and have a very great tendency to respond to like motions. This is a fact well known to all familiar with the operation of vibrations. "But," it may be objected, "the eye is also an organ capable of discerning motion." To this I answer, that we are at present concerned only about the *beautiful* in motion, and that this is chiefly the province of the ear. The *beauties of motion* open to the perception of the eye are of an inferior kind. The pleasures in viewing dancing or marching do not really come within the range of those caused by the contemplation of the beautiful in the highest sense. In the motion of the waves of the sea, we are more impressed by the natural association of ideas than by the beauty of the motion itself.

There is, indeed, another, higher kind of motion in visible Nature—the motions of the heavenly bodies. Daily the sun rises, and tranquilly and majestically pursues its course in the firmament, to set amid splendor and glory. Then the stars appear, and with equal majesty traverse the skies, set and rise until the king of day again ascends from the horizon and eclipses them by the exceeding power of his light. Nor is this motion limited to the day and night; as the year moves on, sun and stars move with it. This month the sun rises in one sign, the next month in another, until he has traversed the whole circle of the zodiac. This month Arcturus is the proudest of the starry host shining high above; in the next he is already dethroned, and bright Antares for a brief time assumes his honor. But the beautiful imperial Lyra follows in the wake, and in her turn claims homage as chief of the stars. Less steady wanderers, too, are there in the heavens, the planets moving unconcerned in their orbits, now visible here, now there. The lovely moon, queen of the night, pursues her tranquil course. Now seen but as a silver thread in the west, she waxes lovelier and prouder as she approaches the east, until she almost rivals the sun in the refulgency of her light—but it is only to wane and wane again, until she is seen no more.

There is, too, the subtile motion of the seasons. Now the forest is in the garb of a beautiful green, the garden is fragrant with flowers, the trees are loaded with fruits, the fields teem with the heaving corn. Soon the green changes into numerous varieties of color, the leaves fall and strew the ground, the flowers are plucked from the garden, the corn is gathered from the fields. Then comes winter; snow covers the ground, the water-courses of the mountains and the rivers of the valleys are turned into ice, cold and bright, the mild breezes give way to the fierce blasts of the storm. But spring follows close behind, and wafts the breath of life before him. The snow melts, the ice thaws, the mountain-torrents tear on with renewed and tenfold increased vigor; the pulse of Nature throbs with the freshness of youth. Soon all is again in bloom, the trees are white, the plants begin to shoot forth. Then summer is here once more, and the course of the year begins anew.

The motions of the spheres and of the seasons are, indeed, full of sublimity. The ancient philosophers and their followers unto recent times, however, saw in them yet the workings of music. All these motions were to them but the visible manifestations of a transcendental harmony. Therefore does the Pythagorean say, "It is the business of music, not only to preside over the voice and musical instruments, but even to harmonize *all things* contained in the universe." Therefore does the Scholastic exclaim, "The music of the universe is a great unity, and by command of God it governs all things in motion—all things that move in heaven, or on earth, or in the sea, all that which sounds in the voices of men and animals—*it is the regulator of days and years.*"

A similar sentiment inspired Shakespeare when he wrote:

> . . . Look how the floor of heaven
> Is thick inlaid with patines of bright gold;
> There's not the smallest orb, which thou beholdst,
> But in his motion like an angel sings,
> Still quiring to the young-eyed cherubins.
> Such harmony is in immortal souls;
> But, whilst this muddy vesture of decay
> Doth grossly close it in, we cannot hear it.
> *(The Merchant of Venice* V. 1)

If I were asked to give my own views on these motions, I should say that I do not believe that they properly come under the head of the beautiful in motion. If we admire the landscape, or look up to the starry vault of heaven, our purely aesthetical pleasure does not take in the factor of motion. It is continuous and imperceptible to the eye—it is only by the aid of memory that we know that it exists, and the feeling of awe connected with it arises solely from the *association of ideas.* Indeed, if we wish thoroughly to examine an object in Space, we require it to be perfectly at rest—its motion as a rule has a tendency to confuse us; and, if the motion be rapid, the object becomes blurred. The very life of the beautiful in audible Nature, however, is motion—it exists in Time and not in Space.

As Time and Space—visible and audible Nature—are the counterparts of each other, there must be great analogies in the manner in which beauty is *perceived* and *produced* in either. Of the analogies of perception I shall now treat, under the heading of vibrations.

Vibrations

We perceive things in visible Nature by means of light—things in audible Nature, by means of sound. To one unacquainted with physics, light and sound are entirely distinct phenomena, having no connection whatever with each other, and yet they are intrinsically very nearly related to each other, being but different manifestations of the same cause. Vibrations of a certain rapidity are perceived by the instrument constructed to respond to them—the ear, as sound; vibrations of greater rapidity are perceived by the instrument constructed to respond to them—the eye, as light. And not only are they produced by the same cause, they are also propagated by the same cause, they are also propagated by the same means—undulations. Entering into details, we find the analogies between the two phenomena in almost all of the principal manifestations. Some bodies are transparent, others translucent, others again opaque to light; in like manner some bodies permit sound to pass on through them without practically enfeebling it; others, like thick walls, transmit it much weakened; while others again do not transmit it to any appreciable degree. An instance of the last case is a tunnel. To any one standing at a distance, the roar of a train entering it is hushed, and remains so until the cars emerge, when it is immediately renewed.

Some of the principal properties of light are absorption, reflection, refraction, and diffraction. These are also the properties of sound. That it may be absorbed can easily be tested, by comparing the sound of a musical instrument in a carpeted and furnished room with that of one heard in an empty room. The echo is a familiar illustration of reflection of sound. The experiments of Sondhauss and Hajech[2] prove conclusively that it is refracted when it enters a medium whose density differs from the one it leaves, in the same manner and under the same conditions as is light. The diffraction of sound has been demonstrated by Seebeck.[3]

It is, however, not only in the physical manifestations of sound and light that we discover great analogies; the construction of the instruments for their perception—the eye and ear—is essentially based on analogous plans. Like the ear, the eye is a membranous structure. The ear is composed of three parts—the auditory canal, with the tympanum, the tympanic cavity, and the labyrinth. The corresponding parts of the eye are the sclerotic coat, the choroid coat, and the iris. The aqueous and vitreous humors present strong points of resemblance with the water of the labyrinth. The difference between light and sound is not in kind, but in degree. Extremely rapid vibrations produce light; slower ones, sound. The rapid vibrations have, however, a proportionately small amplitude; slower vibrations a proportionately large amplitude. Hence the difference in the anatomy of the eye and the ear. The first is prepared to receive and respond to vibrations of enormous rapidity and small amplitude; the latter to receive and respond to vibrations of comparative slowness, but with a relatively large amplitude.

Tones and colors are essentially the *same* things. Colors are tones of tremendous height of pitch. Tones are colors of tremendous depth of pitch. The ear perceives as tones from 8 (Savart) to 38,016 (Dupretz) vibrations in a second. The eye perceives as light from 458,000,000,000,000 (extreme red) to 727,000,000,000,000 (extreme violet) vibrations per second. From the most acute tone capable of being perceived by the ear to the extreme red color there is, therefore, an interval of about thirty-four octaves. To give an illustration of the enormity of such an interval, let us take the length of the string of the highest C of a seven and a quarter octave piano-forte, which is about 1¼ inch, and it will be easy to calculate that a string of the same material and thickness, in order to produce the extreme red light, would have to be cut down to about 1/10,000,000,000 of an inch! The rapidity of the vibrations defines the length of the undulations; the length of a sonorous wave produced by 8 vibrations per second is 140 ft.; the length of a luminiferous undulation in the extreme violet ray is 167/10,000,000 of an inch; otherwise expressed, while of the former there would be but 37½ in a mile, in the latter there are 59,150 in an inch!

Rapidity of the vibrations is, however, the means of distinguishing tones from tones, and colors from colors, as well as tones from colors; and consequently, difference in *rapidity* of vibrations solely, cannot be considered an intrinsic difference.

The principal phenomena connected with colors—analysis and interference—are also proper to tones.

For colors the triangular prism acts as analyzer; for tones that office is performed by *resonators*. Professor Helmholtz has constructed a series of the latter, which serve as analyzers for isolated tones—by resolving them into the fundamental and overtones—as well as for those tones of combination produced by the simultaneous existence of two or more independent tones. Interference in sonorous waves has been demonstrated ocularly as well as auricularly by numerous apparatus.

Having now sufficiently illustrated the identity of the manner of *perception* of the beautiful in visible and audible Nature, I shall proceed to the consideration of the fundamental analogies regarding the *production* of the beautiful in Space and Time.

Colors and Forms

The elements of the beautiful in Space are colors and forms. The counterpart of colors having been found in tones, there remains but the question, "Is there also a counterpart of *forms* to be found in music?" This question I answer categorically in the affirmative: *Rhythm is the shape, form, or proportion of things in Time; and shape, form, or proportion, is the rhythm of things in Space.* And this answer is not based on any arbitrary ideas, but on incontestable facts—facts as indisputable as is the theorem that colors are the tones of Space. Time is but the Space of motion, and rhythm defines that space in the same manner that the

Space of rest is defined by forms. The lines of Space are translated, as it were, into Time, by its means. On entering into an investigation of the prime principles of morphology, we find that the *straight line* and the *curve* are the *fundamental types* of form. In like manner, the fundamental types of rhythm are found in the *dual*, and in the *triple* meter. The geometrical *point* is an impossibility—so is *single* meter. The reason is plain. Rhythm, like form, is based on proportion; in other words, on relativity. We have no perception of rhythm on hearing a single beat. A beat must be defined and bounded by a second one to become a meter, i.e., a measure of Time. The analogy between dual meter and the straight line, and triple meter and the curved line, is by no means a fanciful conception[4]—it has been intuitively felt by musical composers in all times; *and tones spread over rhythms as colors do over forms.*

We cannot, however, overlook the remarkable fact that, while in visible Nature colors play the subordinate and forms the principal part, the order is reversed in audible Nature, where rhythm is subordinate to tones. Well, this is *necessitated* by the fundamental characteristics of Space and Time, rest and motion. In Space, things may remain at rest; our eyes can take in a great variety of forms simultaneously. They have time to examine beauty—extend comparisons over a wide field. Forms and proportions may establish themselves in unlimited variety; for we have *coexistence* on a large scale. Time, however, is motion. In it, proportions and forms are perceptible by their very motion, and *only* by motion; one tone vanishes as the next comes on. Here there is no room for such an extreme variety of forms; rhythms (though they may yet be infinitely varied and complicated) must be much simpler than the forms of visible Nature.

On the other hand, the tones, which constitute the material of melody, embrace about seven and a half octaves, good for practical purposes; while the colors do not extend over more than one octave. This octave, even, is not entirely visible under ordinary circumstances, its eighth degree being that which is called the lavender light of Herschel,[5] and only produced by concentration. Practically, the whole combination of colors does not exceed the interval of a seventh. There are, therefore, conclusive reasons why the chief riches of visible Nature lie in forms, while the chief riches of audible Nature are in tones.

There is, however, another factor besides melody and rhythm that enters into the composition of music—*harmony;* and it may be asked whether any analogy for it can be found in visible Nature. To this question I reply that the fundamental theory of musical harmony lies in the very nature of Time. Each particle of Space is infinitely small, consequently no two things can occupy the same space. Each particle of Time, on the contrary, is infinitely large—embracing the whole cosmos—consequently an infinite number of things can occur at the same time. Space, however, is rest, and the mind can therefore take in a large number of particles of Space at once. Time being motion, does not admit of perceiving more than *one* of its particles at once; and, therefore, the simultaneity, compatible with it, acts as a certain compensation for those advantages which, by its definition, rest has over it. I am, of course, using the word harmony in its narrow signification, in the sense of counterpoint, and not in its spiritual meaning. In the latter higher sense, it pervades the whole universe, existing both in Space and in Time; the soul of the cosmos, says Plato, is musical harmony. The whole topic of colors, and forms, and rhythms, and tones, and harmony, together with the many analogies of detail in these matters, is a tempting and prolific subject for speculation. I shall, however, resist the temptation of going any further into the matter, for it is beyond my scope in this little work to introduce any but plainly demonstrable facts. One thing only I must yet allude to; and this is, that, in instituting comparisons between the beautiful in Space and in Time, we should never forget that in the former case the task of beautifying has been undertaken by Nature itself with the unbounded resources at its command, while in the latter it is left to the limited means of man. Were Nature to beautify Time as it does Space—could we hear, for instance, such a thing as the harmony of the spheres—the sublimity of such music might transcend all possible conceptions.

And now we leave the field of the material analogies, respecting the perception and production of the beautiful in Space and Time, to enter into that of (what may be termed) the spiritual analogies. Those of production I shall class under *Internal Government*, those of perception under *States of Mind*.

Internal Government

Several of the great forces which we see manifested in visible Nature have their counterparts in audible Nature, and prime among these are the forces of *gravity* and *attraction* and the *centrifugal force*.[6] I am not aware that any writer has ever had the boldness to make this assertion in so positive a manner, but certainly the influence of the first over music has been instinctively felt in all times and among all nations, while that of the last two was discovered as soon as it could have been, namely, in the first stages of the development of the science of harmony.

The *center of gravity* of the musical scale is the *tonic*. The whole history of music tends to confirm this in an unequivocal manner. I have before me, as I write, a volume of August Wilhelm Ambros, wherein I find scraps of melodies from the land of the Esquimaux and from the

Friendly Islands, from New Zealand and from Abyssinia, from Gorea and from Senegal. Besides these, a number of finished Chinese melodies, a number of beautiful songs of Hindostan, together with Arabian, Persian, and Turkish airs, and two of the three ancient Greek *nomoi* that have come down to us; and in each one of them, from tutored and from untutored peoples, the audible manifestation of the principle of gravity is unmistakably discernible; all the tones gravitate toward their common center—the tonic. A characteristic passage from the writings of Aristotle proves indeed, beyond doubt, that the consciousness of the force of the tonic was not only apparent in the practice of music of the ancients, but that they were also aware of its spiritual relation to the other degrees of the scale, and attempted to account for it philosophically. "Why is it," he asks, "that, when the tonic *(mese)* is changed (sharpened or flattened), all the other strings sound out of tune, but, if the tonic is in tune, and one of the other strings is changed, only the changed string sounds out of tune? Is it because not only all the strings are tuned, *but also that they are tuned with respect to the tonic,* and that the latter defines the order in which they appear? But when the basis of the tuning and that which keeps (the melody) together is taken away, there can no longer be the same kind of order." But should the reader be disinclined to accept the testimony of a single person, no matter of what importance it may be, I have still another powerful proof in support of my argument to bring forward. Let us cast a glance into pre-Ptolemaic astronomy, and what do we find? That the prime principles of modern astronomy, those contained in the Copernican system, were essentially known and taught in the sixth century B.C. by Pythagoras.[7] His doctrine was, that the sun is the center of the universe, and that the earth has a diurnal motion around its axis, and an annual motion around the sun! Now, we have already seen (in an earlier part of this little work) that the planets and the sun were compared and considered mysteriously related to the tones of the scale. And the sun, the central sphere, was suposed to be the *mese* of the scale—the manifestation of the principle embodied in the tonic. Cicero,[8] however, did not believe in the Pythagorean doctrine of the revolution of the planets around the sun. He was of the opinion that the sun and planets revolve around the earth, which remains stationary. And, in consequence, too, he changed the Pythagorean division of the scale among the heavenly bodies: the sun was no longer *mese*, it became simply the *lichanos hypaton.* But what did he do with the *mese?* It could not be given to the earth, because she, being stationary, represented silence; so he made *mese* symbolical of *the whole expanse of the firmament!*[9]

In the ninth century, Hucbald de St. Amand worked out his "Organum," the first step toward modern harmony. One of the first results of his discovery was the introduction of the *leading note* into the musical scale—the forced recognition of the second great governing principle of the scale—*attraction*. By-and-by *retards* came into use, and showed that this attraction acted in a dual manner—upward and downward. The analogy between it and magnetic attraction with its two poles must I think, strike every one—the attraction of one pole manifested in the retard and the resolution of the seventh, that of the other in the *alteration* and *leading-note*.

The *dominant*, which is in all respects the contrary of the tonic, *is the audible manifestation of the centrifugal force*. In our modern system of harmony, where the principle of tonality is fully understood and recognized, the intensity of this force is even considerably increased by the fact that the dominant is likewise the tonic of the next related key, and consequently as such exercises an extraneous attraction, tending to oppose, by a secondary gravity, the force of gravity in the tonic.

And now, remembering that the emotions are motions, and consequently in sympathy with like motions, we cannot but be convinced that the foregoing facts serve as explanation for many of the characteristic effects which music has on our emotions. The center of gravity, manifested in the tonic of the musical scale, is likewise manifested in the *emotions expressive of satisfaction*. I am using this word, not in the sense of contentment, but in contradistinction to the term suspense; this satisfaction need, of course, not have any gay or even cheerful sentiments connected with it; it may in fact be accompanied by extreme despondency: it is but the relief from suspense, or typifying suspense as the question—it is the answer. The centrifugal force is manifested in the dominant, and likewise in the *emotions expressive of suspense*; it is typified in the question. The perfect cadence is universally, and I may say intuitively, recognized as the only manner in which a composition can be satisfactorily closed. But what is the perfect cadence? It is a chord built on the dominant of a key followed by one on its tonic. And what is the reason that it is the most satisfactory manner of closing a composition? This question is easily answered by a consideration of the data that have just been adduced. The satisfaction is most intense when we have tasted the suspense to its extreme limit; the answer is most complete when it follows the question directly, taking it in to its fullest extent.

Of course, there are different degrees of satisfaction. If the tonic alone is employed and merely doubled in the higher parts, the satisfaction is perfect; if the tonic occurs in the highest part, as well as the lowest, it is nearly perfect; if the mediant is heard in the highest part, a feeling of vagueness—often charming—is superadded; and if the dominant is sounded in the highest part, the vagueness is considerably augmented:

the mixture of suspense with the ground feeling of satisfaction creates, in fact, a peculiar weird impression, easily and (as I believe) only explicable by considering the forces that I have asserted to be manifested in the tonic and the dominant.

Before closing this part of my argument, let me refer the reader to two powerful passages taken at random from the works of Beethoven, where he will find a remarkable corroboration of my conception of the tonic, dominant, and attraction. The first consists in the closing measures of the "Largo Appassionato" of the Sonata Opus 2, No. 2. It is impossible to describe the apathy expressed in that cadence. Yet, by analyzing it, according to the principles which I have set forth, we can explain its effect on natural grounds. The A in the bass is the dominant of the key—the manifestation of suspense—and the mind expects it to move on to the tonic toward which it gravitates. It does not do so, however; and, though the upper parts have already entered the domains of the tonic, the bass still clings to the dominant as though in complete abstraction. The upper parts attempt to console—to urge it to abandon the dominant. It does so, but only to return to it as though to a forlorn hope. Then the upper parts finally move on to the tonic and remain there, and so nothing is left for the bass but to follow. But it does so reluctantly, tardily, as if awakening from a reverie. The emotions of the hearer respond to all these movements, and hence are affected as they are. The second passage occurs near the close of the first movement of the Seventh Symphony. It is expressive of an intense longing that can never be satisfied, of a passionate yearning for the unattainable, or to use a magnificent figure of a modern German poet, it is like "the love of the sea for the moon." This is due to the conflict between the forces of attraction and of the tonic. A seventh is attracted to the degree below it. Here we find the seventh inverted in the bass and in consequence the chord cannot resolve itself on the tonic, but must do so on the chord of the sixth on the mediant. This chord is, however, unsatisfactory, and can never be the concluding one; and the vain attempts of the inverted seventh to resolve itself satisfactorily, repeated and repeated with obstinate fervor, though warned and entreated by the pleading tones of the upper part, constantly obtaining the same discontenting answer, which it is fated to receive, and which it knows that it must receive, is the picture of fervent hope doomed to eternal disappointment.

51

Saint-Yves d'Alveydre
1842–1909

Born Alexandre Saint-Yves, this French esotericist was also a gifted musician and composer. After an unsuccessful career dedicated to social and political change through his own scheme of "Synarchy," Saint-Yves went into semiretirement where he elaborated, step by step, a system of universal correspondence called the *Archéomètre*. He was convinced that he had rediscovered in it the lost science of Antiquity, and since the system pivoted around the names of Jesus and Mary, this supported Saint-Yves' conviction that a premonition of Christ had been an essential part of this sacred science since prehistoric times.

The Archéomètre is based on a circle divided twelvefold. Each compartment has a sign of the Zodiac, a letter of the Hebrew alphabet, a number, a color, a planet, and a tone. One of Saint-Yves' practical realizations based on the scheme was the *Archéomètre musical* (Paris, 1909), a series of over two hundred short piano pieces in the seven planetary modes, most of them perfectly diatonic and using melodies based on the different intervals (seconds, thirds, etc.[1]). While Saint-Yves' free compositions are more or less Lisztian, the *Archéomètre musical* resembles no music of its own or any other period.

Saint-Yves' notes and papers relative to the Archéomètre were collected after his death and published in 1912 by a committee of "Friends of Saint-Yves" led by Papus (Gérard Encausse), but they did not include this short essay from one of his notebooks. Although some details of it are cryptic, necessitating a familiarity with Saint-Yves' personal universe, the main points are clear enough. The author is a Christian Hermetist of the traditional type, the same as Ficino, Giorgi, Saint-Martin, and many others in this collection. He believes in Christ as the creative Word, responsible for a cosmos whose every level is linked by correspondences, and of which the material world is the lowest. Therefore our audible music is but a pale reflection of the true music of higher spheres. A key to the latter is to be sought in number, which alone allows us to correct our musical system by putting it quite literally in tune with the universe. Saint-Yves insists on the Ptolemaic scale-tuning (also adopted by Zarlino), with its harmonic thirds and fifths, because the numbers that enable one to calculate it as lengths of a monochord string are numbers that occur with a sacred meaning in the Archéomètre.

Source: Saint-Yves d'Alveydre, untitled essay in a notebook, Paris, Bibliothèque de la Sorbonne, Ms. 1823 E, folios 214–216. Translated by the Editor.

Music and Archeometry[2]

[214] Music is one of the archeometric equivalents of the living Word. It is the language of Numbers and Forms. Man does not create it: he discovers it; and this discovery is part of the Revelation of the divine, and of divine Love, to human intelligence and consciousness. It is thus because man, having only one mode of reproduction, has only one reflective mind. Man cannot change the fact that the diatonic scale has seven notes, the chromatic (sharp or flat) twelve, the enharmonic (sharp or flat) twenty-two. Man can misunderstand or alter these numbers and their laws; but the Science which reveals them to him, or recalls him to them, simply establishes them, and nothing more. God alone is the expert, just as He alone is wisdom and life.

Twenty-two letters of the solar Alphabets of the Word,[3] twelve zodiacal letters, seven planetary letters; twenty-two enharmonic notes, twelve chromatic, seven diatonic: such is the correspondence of music and of the Word in the living Principle. This Principle is the Word.

This speech and this music are sacred. [214'] Man is merely its receiver, and its unconscious profaner until this science imposes on him humility, knowing Him who alone is the Artist as He alone is the Expert and the Sage.

Music is thus an empirical art only among the barbarians and semibarbarians who demonstrate their knowledge of it without understanding it.[4] Music is a sacred science, and consequently an exact one, and the harmonic pivot of all the sciences, and consequently of all the arts.[5] The artist who lacks this science speaks the language of music like a sleep-talker, when he does not actually blaspheme it by polluting usage. No science is experimental because there is nothing abstract in God: all in Him is living and real. The Fact, the Law, the Principle are indivisible. The Law shows itself in the Fact, the Principle in the Law, and the Fact is experimental.

The sounding string is to audible sound what the prism is to visible light, for there is unheard sound, cause of the audible, just as there is the light of glory invisible to the flesh; but the Law is identical, just as the Principle is. Thus there is correspondence between the unheard sound and the invisible light, both vibrating in creative waves in the pure ether, and the audible sound and visible light [215] reflected by the prism of the astral atmosphere. The difference is that the sounding string and the prism demonstrate to our world of physical reversals and mirages in a dead, rectilineal band that which moves living in sonorous and luminous waves in the real and eternal world of the Principles, whose only atmosphere is the pure ether.

The Archéomètre allows one to establish and experiment with these correspondences, which would otherwise have remained forever a sealed book. The seven seals of planets, tones, and diatonic radii imprinted on the whole astral axis can be opened in no other way.[6]

It is for this reason that the Celts, Indians, Mongols, Chinese, Arabs, Egyptians, Jews, and Greeks offered to advanced students nothing but the confused fragments of a lost musical system, based on the melodic septenary but without exact harmonic correspondence.[7]

It therefore fell to the worshipers of Jesus incarnate to [215'] establish here, as everywhere else, the universal Unity revealed by Him as creative Word to the most ancient patriarchal universities.[8]

The tempered scale must be rejected as false, and that of Pythagoras for the same reason. The scale of Ptolemy adopted by the physicists is the only correct one. But one must verify scientifically why and how this is so. One will then see that this Egyptian scale, on the C string, is only a section of the Sonometer of the pre-Hebraic Patriarchs, worshipers of Jesus the creative Word and King of Heaven.[9]

In the first place one must put aside the traditions of the Greek

philosophers and of the ancient Jewish rabbis, each as unscientific as
the other, not even knowing the true value of their alphabet, much less
the Archéomètre of the prophets and patriarchs, known by the ancient
ones under the name of Wisdom. [216] For if they had known it, as
Saint Paul says—who had a complete knowledge of Judaism and of
Hellenism—they would have understood Moses, they would not have
slain their prophets nor crucified IEVE in his incarnate Word, Jesus.[10]

52

Azbel
1855–after 1917

Azbel was one of the pseudonyms of Emile Abel Chizat, a pupil of Jules Massenet and composer of popular works. In the 1890s he produced in Paris a series of *auditions voilées*, "veiled concerts," in which the auditorium was dimly lit, filled with plants, and the audience sat on sofas listening to invisible musicians and recitations (Chizat wrote the poems as well as the music). Under the name "Hizcat" he published *La légende de l'être Althéus*, a work of mythological history. As "Athénius" he wrote some very pointed articles on the Great War and its roots in the German character. Finally—unless further pseudonyms are discovered—he took the name "Azbel" for his aesthetic and theoretical works, of which the most important is the multidisciplinary *Le Beau et sa loi*, 1899.

This chart of Azbel's system is a suitable conclusion to our collection, for it shows the Harmony of the Spheres come of age and reconciled with modern science. Azbel takes the distances of the planets and asteroids from the sun, both as established by astronomy and as placed according to Bode's Law, and calculates the tones that correspond to them. The coincidences with the harmonic series are too obvious to ignore, and the grand chord that emerges is a symbol of the intelligent harmony that regulates Azbel's universe, as it does Plato's, Kepler's, and perhaps our own.

Source: Azbel, *Harmonie des mondes* (Paris: Hughes Robert, 1903).

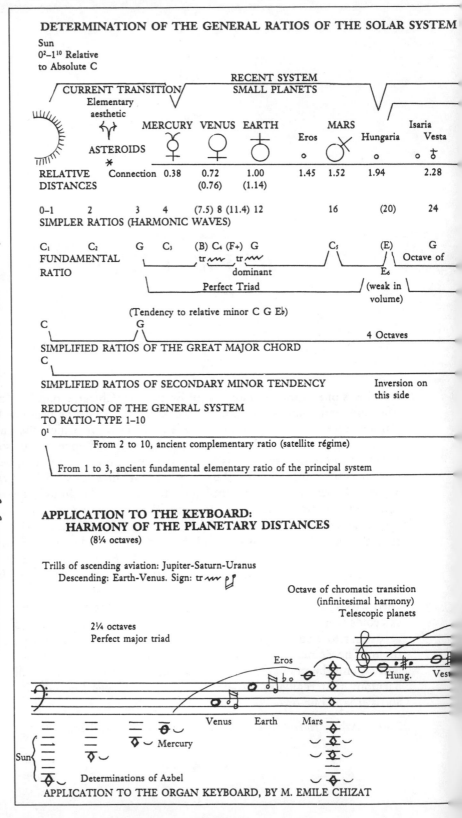

DETERMINATION OF THE GENERAL RATIOS OF THE SOLAR SYSTEM

Sun
0^2–1^{10} Relative
to Absolute C

RECENT SYSTEM

CURRENT TRANSITION / SMALL PLANETS

Elementary
aesthetic

			MERCURY	VENUS	EARTH		MARS		Isaria
						Eros		Hungaria	Vesta
	ASTEROIDS								
RELATIVE DISTANCES	Connection	0.38	0.72 (0.76)	1.00 (1.14)		1.45	1.52	1.94	2.28

0–1	2	3	4	(7.5) 8 (11.4) 12	16	(20)	24

SIMPLER RATIOS (HARMONIC WAVES)

C_1 C_2 G C_3 (B) C_4 (F+) G C_5 (E)   G

FUNDAMENTAL
RATIO   tr~ , tr~   Octave of
                                            dominant E_6

                              Perfect Triad   (weak in
                                                                          volume)

(Tendency to relative minor C G E♭)

C   G   4 Octaves

SIMPLIFIED RATIOS OF THE GREAT MAJOR CHORD
C

SIMPLIFIED RATIOS OF SECONDARY MINOR TENDENCY   Inversion on
                                                                            this side

REDUCTION OF THE GENERAL SYSTEM
TO RATIO-TYPE 1–10
0^1

From 2 to 10, ancient complementary ratio (satellite régime)

From 1 to 3, ancient fundamental elementary ratio of the principal system

APPLICATION TO THE KEYBOARD:
HARMONY OF THE PLANETARY DISTANCES
(8¼ octaves)

Trills of ascending aviation: Jupiter-Saturn-Uranus
Descending: Earth-Venus. Sign: tr ~ ♭

Octave of chromatic transition
(infinitesimal harmony)
Telescopic planets

2¼ octaves
Perfect major triad

Eros

Hung.   Vest

Venus   Earth   Mars

Mercury

Sun

Determinations of Azbel

APPLICATION TO THE ORGAN KEYBOARD, BY M. EMILE CHIZAT

UNIVERSE BEYOND

Tendency ——→ Absolute C

ANCIENT SYSTEM

FIRST SUPERIOR TRANSITION \γ°/ LARGE PLANETS INFINITE 0¹\

TELESCOPIC PLANETS

Emma		Thule			JUPITER	SATURN	URANUS	NEPTUNE
Ceres	PSYCHE	Ismene	CONNECTION					

2.76   3.04   3.84 4.26 4.56   5.20   9.53   19.18   30.00   (73000)
                                        (9.12)   (18.24)

28   32   40   (44) 48   52   96 (100) 192 (202) 320

Theoretical

Bb   C₆   E   (F)   G   A   G₇ (G+) G₈ (G+) E   (G)

transition   /\   Relative   tr〰〰 /\   tr〰 /\ tr〰/\   / of stellar
                    minor of dominant   (Tendency to relative   transition
Modulating chromatics   /   \   minor A E C)   /

Dominant of the waves
of the Great Connection
G   E₉
/\   2¾ octaves   /\
                                                                  C
A₆   E₉
/\   /\

←—— Tonic of Minor Tendency ——→   Minor perfection
                                                 beyond

32   (4.56)   (52) 64 96 128   192 256 320
/1\   /2\ /3\ 4   /6\ 8 / 10
C   C   G   G Complementary
                                   /   \   dominant
                                                        (Uranus)
                              G
                    Fundamental elementary
                    dominant (Saturn)

UNIVERSE BEYOND
Absolute C
8v³0¹

Neptune

Dominant
of the Great
Connection   8v³   Uranus
G₆
                                   Saturn

eres   Psyche Ismene   Jupiter

TOTAL

Notes

Notes by other authors, translators, or commentators are indicated by their initials in brackets.

1 · PLATO

1. A. E. Taylor, *A Commentary on Plato's Timaeus*, p. 106.

2. *Plato's Cosmology*, pp. 59–94.

3. *The Pythagorean Plato*, pp. 57–70.

4. *The Dimensions of Paradise*, pp. 150–169.

5. Soul, throughout this description, is a paradoxical entity, partaking of the incompatibles which it is its task to unite. As Plato says, "two things alone cannot be satisfactorily united without a third; for there must be some bond between them drawing them together." (31b–c) The whole "Timaeus scale" is an extension of this principle of placing bonds or means between things: in the first instance, between the unitary existence of the Gods and the divided existence of the cosmos with all its parts.

6. The three ingredients of the soul are (1) an Essence, midway between indivisibility and divisibility, (2) Sameness, (3) Difference. The latter pair are again incompatibles, stressing the soul's simultaneous possession of its primordial unity and of the multiplicity which it experiences

in the world of time, space, and separated creatures. For a more complicated reading of this passage, see Cornford, pp. 59–61.

7. The Demiurge "ladles out" portions in the measures 1, 2, 3, 4, 9, 8, 27. The figure which sets these out in the form of the Greek letter lambda dates at least from Plato's century:

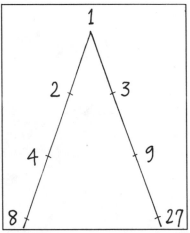

FIGURE 31: *The Timaeus Numbers in Lambda Form*

This figure clarifies the sequence of numbers as being the squares and cubes of 2 and 3, expanding the Pythagorean "first female" and "first male" numbers into the three dimensions required to form a spatial cosmos. But in order to visualize the next process, it is necessary to set them out as a single line:

```
0  1  2  3  4        8 9                                        27
|  |  |  |  |         | |                                        |
```

8. A description of the Harmonic Mean. Thomas Taylor's translation may be illustrated as follows:

$$\text{When } a > b, \ a - \frac{a}{x} = b + \frac{b}{x}$$

(x being the "part" or fraction). The formula for the mean itself is

$$\frac{2\,a\,b}{a+b}$$

It is harmonic because this is the relationship between any three contiguous members of the harmonic series, e.g. between the 2nd, 3rd, and 4th harmonics produced by ½, ⅓, and ¼ of the fundamental string length. Clearing fractions, one has the progression 6, 4, 3, which answers to the formula.

9. The Arithmetic Mean, defined as $a-x = b+x$, found by the formula

$$\frac{a + b}{2}$$

10. By placing the harmonic and arithmetic means between the powers of 2, any pair of which forms an octave, the sesquialter (2:3), sesquitertian (3:4), and sesquioctave (8:9) intervals emerge. For example, between the octave E'=6 and E=12 (Greek scales always going downward), the harmonic mean is B=8 and the arithmetic mean A=9. The four tones 6, 8, 9, 12 will sound E' B A E, creating the sesquialters (perfect fifths) E'–A and B–E; the sesquitertians (perfect fourths) E'–B and A–E, and the sesquioctave (whole tone) B–A. The resulting matrix, 6:8:9:12, is known as the musical *Tetraktys*, a term borrowed from Pythagorean arithmology.

11. Every interval of a fourth is now filled as far as possible with whole tones. Referring to the previous note, this would fill out the matrix E'–B–A–E with the new whole tones E'–D, D–C, A–G, G–F, making a complete diatonic Dorian scale of E' D C B A G F E.

12. When a perfect fourth is filled with two whole-tones, there is a "left-over" (Gk. *leimma*) that we would call a semitone, but which in strict Pythagorean tuning is a rather smaller interval, 256:243. Broadly speaking, the whole range from 1–27 has now filled out with a scale, made from two sorts of contiguous interval: the 9:8 tone, and the leimma. It covers four octaves and a sixth, far exceeding the range of Greek usage. For a complete analysis of the numbers and notes involved, and a tour de force of musical interpretation, see Ernest McClain's *The Pythagorean Plato*, pp. 64–69.

13. The diatonic scale is now imagined as set out on a strip of parchment. The strip is bent round so that its ends meet, and then this circular strip is sliced lengthways into two. The two circles are swiveled relative to one another until, viewed from the side, they have the shape of an X. For a masterful numerical interpretation of this figure, see John Michell's *The Dimensions of Paradise*, pp. 154–169.

14. The circle of the Different is further cut lengthwise into seven narrow rings. At this point we move abruptly into our own cosmic system, in which (assuming the traditional geocentric model) the unchanging sphere of the fixed stars surrounds the seven different spheres of the planets, the latter turning at different and sometimes relatively "contrary" rates. Thomas Taylor's original translation reads: ". . . as

each of them are three, he ordered . . .", which I have amended to the later version found in his *Proclus on the Timaeus of Plato*, vol. II, p. 131. In Proclus' extensive commentary (see no. 12), the Neoplatonist confirms what the visual imagination suggests, namely that each of the seven circles is imprinted with the original three double and three triple intervals (1, 2, 4, 8 and 1, 3, 9, 27), and, presumably, with all their diatonic subdivisions. Proclus says that "each of the seven spheres comprehends a multitude of powers, some of which are more total, but others are more partial" (ed. cit., vol. II, p. 134), i.e., the structure of the World-Soul in its entirety is to be found in every partial soul. See also ed. cit., vol. II, p. 449, for a further commentary on the soul's completeness. Other (and older) commentators have interpreted the intervals as belonging *between* the circles and expressing the relative spacing between the orbits (see A. E. Taylor, op. cit., pp. 154–163). These may simply mean that the relative radii are 1, 2, 3 . . . 27; or that the radii are successive multiples of each preceding radius (as assumed, following Porphyry's lost Commentary, by Macrobius II, 3, 14—see note 5 to Macrobius no. 11). Since Timaeus is still talking about soul, and not about body, Proclus seems to ascribe to him the more plausible intention, but multiple meanings are always a possibility in Plato.

15. The Sun, Mercury, and Venus are the ones that revolve similarly (cf. the Myth of Er from *Republic* X and *Epinomis* 986–987), while the remaining four are the Moon, Mars, Jupiter, and Saturn.

16. Proclus comments on this (ed. cit., vol. II, p. 155) that soul only "participates" in the harmony which preexists in a higher realm than soul, in the Demiurge himself and in his divine creations such as the Muses (cf. Proclus' distinction between Muses and Sirens in his Commentary on the *Republic*—see no. 12). Proclus says here: "Harmony must be conceived to be a threefold substance; so as to be, [1] either harmony itself, or [2] that which is harmonized; being a thing of this kind according to the whole of itself; or [3] that which is secondarily harmonized, and in a certain respect participates of harmony. And the first of these must be assigned to intellect; the second to soul; and the third to body." Of the body, Timaeus later remarks that the senses "agitate and tear in pieces the circulation of the nature distinguished by difference. Hence, they whirl about with every kind of revolution each of the three intervals of the double and triple, together with the mediums and conjoining bonds of the sesquitertian, sesquialter, and sesquioctave ratios, which cannot be dissolved by any one except the artificer by whom they were bound . . . " (*Timaeus* 43d), suggesting that the whole scale is present even in the physical body as a cohesive force.

2 · PLINY THE ELDER

1. Pliny refers to the spherical body of the firmament, which encloses all the other celestial spheres, and which—presuming a static earth—seems to turn around once a day.

2. Pliny gives three different estimates for the relative distances from the earth of the sun, moon, and stars. Taking the distance from the earth to the moon as 1, they are as follows: (1) earth–sun, 20; (2) earth–sun, 3; earth–stars, 6; (3) earth–sun, 3½; earth–stars, 7. See also Alexander of Aphrodisias, *In Metaph.* A5, for another Pythagorean theory.

3. This is exactly half of Eratosthenes' figure for the earth's circumference, 252,000 stades, as reported by Pliny at the end of this book (II, cxii). Pliny's own emendation in his brief final chapter (II, cxiii), where he says that *harmonica ratio* adds 12,000 more stades to Eratosthenes' figure, has been demonstrated by John Michell (*Ancient Metrology*, pp. 3–13) as perfectly accurate, equal to 24,883.2 miles when the correct units are understood.

4. Pliny does not say whether the following scale goes up or down, but implies it in the conceit recorded at the end of the same chapter. Martianus Capella, who reports almost the same scale (*De nuptiis* II, 169–199; see *MM&M*) confirms the upward direction, with the consequent scale of Earth D, Moon E, Mercury F, Venus F#, Sun A, Mars B, Jupiter C, Saturn C#, Fixed Stars E. This is a Dorian scale consisting of a disjunct pair of chromatic tetrachords, plus the earth's added tone. Theodore Reinach (see Bibliography) suggests that the original Pythagorean scale was the enharmonic version of this, implying the following changes: Mercury E quarter-sharp, Venus F, Jupiter B quarter-sharp, Saturn C. Reinach recommends the enharmonic genus because in the fifth century B.C.E. it was the one most used in theory, whereas the chromatic was considered frivolous; but since both were written in the same notation, confusion of the movable notes was easy. Reinach interprets the whole matter as a projection into the sky of the existing musical system as applied to the nine-stringed lyre of the fifth century. Cf. note 5 to Theon of Smyrna (no. 4), who reports a Phrygian version of this as his planetary scale, and the discussion by Glarean (no. 31).

5. *In ea Saturnum Dorio moveri phthongo;* the meaning is that since the Dorian scale is pitched a note higher than the Phrygian, Saturn, Jupiter, etc., may be said to move at successively lower modes. Martianus Capella reproduces this idea, saying that Jupiter rang with a Phrygian, Saturn with a Dorian sound (*De nuptiis* II, 196–197).

3 · NICOMACHUS OF GERASA

1. Flora R. Levin, in *The Harmonics of Nicomachus and the Pythagorean Tradition*, p. 36, summarizes the difficulties in translating this passage and the disagreements of all the experts, and offers the following: "For they say that all bodies that whir when something yields to them and is quite easily set into undulating motion, necessarily produce sounds which differ from one another in intensity, [speed] and pitch, these differences in sound being attributed to the weights of the particular bodies or their individual speeds or the positions in which the orbital swing of the earth takes place. And these positions are either somewhat subject to fluctuation or, conversely, are intractable."

2. Cf. Plato, *Cratylus* 410b.

3. A basic confusion of terminology arises from the fact that the Greek lyre was held, like the guitar, with the lowest-pitched string at the top, and the highest-pitched at the bottom. Nicomachus' scale as described here is generally interpreted as a pair of conjunct tetrachords: Moon D (nete), Venus C (paranete), Mercury Bb (paramese), Sun A (mese), Mars G (hypermese lichanos), Jupiter F (parypate), Saturn E (hypate). Of these the fixed tones are nete, mese, and hypate; the others are movable (cf. the scales in the notes to Pliny, no. 2). In one fragment of Nicomachus (von Jan, *Musici Scriptores Graeci*, p. 272) the positions of Venus and Mercury are corrected by being reversed. He says that his predecessors had assigned the notes to the planets in the reverse order, i.e., with the moon at the lowest pitch. The scale of his *Manual* is certainly unusual in descending from Moon to Saturn. The general assumption of classical natural philosophers was that high pitch is caused by rapid, low by slow speed; and since the fastest of all bodies was assumed to be the sphere of the fixed stars, whirling around once in twenty-four hours, that would naturally be assigned the highest pitch. The seven planets all move in a contrary direction to it, gradually slipping counter-clockwise, as it were, while they are carried along clockwise. Saturn completes a whole contrary circle once in 30 years, the moon in a month. Hence in an absolute sense, Saturn moves faster than the moon, and hence the pitches might best be made lower as one approaches the earth. This they do in all other classical sources except Nicomachus (and Boethius, where he reports Nicomachus). The earlier Greek astronomers, however, such as Anaximander, believed in a rotating earth (see A. E. Taylor, *A Commentary on Plato's Timaeus*, p. 162n.), and therefore approached closer to modern cosmology by leaving the stars at rest. Like the static earth of Cicero, it would then be the stars that would remain silent—as they do in Nicomachus—and the music would

begin with Saturn, the slowest-moving planet, at the lowest pitch. It is therefore probable that Nicomachus transmits the original planet-scale of the Greeks, whose authenticity is supported by its restriction to the sacred number seven, the number of the Chaldaean planets, the different scale-tones, the days of the week, etc.

Roger Bragard, in "L'harmonie des sphères selon Boèce" (see Bibliography) reinterprets the scale as follows: he says that the mese in Pythagoras' time was the middle string but no longer the middle pitch, the tetrachords having become disjunct; and therefore gives the following arrangement in disjunct tetrachords, one of them defective: E' D B, A G F E. See also note 6 below.

4. *Logos:* the meaning is probably the Pythagorean tradition, but it might simply mean "as reason suggests."

5. In this chapter, which is of great interest to the historian of acoustical theory, Nicomachus brings into his Pythagorean exposition the Aristoxenian concept of *kinesis*. See Levin, op. cit., pp. 51–63, for an analysis of these issues. Her point, in short, is that Nicomachus was determined to attribute everything worthwhile in harmonics to Pythagoras.

6. The earlier scale is the first one given in note 3. The expansion with which Pythagoras is credited leads to the scale of two disjunct tetrachords: E' D C B, A G F E. Levin, op. cit., pp. 78ff., believes that here as elsewhere Nicomachus, as an excessively loyal Pythagorean, is claiming every invention as his master's, including this expansion of the lyre's range from a seventh to an octave, which had been done by Terpander a century before Pythagoras. Terpander's scale was E' D C, A G F E (Pseudo-Aristotle, *Problemata* 19–32).

7. Nicomachus is the earliest source for this famous legend. The story as it stands is fallacious, because (1) the different tones of hammering depend more on the weight and shape of the anvil than of the hammer (cf. a xylophone's constant pitch under different mallets), and (2) the experiment of stretching the strings by suspended weights reports a variation of pitch directly proportional to the weights, whereas the actual proportion is to the square roots of the weights. Ruelle (ed. cit., p. 20) cites skeptically A.-J.-H. Vincent in *Notices et extraits des mss.*, vol. XIX, pt. 2, p. 274, as proposing the replacement of *sphura* ("hammer") by *sphaira* ("sphere"), which would make more sense; the workers might have been hammering out hollow globes on the anvils, which would have rung with different pitches.

An interesting sidelight on this is shed by a passage in Iamblichus' *Life of Pythagoras*, Ch. 28: "[Pythagoras] honored the Gods in a way similar to that of Orpheus, placing them in images and in brass, not conjoined to our forms, but to divine receptacles; because they comprehend and provide for all things, and have a nature and *morphe* [form] similar to the universe." (Thomas Taylor's translation, London, 1818, pp. 109–110) Taylor adds to the word "receptacles" this footnote: "i.e. to spheres; Iamblichus indicating by this, that Pythagoras as well as Orpheus considered a spherical figure as the most appropriate image of divinity. For the universe is spherical; and, as Iamblichus afterwards observes, the Gods have a nature and *morphe* similar to the universe; *morphe*, as we learn from Simplicius, pertaining to the color, figure, and magnitude of superficies." Nothing would be more Pythagorean than that such metal spheres, like the resonators of Vitruvius' theater (*De arch.* V, 5) should have been tuned to the perfect consonances. See also the reference to "instruments consisting of concave brass" in Boethius (transmitting a lost passage of Nicomachus), *De Inst. Mus.* I, ii; and on musical brass disks, the sources assembled in Burkert, *Lore and Science in Ancient Pythagoreanism*, p. 377. Isaac Newton (see no. 40) believed that the second fallacy concealed the Pythagorean knowledge of the inverse-square law of gravitation.

8. A surveying instrument like a theodolite.

9. It is surely no coincidence that Pythagoras is imitating the procedure of the Demiurge in Plato's *Timaeus* (see no. 1), filling out the matrix to make a diatonic scale.

10. This reservation is due to the Pythagorean rejection of the term "semitone" as an inaccurate, Aristoxenian substitute for the leimma.

11. Condensed and adapted from *Timaeus* 36a–b. For some reason, Nicomachus changes Plato's statement that all the fourths were filled out with whole-tones and leimmas, making him say only that the distance between the fourths and the fifths was a whole-tone. The remainder of our extract shows Nicomachus demonstrating that the 6:8:9:12 proportion contains the three means: harmonic, arithmetic, and geometric. The latter is a little forced, since a true analogy would place it at the square root of 72—an irrational number, hence foreign to Pythagorean mathematics. See also note 30 to Proclus (no. 12).

4 · THEON OF SMYRNA

1. Plato does not say this explicitly: he recommends the investigation of which numbers are harmonious, and why (*Republic* 531c). But he has already said that Harmonics is a sister science to Astronomy (530d), and that the actual movements of the heavenly bodies are not the concern of the latter, but only "problems" (530b).

2. Theon's books on geometry and on stereometry are lost.

3. Theon proceeds after this introduction by expounding arithmetic, beginning with the Monad.

4. Calcidius (see no. 10) cites a Latin version of the same lines, but attributes them to Alexander *Miletus*. It is generally assumed that Theon was mistaken, and that the author was Alexander of Ephesus or Alexander Polyhistor (both first century B.C.E.). See A. E. Taylor, *A Commentary on Plato's Timaeus*, p. 166n.

5. The relative pitches give the following pattern, also that of Censorinus (no. 6): Earth D, Moon E, Mercury F, Venus F#, Sun A, Mars B, Jupiter C, Saturn C#, Fixed Stars D'. Achilles Tatios, in his commentary on Aratus (p. 136 Petau), gives a similar scale: Earth D, Moon E, Sun F#, Venus G, Mercury A, Mars B, Jupiter C, Saturn C#, Stars D'. Plutarch (*De anim. procr. in Tim.* 31) cites a scale that has the sun as *mese*, earth as *proslambanomenos* a fifth lower, and the stars a fourth above the *mese:* Earth D, Sun A, Stars D'. He adds enigmatically that Mercury and Venus move *en diatonois kai lichanois*, which cannot be explained. But his framework may be that of a similar scale with the sun in the central position. Theodore Reinach (op. cit; pp. 441–443—see note 4·to Pliny, no. 2) explains this scale-type as a chromaticized version of the nine-note Phrygian enharmonic scale of the Pythagoreans, given by Aristeides Quintilianus as: D, E, E quarter-sharp, F, A, B, B quarter-sharp, C, D'. *Musica Mundana*, the unpublished dissertation of James Haar (see Bibliography), is an unrivaled guide through this labyrinth.

6. Theon's source is evidently the lost *Platonicus* of the great astronomer Eratosthenes of Cyrene (circa 275–194 B.C.E.), which he cites in his introduction and elsewhere. Eratosthenes' order of planets begins Moon, Sun, Mercury, Venus (see also Calcidius 73), a slight variation on the Platonic one and in fact the only one in agreement with the traditional order of astrological rulerships: Cancer-Moon, Leo-Sun, Virgo-Mercury, Libra-Venus, etc. Theon quotes Eratosthenes as saying: "These eight spheres also harmonize together while making their revolutions around the earth." (II, xlvii)

5 · PTOLEMY

1. Ptolemy defines the *emmeleis* as "those sounds which are easily accepted by the ear" (*Harmonics* I, 4), giving them (I, 7) as the semitone 16:15, the major and minor tones 9:8 and 10:9, and the major and minor thirds 5:4 and 6:5. These are later (III, 5) distinguished in their symbolism from the perfect consonances of fourth, fifth, and octave. Later he admits such intervals as 12:11 (see Chapter 9 of our selection). Düring explains (ed. cit., p. 174) that they include any intervals usable

for melodic purposes, irrespective of their intervals as such.

2. Ptolemy is neither a doctrinaire Pythagorean nor an Aristoxenian: he allows both theory and the practical judgment of the ear to govern the rules of harmony. On the issues raised by Aristoxenus, see note 9 to Censorinus (no. 6).

3. Socrates says of Astronomy and Harmonics: "It is likely that, as the eyes are fixed upon astronomy, so the ears are fixed upon the movements of harmony, and that these sciences are closely akin, as the Pythagoreans say and we agree with them, Glaucon..." (*Republic* 530d).

4. Aristeides Quintilianus (*De musica* III, 14) compares the five senses with five elements (the four material ones plus ether, corresponding to sight) and the five tetrachords of the Greater Perfect System, hence does not have to disallow touch as Ptolemy does.

5. See Düring, ed. cit., pp. 271–272, on these divisions of the power of thought, and parallels in the "philosophasters" (*Afterphilosophen*) Speusippos and Sallustios. Ptolemy's terms are *sophrosune, enchrateia, aidos, praotes, aphobia, andreia, charteria.*

6. These are the three parts of the soul as described in Plato's *Republic* IV. For other English equivalents, see Clement A. Miller's translation of Gafori (no. 29) and our version of Zarlino (no. 32).

7. Ptolemy's terms for the seven virtues of Reason are: *oxutes, euphuia, anchinoia, euboulia, sophia, phronesis, empeiria.*

8. In the chromatic and enharmonic tetrachords, *pyknon* is the sum of the two smaller intervals (see note 19 below), and *apyknon* is the larger interval.

9. See the Greek note-names in note 12 below.

10. Düring gives in a note to this passage a list of sources on the power of music to calm the soul: Plutarch *De anim. proc.* 441e; Aelius *Var. hist.* X, 23; Porphyry *Vit. Pyth.* 30; Cicero *Tusc. disp.* IV, 2; and, as an antidote to intoxication, Quintilian I, 10, 32; Sextus Empiricus 749, 23; Iamblichus *Vit. Pyth.* 112 and 195. The numerous passages in Plato are listed in Abert, *Lehre vom Ethos*, pp. 9–13nn, the most important being *Republic* II, 398c–401e; *Laws* II, passim.

11. On Pythagoras' use of music, see Guthrie's *The Pythagorean Sourcebook* with its accounts by Iamblichus, Porphyry, and others.

12. Ptolemy is aware that the range of possible pitches, like the circle of the Zodiac, is a continous one; but that for practical purposes both are divided into discrete steps: pitch into scale steps, the Zodiac into 360 degrees and twelve signs. Essential to the understanding of his system is the *Systema teleion* or Greater Perfect System of Greek music:

Greek pitch-names	Approximate relative pitch in the diatonic genus	
nete hyperbolaion[a]	a'	
paranete hyperbolaion[a]	g'	
trite hyperbolaion[a]	f'	
nete diezeugmenon[a,b]	e'	
paranete diezeugmenon[b]	d'	d'*
trite diezeugmenon[b]	c'	c'*
paramese[b]	b	b♭*
mese[c]	a	a*
lichanos meson[c]	g	
parhypate meson[c]	f	
hypate meson[c,d]	e	
lichanos hypaton[d]	d	
parhypate hypaton[d]	C	
hypate hypaton[d]	B	
proslambanomenos	A	

[a]Tetrachord hyperbolaion
[b]Tetrachord diezeugmenon
[c]Tetrachord meson
[d]Tetrachord hypaton
*Tetrachord synemmenon, from Lesser Perfect System

13. Ptolemy is bending the Greater Perfect System into a circle. Since it spans two octaves in all, any points diametrically opposed on the tone-circle will be an octave apart, and the extremes which we (arbitrarily) translate as a' and A will meet at the beginning/end. A diagram of Ptolemy's tone-zodiac will be found in Godwin, *Harmonies of Heaven and Earth*, p. 153.

14. This is the justification for a comparison which seems, at best, artificial; for the Zodiac is a closed circle, whereas the two-octave system is only a small part of the indefinite range of pitches. This indefinite range is articulated, however, by the recurrence of the 1:2 octave ratio and its audible identity. Might it not, then, be better envisaged as a helix in which a line parallel to the axis joining adjacent levels would unite pitches an octave apart? The Zodiac circle would then correspond to a single octave, making the astrological opposition a tritone rather than an octave, and each sign a semitone. Unfortunately such a system leaves the fourth and fifth unsuitably matched, both corresponding to the astrologically weak aspect of the quincunx (150°). Hans Kayser's solution (*Lehrbuch der Harmonik*, pp. 167ff.) was to place the perfect fifth at the opposition point, the major third and flat seventh (seventh

harmonic) at the squares. Ptolemy's choice of a two-octave circle is due in part to the current Greek tonal system, but also to reverence for the Pythagorean *tetraktys* (1,2,3,4) which when musically translated covers a double octave (1:4).

15. To summarize this paragraph and figure 3:

			Octave			**Twelfth**	
ABCD	=	12	ABCD	:	AB	ABCD : AC
ABD	=	9	ABC	:	AC	ABD : AD
ABC	=	8	AB	:	AD	
AB	=	6	**Fifth**			**Double Octave**
ADC	=	4	ABCD	:	ABC	ABCD : AD
AD	=	3	ABD	:	AB	
CB	=	2	AB	:	AC	**Eleventh**
CD	=	1	**Fourth**			ABC : AD
			ABCD	:	ABD	
			ABC	:	AB	**Large whole-tone**
			AC	:	AD	ABD : ABC

16. The correspondence is even more perfect with the equal temperament system, in which the double octave does contain twelve identical whole-tones. Several tone-zodiacs based on this are given in Godwin, *Harmonies of Heaven and Earth*, pp. 148–166.

17. Taking C as a string equal in length to the circumference, the intervals would be as follows:

12:1	C – g''
12:2	C – g'
12:3	C – c'
12:4	C – g
12:5	C – eb
12:6	C – c
12:7	C – A slightly sharp
12:8	C – G
12:9	C – F
12:10	C – Eb
12:11	C – D slightly flat

18. None of these motions has anything to do with the tone-zodiac already explained. The first motion is the daily orbit of the fixed stars and planets around the earth, exemplified by the Sun's journey from east to west: it is imagined to rise in pitch to its culmination or zenith, and to fall at sunset. The second motion (see Ptolemy's Ch. 11 of our

selection) considers the epicyclic rotations within the planetary orbits, elaborately calculated by Ptolemy and his predecessors to explain variations inherent in a geocentric system. In each epicycle, the planet comes nearer to the earth and then recedes. The third motion (see Ch. 12) is more contrived: it is of the planet's declinations as they move through the Zodiac from the most "northerly" sign (Cancer) to the most "southerly" (Capricorn), and back.

19. The parallel is as follows:

> Dorian tetrachord in diatonic genus
> $\quad$ E D* C B $\qquad\qquad\qquad$ = widest epicycles
> Dorian tetrachord in chromatic genus
> $\quad$ E C#* C B $\qquad\qquad\qquad$ = medium epicycles
> Dorian tetrachord in enharmonic genus
> $\quad$ E C* B-quarter sharp B $\quad$ = narrowest epicycles
>
> * = the *lichanos* note in each tetrachord

20. I.e., to the radius of the epicycle which encircles the major orbit.

21. Imagine the intervals as follows:

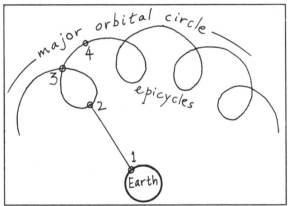

FIGURE 32: *The Genera and the Epicycles*

In the chromatic genus, 1 = E, 2 = C#, 3 = C, 4 = B. The bigger the epicycle, the larger the intervals 2–3 and 3–4 relative to the whole.

22. Each planet, maintaining the same "genus" of epicycle, passes through the signs of the Zodiac, which cause it to appear variously to the north or the south. The sun, for instance, reaches its northernmost points at the summer solstice, when it is at the first degree of Cancer. Viewed from a point on the Tropic of Cancer, it is directly overhead at noon. This would correspond to the "highest" mode, the Mixolydian.

The other six parallels are:

Gemini/Leo	Lydian
Taurus/Virgo	Phrygian
Aries/Libra	Dorian
Pisces/Scorpio	Hypolydian
Aquarius/Sagittarius	Hypophrygian
Capricorn	Hypodorian

With this scheme Ptolemy supports, as he does throughout the *Harmonics*, the central position of the Dorian mode, putting it at the equinoctial signs.

23. We now return to the two-octave circle of Chapter 8, each twelfth of it equivalent to a whole-tone and to one 30° sign of the Zodiac. The diagram in the text shows the course of the moon, exemplifying that of all the planets, from its conjunction with the sun at the dark of the moon. Its heliacal rising is the point at which it first becomes visible ("new moon"), toward the first octant, while the full moon is when it is in opposition to the sun. The other planets are similarly obscured when they are close to the sun, and gradually become more visible as they and the sun move apart through the Zodiac. Their heliacal setting is when they again disappear into the sun's proximity.

24. Seen from the earth, Venus and Mercury can never reach opposition with the sun. That is why Cicero calls them "the sun's companions" (*Somn. Scip.* IV, 2). This chapter survives only in very corrupt versions, and the placement of the planetary symbols on the diagram which I have redrawn from Düring's edition, p. 133, is, as he admits on p. 276, arbitrary. I have therefore omitted them.

25. From here to the end of the treatise is missing, presumed unfinished at Ptolemy's death. What follows is a summary, absent from the best manuscripts but found in the hand of Nicephoras Gregoras, fourteenth century, which makes a feeble attempt at filling out Ptolemy's proposed chapter titles (see von Jan, "Die Harmonie der Sphären," p. 33).

26. See our extract from Kepler's *Mysterium Cosmographicum* (no. 34) for an attempt to correlate regular geometrical figures, aspects, and intervals. Kepler admits that Ptolemy had almost hit on Kepler's own ideas, but was prevented by the geocentric system (*Harmonices Mundi* V, 5).

27. This chapter is thought to be more authentically Ptolemaic than the preceding two. Otto Neugebauer, in *A History of Ancient Mathematical Astronomy*, Pt. 2, pp. 913–917 and 934, describes the contents of the

"Canobic Inscription" attributed to Ptolemy, which contains astronomical data and the following correspondences, very similar to those which end this chapter:

Hyperhyperbolaia	Stars	36	b'
Nete hyperbolaion	Saturn	32	a'
Nete diezeugmenon	Jupiter	24	e'
Nete synemmenon	Mars	21⅓	d'
Paramese	Sun	18	b
Mese	Venus, Mercury	16	a
Hypate meson	Moon	12	e
Hypate hypaton	fire, air	9	B
Proslambanomenos	water, earth	8	A

According to Neugebauer (p. 934n.) the original sequence was "obviously" 36, 32, 28, 24, 20, 16, 12, 8, 4; he says that "the ⅓ is a misinterpretation of the 20 as a sexagesimal fraction." Neugebauer's sequence of numbers, however, though it gives a harmonic series covering three octaves and a whole-tone, bears no relation to the Greek note-names. His book is a monument of scientific history, but his judgment is occasionally impaired by an impatient contempt for anything that does not foreshadow the quantitative exactitude of modern science (see, for example, his unscholarly dismissal of Cicero and Martianus, p. 1030). Von Jan found the same scale as the Canobus one in three Naples manuscripts of the late Byzantine period, but thought that the Canobic Inscription was a forgery because of its lack of fit with the Greek system. As an amendment, von Jan suggested that the scale should have excluded the highest note, the sphere of the stars, and the four elements, and reversed the order of numbers so as to correspond to string-lengths. That would give:

Proslambanomenos	Moon	36	A
Hypate hypaton	Mercury	32	B
Hypate meson	Venus	24	e
Mese	Sun	18	a
Paramese	Mars	16	b
Nete diezeugmenon	Jupiter	12	e'
Nete hyperbolaion	Saturn	9	a'

This certainly fits well the tetrachords of the Greater Perfect System, and also works much better with the zodiacal circle of 360°:

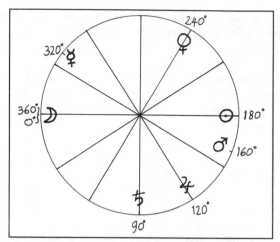

FIGURE 33: *The Aspects of the Planets*

Düring, on the other hand, argues strenuously (ed. cit., pp. 281–284) for the authenticity of the Canobic Inscription as the work in which the mature Ptolemy wished to record his most profound discoveries for posterity. As such, they entered the Islamic as well as the Byzantine domain (see note 8 to the Brethren of Purity, no. 18). None of these versions succeed, however, in making the good and bad aspects work astrologically as well as musically, thanks to the basic fallacy of applying linear arithmetic to the circle. Thus the 9:12 fourth is reflected in the Zodiac as a 90°–120° semi–sextile, but the 18:24 fourth as a 180°–240° sextile.

28. More correctly *hypate meson*, if the moon is a fourth from Venus. The modern pitches are conjectured from the description.

6 · CENSORINUS

1. Censorinus has just given a summary of authorities on the question of which months of pregnancy can give live births. Of these, only Diocles Carystius and Aristotle (*Hist. an.* VII, 4) believe that the eighth month is possible, while Aristotle alone admits the eleventh as well.

2. The assumption is that only the astrologically effective aspects (sextile, square, trine, and opposition) can bring about live births. These are calculated as movements of the sun, at the rate of 30° a month, from the time of conception. (See figure 34.)

3. The "births" so far, occurring in the third, fourth, and fifth months of pregnancy, must be miscarriages. The live births that follow, in the seventh, ninth, and tenth months, are brought about by the aspects beyond 180°.

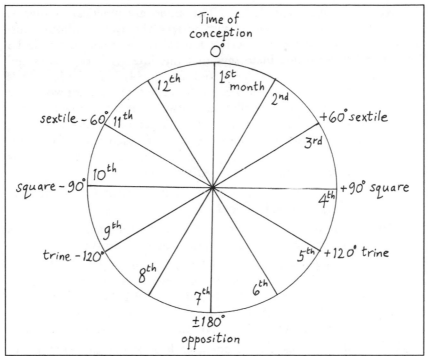

FIGURE 34: *The Aspects Favorable to Birth*

4. Marcus Terentius Varro (116–127 B.C.E.), librarian and the greatest Roman encyclopedist. His *Tubero de origine humana* is a lost work.

5. Diogenes Apolloniates: eclectic philosopher of the late fifth century B.C.E.; his works lost.

6. Hippon or Hipponax, natural philosopher of the fifth century B.C.E., much concerned with embryology; his works lost.

7. Pythagoras' teachings were oral. This must be the opinion of a Pythagorean, probably reported by Varro. That the embryo develops purely from the male substance, making the mother a mere receptacle ("matrix") for the seed, was the common Greek view before Aristotle, for whom the menstrual blood was the matter, semen the formative element. See Needham's *History of Embryology*, p. 43.

8. *Musica est scientia bene modulandi*, a definition repeated by Isidore of Seville and often thereafter.

9. Aristoxenus of Tarentum, born between 375 and 360 B.C.E., philosopher and musical theorist, pupil of the Pythagoreans and of Aristotle. In his *Harmonic Elements* he criticizes the geometers and speculative music theorists and advocates the primacy of the ear's judgment over mathematical hairsplitting (ii, 33–34): hence his refusal to admit differ-

ent sizes of semitones (ii, 46–57). The *musici* are practical musicians. Theon of Smyrna (see no. 4) summarizes the two positions thus: "Aristoxenus says that it [the perfect fourth] is composed of two and a half perfect tones, while Plato says that this interval is two tones and a remainder (*leimma*)...of 256:243." (ed. cit., p. 44) Now the leimma is less than half of the wholetone 9:8: a difference that does not worry the ear—hence Aristoxenus' assimilation of two of them to make a sixth whole-tone for the octave, as in modern equal temperament—but which is mathematically significant. Hence Censorinus' stricture on Plato's terminology at the end of this section. See Boethius *De Inst. Mus.* III, i–iv for a mathematician's refutation of Aristoxenus.

10. Plato never used the term; *Timaeus* 36b has *leimma*. Censorinus was presumably using a secondary source.

11. Censorinus only reports the second part of Pythagoras' famous experiment, which we have here in Nicomachus' version (see no. 3).

12. The following passage, omitted here, is translated in *MM&M*. It explains how the development of the embryo, counting days of pregnancy, is in stages proportioned in the same 6:8:9:12 ratios. Censorinus concludes (XII, 5) that "if there is a harmony in the motion both of the body and the soul, music is definitely not foreign to our nativities."

13. The fixed stars must be intended. Only Ptolemy (no. 5) writes of their harmony. Censorinus is here referring to much lost material (Dorylaus, etc.).

7 · ORPHEUS

1. The poet has already referred to Apollo's journey, as the sun, beneath the earth every night. He now alludes to the god's yearly journey from winter solstice (lowest note) through spring equinox (middle) to summer solstice (highest), tending alternately toward the north and south poles. The Dorian, the centrally pitched mode, is given the middle position here as in Ptolemy *De Musica* III, 12 (see no. 5), where it represents the equator.

2. This refers to the five zones (= "belts") into which ancient geography divided the earth: two frigid ones at the poles, two temperate ones (the northern including all the known world), and the torrid equatorial zone. The only two habitable, temperate zones "have no communication with each other because of the fiery heat of the heavenly body." (Pliny, *Hist. nat.* II, lxviii, 172; see also Cicero *Somn. Scip.* VI, 1–3). All of this imagery recognizes the earth as in every sense coeval with the sun.

3. In later Antiquity the rustic god Pan developed into a universal god of the All (Gk. *to pan*), cf. Orphic Hymn no. 11, "To Pan": "You weave your playful song into cosmic harmony . . . a veritable horned Zeus"

who also supports the earth and separates its zones. Cf. Aratus' reference to the East and West extremes of the horizon, where the stars rise and set: "Ocean will give you signs at either horn" (565–566). Pan's seven-piped syrinx, source of "whistling winds," is his equivalent of Apollo's seven-stringed lyre. Cf. Johannes Galenos (twelfth century C.E.) on Hesiod's *Theogony*, who quotes from Orpheus: "Jupiter is the god of the all, and the mingler of things, emitting shrill sounds from winds and air-mingled voices." (Quoted among the extensive notes of Thomas Taylor to his translation of this hymn in *The Mystical Hymns of Orpheus*.) Cf. also the "whistling" descent of the soul as it fills with breath in Aristeides Quintilianus II, 17 (no. 9).

8 · SAINT ATHANASIUS

1. This section comes in a context of the proof of God through the revelation of the natural order and its "concordant harmony" of opposites (Ch. 36). Parallels occur in the pseudo-Aristotelian *De mundo*, 399a, 400b, which also compares God to a city-ruler, a helmsman (as Athanasius in Ch. 39), and a conductor (see next note). For further comparisons and sources, see Meijering's *Orthodoxy and Platonism in Athanasius*, pp. 5–40.

2. The simile of the lyre goes back to Heracleitus (Fragment 51) and is also found in Irenaeus, *Adv. Haer.* II, 37, 2. Athanasius is fond of the image; in Ch. 31, 4 he applies it to the relationship of reason to sense: "In like manner [to a musician], the senses being disposed to the body like a lyre, when the skilled intelligence presides over them, then too the soul distinguishes and knows what it is doing and how it is acting." In 42, 3: "For just as though some musician, having tuned a lyre, and by his act adjusted the high notes to the low, and the intermediate notes to the rest, were to produce a single tune as the result, so also the wisdom of God, handling the universe as a lyre, and adjusting things in the air to things on the earth, and things in the heavens to things in the air, and combining parts into wholes . . . produces . . . the unity of the universe and its order." This wisdom is defined (42, 2) as the Johannine Word of the Father, and cast in a demiurgic role reminiscent of the harmonization of the elements in *Timaeus* 32b–c. A further musical simile, likening the Word's relation to the universe to that of conductor and chorus, occurs in 43, 1; cf. Plotinus, *Enneads* VI, 9, 8 (in *MM&M*).

3. In the next chapter, Athanasius argues that a plurality of gods would imply a plurality of universes; which was certainly possible to God (Origen thought it actually was so), but from which God refrained "lest by the coexistence of more than one a plurality of makers should be supposed" (39, 6)!

9 · ARISTEIDES QUINTILIANUS

1. See also our extract from Gafori (no. 29, Ch. 14) for a summary of another chapter of Aristeides.

2. That is, in Book III. The harmonia of the soul is discussed in Book III, section 24 (based on Plato *Timaeus*). Since this harmonia is composed of the numbers 1, 2, 3, 4, 8, 9, and 27 (which will be associated in Book III with specific passions and virtues) and since musical harmonia involves these same numbers for its intervallic ratios, music has special mimetic power. Cf. Plutarch *De mus.* 44 and Sextus Empiricus *Adv. musicos* 3637. [TJM]

3. Explaining the presuppositions of this division between the original home of the soul and its time spent in the body, Thomas Mathiesen writes: "'That place' *(ekeithi)* is the original perfect dwelling place of the soul, and Aristeides Quintilianus will use the term consistently in this sense. The term is in contrast to 'the things in this world' *(ta enthadi)* and 'the things of earth' *(ta tede)*. Aristeides Quintilianus does not assign any other name to this region, but it appears to be Plotinus' Intelligible Realm (see *Enneads* II.4 [12], Ch. 5)" (ed. cit., p. 116n).

4. I.e., it makes a perfect circular movement, along with the Unmoved Mover of the Eighth Sphere, and is expanded to the dimensions of the universe itself (see below, "mentally coextensive with the universe"). This needs to be visualized in order to appreciate the soul's subsequent adventures.

5. The soul contracts now, diminishing successively to the sizes of the different planetary spheres.

6. The highest sublunary sphere, consisting of fire and light.

7. These parts will later be associated with musical instruments: lines and sinews with the string instruments, the breath with the wind instruments. Festugière, pp. 66–67 and 69–73, discusses these bodily parts and the sources of the doctrine. [TJM]

8. This allegorical interpretation is based on the similarities of *hermis* (bedpost) and *Hermes* (Hermes); and *logios* (eloquent)—Hermes' epithet, *logos* (ratio), and *analogia* (proportion). As a god, Hermes was commonly associated with Aphrodite. [TJM]

9. This separation reflects the structure of the soul—as Aristeides Quintilianus proposes—because Ares is associated with the irrational part while Aphrodite is consecrated and withdraws from inferior things, reflecting the higher, rational part. [TJM]

10. These two passages anticipate the later association of the string instruments (which are strung with dried gut) with the higher, ethereal

region, and the wind instruments (which use air and condense vapor) with the lower regions. [TJM]

11. The tissues parallel the "membranaceous surfaces" and the arteries, the "sinuouslike lines." The breath is the third part of the bodily constitution described earlier in this section. In Greek medical thought, it was held that the arteries contained breath rather than blood and that the soul moved through the arteries. The soul did not, however, extend and collapse with the physical development of the body. [TJM]

12. The notion of sympathetic movement in musical instruments and the soul is explored in an important article by Evanghélos Moutsopoulos, *"Mousike kinesis . . ."* [see Bibliography] (p. 201 deals with this particular passage). See also idem, "Sur la 'participation' musicale chez Plotin," especially pp. 383–386. [TJM]

13. Cf. Porphyrius *Ad Gaurum* 11.4. For a survey of sympathetic resonance, see Helmholtz, pp. 36–49; or Willi Apel, "Resonance," in *The Harvard Dictionary of Music*, 2nd ed., p. 726. [TJM]

14. Apollo is also the god who has inspired this treatise (see Book I, proem, and Book III, section 27). The legend of Apollo and Marsyas is preserved in Diodorus Siculus 3.59. Marsyas was suspended over the river after losing his contest with Apollo. The contest was judged by the Muses. The inferiority of Marsyas to Apollo (and concomitantly, of the aulos and its music to the kithara and its music) is developed in the following section. Cf. Plato *Rep.* 3.10 (399c–e) and *Symposium* 215–216); and Aristotle *Pol.* 8.6. [TJM]

15. Cf. the interpretation of Odysseus and the Sirens in Proclus, (no. 12).

16. This characterization of Polyhymnia follows Plato *Symposium* 187. The kithara is associated with Apollo in Plutarch *de mus.* 14 (1135F). [TJM] See also Dalberg (no. 44).

17. Hermes was the inventor of the lyre (*h. Merc* 24ff); since he is also the guide of souls and the patron of young men, it is natural to also include the association with paideia. [TJM]

18. Euterpe, whose name means "Well-pleasing," is associated with the aulos (*Anth. Pal.* 9.504–505). Athena is feminine in genus, of course, but because she also exhibits discretion and is associated with war, she is masculine in ethos. Athena invented the aulos but threw it away because playing it distorted her face (Plutarch *De cohibenda ira* 6 [456b]; Athenaeus *Deip.* 616e–617b). [TJM]

19. Cf. Plutarch *De mus.* 37 and Sextus Empiricus *Adv. musicos* 7–9 and 23. Since breath is associated with the lower regions (Book II, sections 17–18), it would defile the hearing of the Pythagoreans, who were seeking transcendence. The aulos would, of course, excite the irrational im-

pulses because of its character, but the lyre, which is dry and associated with the ethereal region (Book II, sections 17–18), would raise and purify the soul. [TJM]

20. That is, 1, 2, 4, 8 and 1, 3, 9, 27. [TJM]

21. This controversy of the soul and number is taken up in Plutarch *De anima pro.* 1013C–D, 1016E–1017B, 1023B–1024B. Cf. Plotinus *Enneads* VI.6 [34], especially Ch. 16. The "more precise" would be Plato and Plotinus; the other group would probably by Xenocrates and Posidonius and their followers. [TJM]

22. These are the various numerical diagrams demonstrating the pattern described above. One such display appears later in the section. [TJM]

23. Because number belongs to the higher, ethereal region. [TJM]

24. In the One, the creative cause. [TJM]

25. For a discussion of these numerical patterns, see Cornford, pp. 66–71. [TJM]

26. Incorporeal things are inseparable because they are simple. On this matter of indivisibility of the incorporeal, see Cornford, pp. 60–66. [TJM]

27. On this entire paragraph, cf. Plutarch *De animae pro.* 1027D–F, 1017C–1019B, which associates this numerical pattern with the plane figures [given by Aristeides in] Book III, section 4. [TJM]

28. That is, 2 x 2 = 4. [TJM]

29. Here the virtues are associated with human conditions, which are concomitantly associated with the numbers of the psychogony. [TJM]

30. Aristeides has associated a "healthy condition" with the power of music in Book II, sections 6 (para. 8) and 15 (para. 2). [TJM, adapted]

31. Because this is the transcendent virtue, associated with the One. [TJM]

32. The duple pattern (1, 2, 4, 8) was assigned to the material (or the affective) and the triple (1, 3, 9, 27) to the incorporeal in the previous paragraph. [TJM]

33. The following diagram is not preserved in the manuscripts. [TJM]

34. This sentence refers to the nature of the Same and the Other in the psychogony. Although the circles are not mentioned until the next paragraph, it appears that they are implied here because the circle of the Same is uninclined and senior (*Timaeus* 36c) and it exhibits concavity and convexity together in respect to a central diameter and because the circle is not further subdivided. The circle of the Other, by con-

trast, is set on the diagonal (crooked in repect to the axis) and is subordinate since it is further divided into seven unequal circles. On this, see Cornford, pp. 60–93. [TJM]

35. The crook demonstrates the straight and the circular in its straight shaft and curved hook. [TJM]

36. At the end of section 1 of Book III, Aristeides names the four numbers 192, 216, 243, and 256 (the latter pair mentioned in *Timaeus* 36b), which constitute a tetrachord of the psychogonic scale. He also cites Plato's remark on how "sensible music falls far behind noetic music in precision." Thomas Mathiesen here draws the attention to the important dissertation of C. André Barbara, "The Persistence of Pythagorean Mathematics in Ancient Musical Thought."

37. The reference here is to the division of the intervals produced by the duple and triple cubes (*Timaeus* 36a). Plato actually inserted two means: the harmonic (which produces the sesquitertian ratio of A.Q.'s description) and the arithmetic (which would produce the sesquialteran ratio). [TJM]

38. This sentence may refer to the splitting of the circle of the Other into seven circles, where three move at the same speed and four move at speeds differing from each other and from the three; or it may refer to the squares in the seris (4 and 9), representing planes—which have length and breadth, and the cubes (8 and 27), representing depth. [TJM]

39. This refers to the movements of the seven planets and the zodiac, which were related to the voice in Book III, sections 21–23 [not given here]. See Ptolemy *Tetr.* 1.12–14 and especially Plutarch *De animae pro.* 1028A–1030C. [TJM]

40. This is associated with the circle of the Other because the eventimes tetraktys used the curved line. The circle of the Other is defined as affective and so rightly associated with the material in accord with the second paragraph of this section. [TJM]

41. The odd-numbered tetractys is the circle of the Same, which was associated with the straight and uninclined. This follows the association of the triple in the second paragraph of the section with the incorporeal, better things. This sentence shows how the whole order of the universe provides a paradigm for the two divisions of music [theoretical and practical] introduced in Book I, section 5, and this is the first association with full divisions rather than with classes or subclasses. [TJM]

42. Because judgment has been associated with the One, it is unchangeable, as is the circle of the Same. [TJM]

10 · CALCIDIUS

1. In section 95 Calcidius says that the intervals between the planets are also musical intervals, comparing them to the circles of the Sirens in the *Republic*. On this, and generally on Calcidius' probable source in Adrastus, see Waszink, *Studien zum Timaioskommentar des Calcidius*, vol. I, pp. 34–35. Later (sect. 267) Calcidius says: "Without a doubt music rationally adorns the soul, recalling it to its ancient nature and causing it eventually to be as God the creator made it in the beginning."

2. *Timaeus* 36c, with a large lacuna, also left out of Calcidius' translation.

3. *Timaeus* 37a.

4. Calcidius' argument is that since soul was originally made from two elements, the Same and the Different, and from the dual nature of Essence (see *Timaeus* 35a), it is inherently numerical; that just as all numbers are built on 1 and 2, so all creatures and all knowledge are built on the soul.

5. "For as the rational soul of the world holds the middle position between divinity and nature, so the rational soul of man is between God and the animals" (Waszink's edition, p. 102n).

6. Plato's harmonies are all within a world-soul that is inherently rational, not between different degrees of souls. Calcidius' observation makes a transition to the reference in the next paragraph to the creation story of *Genesis*. The "sacred school" of the next section is Judaism, not Christianity (see Waszink's edition, p. 103n).

11 · MACROBIUS

1. In sections 15–16, Macrobius summarizes the numbers 1, 2, 3, 4, 9, 8, and 27.

2. From *kalos*, "beautiful," and *ops*, "voice." On the interpretation "best voice" see Plato *Phaedrus* 259d. [WHS]

3. Vergil, *Aeneid* IV, 244. The words are said of Mercury's caduceus.

4. Ibid., V, 728–729.

5. Porphyry's lost commentary on the *Timaeus*, felt by many modern scholars to be the chief source of Macrobius' work, is here acknowledged (for the one and only time) as the authority for these radii of the planetary spheres, relatively 1 unit (moon), 2 (sun), 6 (Venus), 24 (Mercury), 216 (Mars), 1728 (Jupiter), and 36,288 (Saturn). Yet Macrobius says elsewhere (I, xix, 4; also below, II, iv, 9) that Venus, Mercury, and the sun are so close together that their periods are ap-

proximately the same. He does not have any clear system of his own. On Porphyry's theory, see A. E. Taylor, op. cit., pp. 162–163. Of course Porphyry may have been reporting it as one speculation among others.

6. The standard explanation of pitch from Aristotle (*De anima* II, viii, 420a) onward made high pitch the result of rapid, low the result of slow, speed. Therefore the fastest planets will have the highest notes. Macrobius has actually said (I, xxi, 6): "It is a fact that no planet moves more swiftly or more slowly than the others," but this is the very theory that does not allow for any cosmic harmony. Of the other two, moon-high and moon-low (see our note 3 to Nicomachus, no. 3), Macrobius embraces the latter, more popular theory, but declines to put actual notes to the planets. For an intelligent discussion of Cicero and Macrobius, see Glarean (no. 31).

7. Cf. Nicomachus 4, 22 (see no. 3). The motive power of all the spheres is assumed to be the Primum Mobile, beyond the fixed stars.

8. The diatonic genus alone was in use in the fourth century C.E., but the reason that it is the one used for the Timaeus scale is its mathematical simplicity, not the reasons Macrobius adduces. Nevertheless, the prestige given to the diatonic genus by Plato's use of it in the *Timaeus* and his recommendation of it for paideia may well have led theorists of cosmic harmony to prefer it.

12 · PROCLUS

1. While suggesting that the Sirens sing the eight notes of a scale, Proclus does not specify these notes. The "harmony" of a scale sounded all at once would of course be an extremely discordant "tone-cluster," and Proclus could not allow that. He therefore takes the sound as a metaphor for harmonious activity.

2. The Muses inhabit Olympus, which is beyond the visible cosmos. The "Circle of the rings" is the whole complex of stellar and planetary orbits.

3. Because Soul is prior to Body, and so the Sirens, or planetary souls, are the framework within which the bodies of the planets are fitted, not vice versa.

4. *logikai*. Proclus contrasts the simplicity of these planetary souls who know the Real through direct intellection with the clumsy complexity of our own means of knowledge.

5. Cf. Plotinus, *Enneads* VI, 9, 8 (see *MM&M*).

6. Earlier interpretations of the *Odyssey* make the Sirens represent the

allurements of sensual life which drag the soul down into generation. See Thomas Taylor, *Selected Writings*, p. 329. For the later Neoplatonic attitude, see my note 11 to the *Republic* in *MM&M*.

7. The Greek dual is used, untranslatable as such.

8. The celestial monad, or the principle of unity, proceeds directly into a sevenfold archetype. The dyad, on the other hand, already represents the splitting of the primordial unity into duality and opposition, characteristic of the material but not of the intelligible world. Its proceeding must, Proclus argues, be into two opposing sets of seven, each archetype now having a positive and a negative quality. See also notes 41 and 42, below.

9. The Sirens' work is to attract and to harmonize, wherever they are. Thus the celestial ones make the planets move in mutual harmony; the terrestrial ones bring about the harmonious development of the physical body; and "in the kingdom of Hades there are gods, daemons, and souls, who dance as it were around Pluto, allured by the Sirens that dwell there." (Thomas Taylor, "Additional Notes on the *Cratylus*," in *The Works of Plato*, vol. V, p. 702).

10. I.e., the Muses are the models in the Intellectual World to which the Sirens of all types aspire; for intellectual harmony is the supreme harmony. See also note 16 to Plato (no. 1).

11. ". . . the Seirens, having ventured upon a contest of song with the nine sisters, were deprived of the feathers of their wings, which the Muses subsequently wore as an ornament (Eustathius, *ad Hom.*, p. 85)": Quoted thus in G. R. S. Mead's *Orpheus*, p. 95, who refers this to the first opening and subsequent control of the psychic senses.

12. I.e., the division of the soul, as described in *Timaeus* 35b.

13. Proclus, on beginning his discussion of the Timaeus harmonies, immediately dissociates himself from any confusion of levels that might suppose the harmonies of Life and Intellect to be audible to the bodily ears. He has previously expounded the "mathematical theory" at length.

14. I.e., metaphorically or symbolically.

15. Quoted from *Timaeus* 41b.

16. I.e., the harmonic division restores a new wholeness to the divided soul, just as a musical scale or chord has a wholeness which arises out of the combination of its notes, yet transcends their separateness.

17. I.e., the ratios or intervals, present in every subdivided circle. See note 14 to Plato (no. 1).

18. (1) Soul's undivided wholeness as a blend of Essence, Sameness, and Difference; (2) its wholeness as animator of the entire universe; (3) its

wholeness as animator of every separate creature, e.g., as giver of one's sense of individuality that transcends time and bodily change.

19. Existence or subsistence, in a primordial sense.

20. I.e., in the geometrical middle or proportion, which comprehends arithmetical and harmonical proportion. For if to any three numbers in arithmetical proportion, a fourth number is added, so as to produce geometrical proportion [sc. two pairs of numbers expressing the same geometrical proportion], then this proportion will comprehend both that which is arithmetical, and that which is harmonical. Thus if to the terms 1, 2, 3 a fourth term is added viz. 6, so that it may be 1:2:3:6, then 1, 2, and 3 are in arithmetical, and 2, 3, and 6 in harmonic proportion. [TT]

21. "Monadic," "impartible," "indivisible," "intellect," "the bound," are all expressions of wholeness. "Duad," "dyadic," "partible," "divisible," "infinity," pertain to multiplicity. To follow this exposition it is essential to know the Neoplatonic series of hypostases in their hierarchical order:

One
(Being)
Intellect
Soul
(Nature)
Body

Those in parentheses are not discussed here: hence Soul alone mediates between Intellect and Body.

22. *Timaeus* 31c.

23. A reference to the Pythagorean dichotomy of Bound = the primordial Oneness, and Infinity = the multiple universe.

24. Another Pythagorean dichotomy: that of the Monad and the indefinite or indeterminate Duad (dyad).

25. Cf. our extract from Giorgi (no. 30).

26. The double octave (1:4) first arises in the double progression 1, 2, 4; and $1 + 2 + 4 = 7$.

27. Three is the number characteristic of the very highest members of the Neoplatonic hierarchy, the Intelligibles (cf. the Christian Trinity); seven, of a lower rank, the Intellectual Gods.

28. Viz. in the numbers 1.2.3.4.6. the hebdomad, as Proclus observes, consisting of 1.2. and 4. For 1.2. and 3. are in arithmetical proportion. The numbers 2.3. and 6. and also 3.4. and 6 are in harmonical proportion. And the numbers 1.2.3. and 6. are in geometrical proportion. [TT]

29. *"Horai, daughters of Themis and of Lord Zeus—*
  Eunomie and Dike and thrice-blessed Eirena—"
(Orphic Hymn no. 43) The names of these three Seasons mean "Law-abiding," "Justice," and "Peace."

30. Because a pair of ratios in geometric proportion can be constructed so as to express the other two proportions as well; see note 20. As we have pointed out in note 11 to Nicomachus (no. 3), this is a specious argument, since the placement of a geometric mean, just like the other two means, properly requires only three terms of which the middle is the square root of the product of the other two, e.g., 3:9:27. It is true that the geometric mean between two terms is also the geometric mean between their harmonic and arithmetic means, e.g., the geometric mean of 6 and 12, the square root of 72, is also that of 8 and 9.

31. Proclus now aligns the three means with the three ingredients of Soul: (1) Essence = geometric mean; (2) Sameness = harmonic mean; (3) Difference = arithmetic mean.

32. Proclus explains the Timaeus scale as consisting of 34 notes (four diatonic octaves plus a sixth). Each octave (2:1) and twelfth (3:1) has been divided with an harmonic, an arithmetic, and a pair of geometric means, and the larger intervals that remain have been filled in with tones and leimmas.

33. As, for instance, the contrary divisions of the octave C-C' by harmonic mean into C G C, and by arithmetic mean into C F C', are resolved into a pair of adjacent fourths CF, GC', or interlocking fifths CG, FC'.

34. E.g., in the progression 6:8:12, 6/8 < 8/12.

35. E.g., in the progression 6:9:12, 6/9 > 9/12.

36. E.g., in the progressions 6:8, 9:12, 6/8 = 9/12.

37. Simplicity tends toward the original unity, hence is a characteristic of the Intelligibles. The sensible world, on the other hand, is composed of multiple bodies.

38. The meanings of these ratios are as follows:

Equal	as 10:10
Multiple	as 10:20
Submultiple	as 10:5
Superparticular	as 10:11
Subsuperparticular	as 10:9
Superpartient	as 10:13
Subsuperpartient	as 10:7

For further examples and explanations, see Thomas Taylor, *The Theoretic Arithmetic of the Pythagoreans*, Book I, Chs. 10–20.

39. Because all multiple ratios are reducible to x:1, e.g., 10:5 = 2:1.

40. Because all superparticular ratios are reducible to x:x+1.

41. Because the difference in superpartient ratios is always >1.

42. Because all equalities are reducible to 1:1.

43. Proclus is making the following correspondences:
Point (the impartible)—equality
Line—multiple, submultiple
Surface—superparticular, subsuperparticular
Solid (corporeal)—superpartient, subsuperpartient
This enables any musical interval to be related to a particular stage of manifestation, e.g.:
Point—unison (1:1)
Line—octave (2:1), twelfth (3:1)
Surface—fifth (3:2), fourth (4:3), major third (5:4), minor third (6:5), whole-tone (9:8), semitone (16:15)
Solid—major sixth (5:3), minor sixth (8:5), leimma (256:243)

44. The multiple and submultiple ratios (which, taken in sequence, form geometrical progressions, e.g., 1:3:9:27) show archetypal forms mirrored in descending series through the different levels of being, according to the Hermetic principle of the reflection of the greater world in the lesser.

45. Some reflections of the "solar form" in the various realms might be as follows:

divine soul—Apollo
daemoniacal soul—the god or daemon of Socrates
human soul—Orpheus or another son of Apollo
irrational animal—lion
plant—sunflower
stone or mineral—gold

46. This difficult passage may be clarified as follows: Man:animal is as 2:3 because man is wholly an animal (2), plus one certain animal type, viz. the human (2+1=3). On the other hand, Mule:animal is as 5:3 because it is wholly an animal (3), plus two special types, horse and donkey (3+1+1=5).

47. I.e., every creature relates to its species either in an individual way (as man or horse) or in a multiple way (as mule or other hybrids). Further examples follow in the text.

48. The seven parts are 1, 2, 3, 4, 9, 8, and 27; the seven ratios, the types enumerated above (see note 38 above); the seven circles, those of the Same plus the six of the Different (see *Timaeus* 36d).

49. This refers to the primal division of Intellect into Soul being sevenfold. Seven is paternal, being a "male" (odd) number. It is motherless as a prime number. A recapitulation of the hierarchy of ratios follows, Proclus now interpreting them as types of relationship between things.

50. Refer to the table in note 21 above.

51. The "first monads" are the doublings of unity contained in the Tetraktys: 1, 2, and 4. The ogdoad (8) belongs to the doubling, or "diminishing," of these. The triad (3) belongs to a third order, in which the imperfection of even numbers is "led around" to a semblance of the original unity. This paragraph is comprehensible if one keeps in mind the two different modes of development: (A) that initiated by the number 2, aptly called the "prolific duad" because it begins a splitting process that leads, like the divisions of amoeba, to indefinite multitude; and (B) that begun by the number 3, which is superior to the dyad because it does not only proceed away from unity, but also returns again (2 + 1). This strange Pythagorean arithmology comes to life when expressed in musical terms. Series A is then the series of octaves, fruitless repetitions of a single tone, whereas Series B is the inexhaustible spiral of fifths (a "circle of fifths" only in equal temperament) which brings forth ever new tones. The Neoplatonic mind naturally associates these respectively with the, to them, futile exercise of corporeal procreation, and the truly creative acquisition of knowledge.

52. We have inserted the numbers in brackets to show how Proclus now works in order through the seven numbers of the Timaeus story.

53. "Intellectual conversion" is the process through which one comes, through knowledge of something higher, to resemble that thing. "Generation" is the creation of something lower.

54. I.e., soul is involved in both the degenerative and divisive process of Two, and in the regenerative and unifying process of Three.

55. The intervals under consideration in the preceding paragraph are those of geometrical progressions (2:4:8, 3:9:27), hence symbols of Essence. The relations between their appearance in various states of being are represented by the other means: in some respects they are the same (harmonic means), in others, different (arithmetic means).

56. The circular progress of Soul may be shown as in Figure 35.

57. The meaning of the ensuing passage seems to be that the three types of interval represent different degrees of relationship within the soul.

58. I.e., the two intervals 6:8 (fourth) and 6:9 (fifth) are joined by 8:9 (whole-tone).

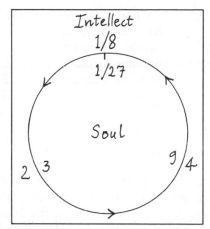

FIGURE 35: *Soul, Intellect, and the Timaeus Numbers*

59. The whole-tones and leimmas represent the penetration of the soul's ordering power into the lowliest manifestations of the world.

60. Proclus now explains the Timaeus scale in its entire range of four octaves plus a sixth as follows:

1 = preexistence of harmony in Intellect (= whole-tone)

4 = soul's participation of harmony in four degrees, 2^0, 2^1, 2^2, 2^3, or 3^0, 3^1, 3^2, 3^3 (four octaves)

10 (1+2+3+4) = physical harmony, derived from the five Platonic or regular Solids (tetrahedron, cube, octahedron, icosahedron, dodecahedron), plus the five centers (North, South, East, West, and the center) = five wholetones

61. This has to be explained via a lost Neoplatonic commentary cited by Ficino (*Theol. Plat.* IV, 128), surmised by Thomas Taylor to be by Proclus or one of the later Platonists (*The Mystical Hymns of Orpheus*, pp. 26–27n). "The professors of the Orphic theology consider a two-fold power in souls, and in the celestial orbs; the one consisting in knowledge, the other in vivifying and governing the orb with which that power is connected. Thus, in the orb of the earth, they call the gnostic power Pluto, but the other Proserpine . . . " The remainder is better summarized in tabular form:

Sphere no. 10	One Invisible Sun	Apollo
	Gnostic Powers	*Vivifying Powers*
9 (World-Soul)	Bacchus Eribromus	Calliope
8 (Stars)	Bacchus Pericionius	Urania
7 (Saturn)	Bacchus Amphietus	Polymnia
6 (Jupiter)	Bacchus Sebazius	Terpsichore

	Gnostic Powers	*Vivifying Powers*
5 (Mars)	Bacchus Bassareus	Clio
4 (Sun)	Bacchus Trietericus	Melpomene
3 (Venus)	Bacchus Lysius	Erato
2 (Mercury)	Bacchus Silenus	Euterpe
1 (Moon)	Bacchus Liknites	Thalia
Fire	Phanes	Aurora
Air	Jove	Juno
Water	Ocean	Tethys
Earth	Pluto	Proserpine

"From all which the Orphic theologists infer, that the particular epithets of Bacchus are compared with those of the Muses, for the purpose of informing us that the powers of the Muses are, as it were, intoxicated with the nectar of divine knowledge; and in order that we may consider the nine Muses, and nine Bacchuses, as revolving round one Apollo, that is about the splendour of one invisible Sun." See also the commentary in Mead's *Orpheus*, pp. 92–96, and Wind, *Pagan Mysteries in the Renaissance*, pp. 277–278nn).

The sesquioctave (8:9) is therefore the communion of the World-Soul with the eight visible spheres beneath it, excluding the earth—for at this stage of the *Timaeus* the elements have not yet been created.

62. The eight Sirens correspond to levels 1–8 in our tabulation (previous note). Apollo is the decad, the number of perfection and completion in Pythagorean arithmology, raising all to a higher unity. Proclus later says: "Hence, as the world consists of eight parts, and also of nine, the ancients, at one time establish eight Syrens, and at another nine Muses, as presiding over the universe, since the harmony of the whole of things proceeds from these" (*On the Timaeus*, ed. cit., vol. II, p. 88).

63. "One, the celestial genus of Gods; another, winged and air-wandering; a third, the aquatic form; and a fourth, that which is pedestrial and terrene" (*Timaeus* 40a).

64. Earth=cube, Water=icosahedron, Air=octahedron, Fire=tetrahedron, surrounded by Ether=dodecahedron. See *Timaeus* 54d–55d.

13 · BOETHIUS

1. Tetrachords synemmenon and meson; see the Greater Perfect System in note 12 to Ptolemy (no. 5).

2. The resultant scale would go: Moon–D, Mercury–C, Venus–B♭, Sun–A, Mars–G, Jupiter–F, Saturn–E.

3. I.e., Cicero, in the Dream of Scipio. See *MM&M*, and Macrobius' commentary and notes, above.

4. This scale would be: Moon–A, Mercury–B [Venus and Sun unspecified] Mars–E, Jupiter–F, Saturn–G, Stars–A.

14 · HUNAYN

1. The *Encyclopedia of Islam* gives Hunayn's dates as 192 A.H./808 C.E.–260/873 (A.H. = after the Hejira, year 622 of the common era), counting in years of twelve lunar months, hence the discrepancy in his age. Amnon Shiloah, in *The Epistle on Music of the Ikhwan al Safa'*, p. 10, gives the birthdate 803 C.E.

2. Ammonius Hermiae, born later fifth-century, died between 517 and 526 C.E., was the son of Hermias, co-pupil with Proclus and Syrianus at the Academy of Athens (founded by Plato). Ammonius also studied under Proclus, then returned to Alexandria to lead the school of philosophy there. Among his pupils were the Athenians Damascius and Simplicius (see *MM&M*); Olympiodorus, the last pagan head of the Alexandrian School, from 550–563; and John Philoponus, his Christian successor (died c. 570). All were important links in the transmission of Greek learning to the Syrian and Arabic domain. On the end of Alexandrian Platonism, see Westerink, ed. *Anonymous Prolegomena to Platonic Philosophy*, pp. ix–xxv, and on Hunayn's sources and references in general, Werner and Sonne, article cited, pp. 558–563.

3. Thus the translation; presumably a lyre was meant in the Greek original, while Arabic readers would imagine an *oud*, the Middle Eastern ancestor of the lute, still in widespread use today.

4. This seems to refer to the Hermetic doctrine of the descent of the soul from its home in the "superior world" above the stars, through the planetary spheres where it makes its "sublime compositions," until it is lured by the vision of sensuous Nature (symbolized in the mirror of Dionysus and the pool of Narcissus) and falls into incarnation. See Aristeides Quintilianus (no. 9) on this descent and music.

5. Rhetor of the second century C.E., familiar of Plutarch and member of the Emperor Hadrian's circle.

6. King Philip of Macedon. Aristotle, also mentioned, was tutor to the young Alexander. The casual mixture of Aristotelian with Platonic sayings reflects the lack of discord between the two philosophers as perceived by Arabic and early medieval scholars. These Aristotle and Alexander stories come mainly from Athenaeus' *Deipnosophistai*.

7. The comparison of tones to virtues recalls Ptolemy, *De Musica* III, 4–6 (no. 5) and Aristeides Quintilianus, but the correspondences with the humors are intrinsic to the four-stringed oud, and must date from the superseding of the classical lyre by this instrument. As soon as four

strings, rather than the classical seven, became the norm, the way was open for sevenfold symbolism (planets, etc.) to give way to fourfold (elements, etc.). See Isaac ben Haim (no. 25), for the penetration of these ideas into Jewish theory.

15 · AURELIAN OF RÉÔME

1. Job 38:37 reads in the Vulgate: *quis enarrabit caelorum rationem, et concentum coeli quis dormire faciet?* The "concert of the heavens" is a mistranslation derived from St. Jerome's *organa coeli;* that in turn from Theodotion's Greek *organa ouranon* (the Septuagint, partly based on Theodotion, has simply *ouranou*); that in turn from the Hebrew expression "the measures of the sky," which the First Targum (fifth-century Hebrew commentary) explains as "the clouds which resemble the waterskins of heaven." Hence the King James Version's "who can stay the bottles of heaven?" See Dhorme, *A Commentary on the Book of Job*, pp. 593–594.

2. This is an early definition of the Seven Liberal Arts, of which the first three (the Trivium) are more generally given as Grammar, Logic, and Rhetoric. The subsequent four (the Quadrivium) are given in the usual sequence, derived by Boethius from the five mathematical sciences of Plato's *Republic* (see our extract from Theon of Smyrna, no. 4).

3. The literal meaning of "geometria" and taken as such by Martianus Capella, for example, whose exposition (*De Nupt.* VI) is almost wholly devoted to earth-measurement as opposed to the abstract, Euclidean science.

4. *Differentiae* are "those variable terminations to the psalm-verse recitation formulae which are necessitated by the varying beginnings of the antiphons that follow them." Lawrence Gushee, "Aurelian" in *New Grove* I, p. 702–704.

5. In Chapter ix these are explained as being untranslatable Greek exclamations of rejoicing, which contain modal melodies. See Gustav Reese, *Music in the Middle Ages*, p. 173.

6. This was also the object of Heinrich Glarean's *Dodecachordon* (see no. 31), though he began not with the Greek but with the eight church modes.

16 · JOHN SCOTUS ERIUGENA

1. From Eriugena's *Periphyseon* or *De Divisione naturae* (*Patrologia Latina*, vol. 122, col. 966).

2. The heading is original. This part of the commentary is on the passage in Martianus Capella's *The Marriage of Philology with Mercury* I, 11–12, that describes the musical trees in the Grove of Apollo (see *MM&M*, pp. 305–306). On Eriugena's musical thought, see the articles of Handschin and Münxelhaus listed in the Bibliography.

3. The only previous author of a double-octave system for the planetary music was Favonius Eulogius (late fourth century B.C.E.), pupil in rhetoric of Saint Augustine and commentator on the Dream of Scipio. Favonius accepted the "Pythagorean" scale of an octave from earth to stars, following the tuning of Pliny (see no. 2), but having only a semitone between Saturn and the starry sphere. Then he drew the sensible conclusion that "since the universe is formed from two hemispheres, the upper and the lower one, it must comprise in all a double octave or twenty-four semitones" (*Disputatio*, para. 25).

4. Eriugena has two different methods of assigning pitches. One is based on the positions of the fixed orbits. Thus in *Periphyseon* (it 122, col. 722) he estimated "not irrationally" that there must be one octave from the earth to the sun, another from the sun to the starry sphere, using words similar to those of the present Commentary. Earlier in the Commentary (f. 9–10') he has expounded a slightly different version, with the sun bisecting a double octave between the moon and Saturn. Eriugena's second principle is based on relative speeds, which give rise to the ratios now explained. Sometimes the tones of both systems coincide, leading even the author (I believe) into confusion.

5. The relative speeds are, from slowest to fastest: Saturn, Jupiter, Mars, Sun, Venus, Mercury, Moon, Sphere.

6. Eriugena's mathematics is not strong. Here three whole-tones make a perfect fifth, while elsewhere he takes the value of π to be 2 (see Handschin's article, p. 331).

7. Eriugena is avoiding the attribution of tones to the spheres themselves—the idea of Plato's *Republic*—in favor of deriving musical proportions from the distances between them—the principle of the *Timaeus*. The principle of harmonious proportion can then, of course, be applied to all manner of nonmusical situations, as at the beginning of the next paragraph.

8. *Sic non locus siderum sed sonus caelestem componit armoniam.* The *locus* is the orbit of the planet, the *sonus* the particular pitch made by its speed. Eriugena is now back in the *Republic* mode (see preceding note).

9. Because the addition of one-eighth to the length of a string gives the whole-tone 8:9.

10. The two numbers, multiples respectively of 8 and 9, which give a

whole-tone in the middle of the Dorian octave 144:288. In *Periphyseon* (PL 122, col. 722) Eriugena gives his own estimate of the cosmic distances in stadia.

17 · REGINO OF PRÜM

1. Regino's source is Boethius *De Musica* I, 1 and 2.
2. See note 2 to Eriugena (no. 16).
3. This is an adaptation of the chapter from Boethius given in no. 13.

18 · THE IKHWAN AL-SAFA'

1. Al-Kindi (first half of the ninth century) also gives correspondences of the strings to the four elements and humors, as well as of the twelve parts of the oud to the Zodiac. He gives the following relative pitches for the planets, Saturn's being the open lowest string of the oud: Moon-f#, Mercury-f, Venus-e, Sun-c#, Mars-c, Jupiter-B, Saturn-A. This scale most resembles the chromatic one of Pliny, Censorinus and Theon (see nos. 2, 4, and 6), and differs markedly from what Al-Kindi has previously given as the seven notes of the oud's scale, beginning with the lowest: A, B, c or c#, d, e, f or f#, g. See Shiloah, "Un ancien traité sur le 'ud ... "

2. In chapter 15 of the treatise (not given here) there is a long list of the four qualities as manifested in the year, the body, winds, seasons, colors, etc. Here the day and night are also divided into four quarters. The exploitation of these correspondences is a form of sympathetic magic, while it also underlies Galenic medicine as the balancing of the four humors.

3. Werner and Sonne, in "The Philosophy and Theory of Music in Judaeo-Arabic Literature," Part 1, p. 274, make a distinction between the uses of music as a homeopathic and an allopathic form of therapy. They say that the former, following the principle of "like cures like," was favored by Aristotle and Galen, e.g., curing the Corybantes with corybantic tunes, while the Brethren of Purity, like the Pythagoreans, preferred to counter a psychological state with its musical opposite.

4. The ensuing explanation is confusing. First it presents five elemental spheres, each one's diameter related to the next in the proportion 4:3, giving the following plan:

Ether/Fire 4:3 Frigidity 4:3 Air 4:3 Water 4:3 Earth

Then it names only four, related in the same proportion as to their relative subtlety:

Fire 4:3 Air 4:3 Water 4:3 Earth

Although the two systems are not entirely incompatible, one suspects that the fourfold model is supposed to underlie them, expressed first as four intervals, then as four tones. The next paragraph concerns the placement of the lowest string on top, when the lute is held in playing position. Here the comparison is made to yet another set of elements: Ether, Frigidity, Air, Earth. These differences are due to the fact that the Brethren of Purity distinguished only two principal elemental spheres beneath the moon, Air and Earth, but subdivided them in various ways. See Shiloah's notes to the French edition of the treatise (see Bibliography), part 2, p. 160.

5. A number whose factors add up exactly to it, e.g., 6, 28, 496, 8128. These were held in great regard by the Pythagoreans; see Thomas Taylor, *Theoretic Arithmetic*, pp. 130ff.

6. The four numbers of the cube, 24:12:8:6, applied to string lengths, give the first four tones of the harmonic series; the terms are thus all in harmonic proportion to one another.

7. In *Timaeus* 55e the Demiurge assigns the cube to the element of earth, the dodecahedron to the heavens.

8. The following scale emerges from these numbers: Earth (8)-b', Air (9)-a', Moon (12)-e', Mercury (13)-slightly sharp d', Venus (16)-b, Sun (18)-a, Mars (21½)-slightly sharp f, Jupiter (24)-e, Saturn (27⅐)-slightly flat d, Stars (32)-B. It is curious that a very similar set of numbers occurs, with the planets in reverse order, in a sixteenth-century Greek manuscript (Naples, Bibl. Naz. III C 3) containing musical material by Porphyry, Nicomachus, Ptolemy, Plutarch, Theon, and Macrobius. The passage was published by von Jan in "Die Harmonie der Sphären," pp. 29–30, and gives, in numbers and Greek note-names, the relative pitches: [Stars conjectured to be 36-b'], Saturn (32)-a', Jupiter (24)-e', Mars (21⅓)-d', Sun (18)-b, Venus (16)-a, Mercury (12)-e, Moon (9)-B, Earth (8)-A. Von Jan traces this scale to Ptolemy's incomplete final chapter (see Ptolemy, no. 5), and surmises that originally the numbers should have gone in the reverse direction so as to reflect relative string lengths. But if the notes, instead, are reversed, inverting all the intervals, then we have virtually the scale of the Brethren of Purity. No doubt a common Greek source, more or less misunderstood, underlies both.

9. *Sic.* The ratio should be 16:9, i.e., an octave less a whole-tone.

10. It may have been noticed that the proportions expounded are not consistent, either within the chapter (in which the interval from earth to moon as formerly given would be far greater than the 8:9 whole-tone

given here) or elsewhere in the *Encyclopedia* (see the planetary distances as given in Nasr, *Introduction to Islamic Cosmological Doctrines*, p. 78). Perhaps this was inevitable in a book compiled by a group of enthusiasts.

11. See Professor Shiloah's long note on the problems raised by this choice of letters, ed. cit., p. 47n. His conclusion is that behind it lies a classification akin to the number correspondences of Qabala and of Greek gematria.

12. The *Khurramiya* were a Manichean sect; the *Kayyaliya* followers of Ahman ibn al-Kayyal, ninth-century philosopher of Ismaelian and gnostic tendencies [AS, abbreviated].

19 · AL-HASAN AL-KATIB

1. The "noble relationship" is the superparticular ratio; cf. note 38 to Proclus on the *Timaeus* (no. 12).

2. This comparison is taken from Ptolemy's *Harmonics* III, 5 (see no. 5).

3. These are evidently the same faculties as mentioned by Ptolemy. According to Shiloah, their origin is a puzzle; they do not correspond to any lists in Wolfson's "The Internal Senses in Latin, Arabic, and Hebrew Philosophic Texts" (in his *Studies in the History of Philosophy and Religion*, vol. I, pp. 250–314).

4. Dionysos is mentioned again in the penultimate chapter, among the philosophers and theoreticians who have contributed to the development of music. [AS]

5. Literally "notes," but this term, as we have said, also signifies "modes." [AS] They are specified, as is usual in Arabic music theory, by playing positions on the oud. These virtues and notes do not correspond to anything in Ptolemy, nor, so far as I know, to anything else Greek.

6. Cf. Ptolemy's alignment of the Greater Perfect System with the circle of the Zodiac (see no. 5, Ch. 8).

7. The argument is that both the three perfect consonances (2:1, 3:2, 4:3) and the twelvefold Zodiac (2×6, 3×4, 4×3) embody the same numbers.

8. This last sentence, translated literally, is an enigma. After long research we believe that the author is expressing here very poorly a Platonic theory that concerns the "harmonic mean": 21:8::8:6. This theory was repeated and commented on by Nicomachus in his *Manual of Harmony*, pp. 26–27 in the translation of C.-E. Ruelle. We refer the reader to this work in which the question is explained very clearly and in which he can verify our hypothesis. [AS] The passage in question seems to be Nicomachus' eighth chapter (Ruelle, pp. 24–25); see no. 3.

20 · ANONYMOUS OF THE TWELFTH CENTURY

1. I.e., Cicero, in the Dream of Scipio; see Macrobius (no. 11).

2. The scale is evidently the Pythagorean one of the *Timaeus* (see no. 1), composed of tetrachords each containing two 9:8 whole-tones and one 256:243 leimma.

3. It seems strange to call the *mese* of the fixed stars "heavy," but perhaps it is now considered in comparison to the Angelic realms beyond.

4. *Vocum*, also meaning "notes."

5. The seven octave-species are those starting on each diatonic note; the types of fourth and fifth, those brought about by different internal arrangements of two tones and a semitone.

6. *Septem discrimina vocum*, quoted from Vergil, *Aeneid* VI, 645–646, who there describes Orpheus:

> *Nec non Threicius longa cum veste sacerdos*
> *Obloquitur numeris septem discrimina vocum.*

Fabio Paolini, a professor of Greek at Venice, would devote a large book to the exegesis of this phrase, speaking on p. 110 of this Seven as "a vessel in which to voyage through ancient learning." See D. P. Walker, *Spiritual and Demonic Magic from Ficino to Campanella*, pp. 126–144.

7. Incomprehensible word, conjectured by Handschin to resemble "Acelnae."

8. Perhaps the meaning is to be sought in the fifty-six year span of life allotted to Scipio (see Cicero *Somn. Scip.* II, 2), on whose factors Macrobius makes an immense excursus (*De Somn. Scip.* I, v–vii).

9. *Vivitur octavo post septem milia, credo.* One can only guess at what this could mean: eternal life in the eighth sphere of Heaven, when the world ends its seven-thousand-year existence?

10. This line is a virtual quotation from Cicero, *Somn. Scip.* V, on the Seven, *qui numerus rerum omnium fere nodus est.*

21 · ISAAC BEN ABRAHAM IBN LATIF

1. The quotation continues: "And it came to pass, when the minstrel played, that the hand of the Lord came upon him," implying that music could be a means to open the soul to prophetic inspiration. The story will recur in the extract from Isaac ben Haim (no. 25).

2. It is not clear what are meant by these eight modes. Rather than scales or rhythmic modes, they may have been bodies of melody compa-

rable to the Indian ragas. See Werner and Sonne in ed. cit., pp. 295–300, and cf. also the "modes" with strange names given by Aurelian (no. 15).

3. Translated as the "voice" of the Lord, which in turn: (1) "echoes over the waters," (2) "is power," (3) "is majesty," (4) "breaks the cedars," (5) "makes flames of fire to burst forth," (6) "makes the wilderness writhe in travail," (7) "makes the hinds calve and brings kids early to birth." The psalm continues: "and in his temple all cry 'Glory!'" Isaac's reticence after this parallel tempts one to suspect a Kabbalistic interpretation, and the comparison of these seven powers of God's voice with the seven lower Sephiroth is certainly alluring.

4. These categories resemble those of Ptolemy's *Harmonics* III, 10–12 (see no. 5) more closely than any other classical source of astral-musical correspondence.

22 · Jacques de Liège

1. Aegidius Zamorensis (thirteenth century), in *Ars musica* (Gerbert, *Scriptores* II, 370ff.) was the first to add to the Boethian categories *musica coelestis*; but for him, *musica mundana* was the music of the seasons, elements, etc., while *coelestis* was that of the heavenly spheres. For Nicolaus of Capua (*Compendium musicale*, 1415), on the other hand, the fourth division was *musica angelica*, but not with the contemplative and intellectual meaning which Jacques has in mind: like Ugolino of Orvieto (see no. 23), he meant by it the music that the angels actually sing before the throne of God. See Pietzsch, *Klassifikation*, pp. 94, 118.

2. Isidore of Seville, *Sententiae de Musica*, Ch. 3; in Gerbert, *Scriptores* I, p. 20b.

3. On the movers or souls of the spheres, see Wolfson, "The Problem of the Souls of the Spheres . . ."

4. See the introduction to Boethius, no. 13.

5. Simplicity is superior to multiplicity, being a characteristic of the divine unity. Cf. Proclus on the *Timaeus* (no. 12).

6. I.e., the heavens of the seven planets, the fixed stars, and the *primum mobile*, the latter being added as an "unmoved mover" for the eighth sphere, or to explain the precession of the equinoxes.

7. According to Dionysius the Areopagite, *Celestial Hierarchies* (see *MM&M*), there are three hierarchies each containing three orders of angels. We will find these enumerated in Berardi (no. 38).

8. See note 1 to Aurelian (no. 15).

9. Aristotle, *De Gen. et Corr.* II, 6, 333a.

10. The identity of this commentator is unknown.

23 · UGOLINO OF ORVIETO

1. Presumably Johannes Sacrobosco or John Holywood, early thirteenth century.

2. Thebit ben Corat, a ninth-century Sabian of Harran, wrote on magic and astrology. Some of his astronomical treatises were much copied in medieval manuscripts, among them *De motu octave spere*. See Thorndike, *History of Magic and Experimental Science*, vol. I, pp. 661ff.

3. Virtually a quotation from Boethius, *De Inst. Mus.* I, 2.

4. *Saturnalia* I, 21–22, tells of the spheres being symbolized by the syrinx of Pan, God of the Universe.

5. See our extract from Pliny (no. 2).

6. Mentioned in the *Dream of Scipio* II as "Catadupa."

7. Presumably *De Caelo*, Book III, although Aristotle's refutation is in ibid., II, ix, 290b. (See *MM&M*, under Simplicius.)

24 · GIORGIO ANSELMI

1. *Harmonia celestis*, evidently meaning both that of the planets and that of the angels. As we shall discover, they are not in fact distinguishable.

2. Anselmi is summarizing the creation of the World-Soul in Plato's *Timaeus* (see no. 1). He quotes directly from Calcidius' version (see no. 10) in section 138.

3. Cf. note 6 to the Brethren of Purity (no. 18).

4. Anselmi now turns to *musica humana* (though he does not use that term), citing several of the favorite instances, mostly taken from Boethius *De Inst. Mus.* I, 1.

5. In the Galenic medical system still favored in the fifteenth century, the four humors of melancholy, phlegm, sanguine, and choler, corresponding to the elements earth, air, water, and fire (see nos. 18, 19, 25), were supposed to be kept in balance for health to be maintained. Comparison with the harmony of the world's elements referred to by Boethius naturally followed, given the parallel of microcosm with macrocosm; both could be envisaged as measured out in right or wrong proportions.

6. These are the souls or intelligences of the planets.

7. Anselmi makes a distinction between the harmony within the World-Soul, and the harmonies of the separate planetary intelligences. In the following sections a polyphonic effect is described for the first time in our texts.

8. *Phthongoi* are notes; the other terms those of the smallest Greek intervals.

9. These are respectively the Primum Mobile, responsible for the precession of the equinoxes and revolving only once in about 25,000 years, and the Eighth Sphere, carrying the fixed stars around the earth once in twenty-four hours.

10. The planets therefore make one sound or vibration by their journey around the Zodiac, another by their daily revolution around the earth.

11. The relative frequencies of the planets' zodiacal revolutions are given here as: Saturn (30 years)–C_2, Jupiter (12 years)–G_1, Mars (3 years)–g, Sun, Venus and Mercury (1 year)–d^2, Moon (1/12 year)–d^5. Anselmi's arithmetic is not very accurate.

12. This refers to the daily revolution of all the spheres, carried along by the Eighth Sphere.

13. This comparison is even more forced than the last. It comes to a mere comparison of the proportions of various aspects (see Ptolemy *Harmonics* III, 15—our no. 5); but there is nothing enharmonic about these intervals, unless Anselmi has in mind the close approximations to the aspects—the astrologically significant "orbs" of the planets—which could be likened to small deviations from pure tuning.

14. See Proclus' commentaries on Plato's *Republic* and *Timaeus* (no. 12). Anselmi plainly makes the equation: Sirens = souls of the spheres = nine orders of angels. Dante seems to have been responsible for the latter connection (*Convivio* II, vi; *Paradiso* 28). His system is conjectured as: Seraphim-Primum Mobile, Cherubim-Starry heaven, Thrones-Saturn, Dominions-Jupiter, Virtues-Mars, Powers-Sun, Principalities-Venus, Archangels-Mercury, Angels-Moon. See Hutton, "Some English Poems in Praise of Music," for some instances of planet-angel correspondence. Without pretending to any expertise in the matter, I suspect that Dante's source was Islamic. In the system of Ibn Sina ("Avicenna"), for instance, the planetary heavens are "each governed by an intellect or group of angels" in addition to their soul and body (see Nasr, *Introduction to Islamic Cosmological Doctrines*, pp. 236–238). The Christian application of such a system would demand its accommodation to the hierarchies of Dionysius—"our theologian" of the following section.

15. Notably Martianus Capella, *De Nuptiis* I, 28.

16. Anselmi presents himself ("George") as the listener, Petrus as the speaker of this passage.

25 · ISAAC BEN HAIM

1. The term *meshorer* means literally poet, but in view of the context we think it refers to the combination of poet and musician. [AS]

2. The original text of the remainder of this sentence is corrupt, and has been amended with reference to the treatise of the Brethren of Purity, who devote much space to this theory. [AS]

3. The word used in the text as the equivalent of *be-r'Och* literally signifies excrement. This interpretation, like those which follow, is based on a linguistic analysis making the word in question derive from another root.

4. It appears that the musicians, in this case the Levites, must have served the two groups at the same time. [AS]

5. This interpretation comes in the Talmud, *Hagiga* 5b. [AS]

6. This story is told in the Talmud, *Berakhot* 3a. It concerns Rabbi Yosse, who went into a ruin to pray and was protected by the prophet Elijah. The dialogue takes place between the two of them. [AS]

7. It is possible that the reference here is to the seven liberal arts, whose purpose is to prepare the student for the study of perfect wisdom. [AS]

8. Our interpolation, in parentheses, is necessary for the proper understanding of a poorly stated proposition; it merely elucidates the author's theory, without any alteration. [AS]

9. In this phrase, opinion or concept is opposed, and inferior, to intellectual knowledge. The latter, like the music attached to it, cannot be manifested at night, which symbolizes the world of generation and corruption which is subject to permanent transformations. Opinion, on the other hand, together with the sounds, imagination, and music that go with it, belong to the night, i.e., to our lower world.

10. The following commentary is found in the treatise *Ta'anit* 5a. [AS]

11. For purposes of demonstration, the author gives to the word *omer* (word), from the root *amor* (say), a very distant and improbable meaning, making it derive from the root *mrr*; the word *vagitmarmar* appears twice in Daniel 7.8, 11.11. In addition, this passage is entirely based on this sort of metaphorical interpretation, one of whose purposes is to demonstrate the opposition of day and night, with all that follows from that, and to introduce two other concepts opposed to the one that follows in the next paragraph. [AS]

26 · Marsilio Ficino

1. For a more general treatment of musical philosophy in the context of the Platonic "frenzies," see Letter 7 to Peregrino Agli, 1 December 1457 (in *The Letters of Marsilio Ficino*, vol. I, pp. 42–48).

2. One of the three Benivieni brothers, all members of Ficino's Academy.

3. *Republic* X, 591d.

4. *Symposium* 187 d–e. This would inspire the fantasy of Dalberg (no. 44). Further on Ficino and Muse-lore, see note 61 to Proclus (no. 12).

5. *Asclepius* IX, 2.

6. *Musicus absolutissimus.*

7. I cannot find the source for this statement of the Pythagoreans, but the number of correspondences Ficino mentions encourages me to add the cosmogonic parallel of the Orphic World-Egg, surely at the back of any Pythagorean's mind: on which see the enigmatic passage in Fabre d'Olivet, *Music Explained as Science and Art*, pp. 99–100.

8. Cf. notes 14 and 15 to Ptolemy (no. 5). Ficino is alluding to the traditional meanings of the twelve "houses" into which the zodiacal circle is divided.

27 · RAMIS DE PAREJA

1. Ramis calls these four modal divisions simply *protus, deuterus, tritus,* and *tetrardus.* Later he distinguishes two *tonoi* in each *modus.* Although his citations are mainly from classical sources, he is evidently thinking of the eight Church Modes, as numbered here: I=Dorian, II=Hypodorian, III = Phrygian, IV = Hypophrygian, V = Lydian, VI = Hypolydian, VII = Mixolydian, VIII = Hypomixolydian. The sequences of notes in these modes are quite different from those of the Greek modes of the same names, e.g. the Greek Dorian runs from E' to E, whereas the Church Dorian from runs D to D'. This makes nonsense of any attempt to apply the Greek ideas of modal ethos to the later scales.

2. Boethius, *De Inst. Mus.* I, 1, actually says that the Taorminian youth was calmed by "a spondaic rhythm."

3. The notes which Ramis assigns are as suggested by Boethius (see no. 13) to fit Cicero's description. Ramis then treats them each as the bottom note of the appropriate modal scale, e.g., Mercury's B becoming the limit of the Hypophrygian octave BCDEFGAB'.

4. *Theogony* 77–79.

5. The correspondences of planets with Muses are as in Martianus Capella, *De nuptiis* I, 28.

6. Ramis ends his first book with a diagram (reproduced and redrawn in our source, p. 61), attributed to Roger Caperon, in which the two octave gamut has each pair of notes an octave apart linked, in turn, with a spiral band, thus: A-a-B-b-c-c', etc. On the band is written, at the appropriate note, the name of the Muse, that of the corresponding

planet, and Latin hexameters summarizing the characteristics of each. See James Haar, "Roger Caperon and Ramos de Pareja."

28 · PICO DELLA MIRANDOLA

1. These are the three hypostases of the *Timaeus:* the Same (unity), the Different (duality), and Essence (that which is).

2. By "unitary," odd numbers are probably meant, by "generative," even numbers, while "substantial" ones are those produced by both even and odd factors. See Thomas Heath, *Greek Mathematics,* vol. I, pp. 70–71.

3. The first numerical progression, depicted as the apex of the Tetraktys.

4. Cf. Macrobius *In Somn. Scip.* I, v, 11: "The reputation of virginity has so grown about the number seven that it is called Pallas. Indeed, it is regarded as a virgin because, when doubled, it produces no number under ten, the latter being truly the first limit of numbers. It is Pallas because it is born only from the multiplication of the monad, just as Minerva alone is said to have been born of one parent."

5. For the three types of proportion, see notes 8 and 9 to Plato (no. 1), note 11 to Nicomachus (no. 3), and notes 29 and 30 to Proclus (no. 12).

6. This and the following Conclusions give proportions for the three parts of the soul defined in Plato's *Republic* IV, 441a: Rational=4, Irascible=3, Concupiscent=2. This differs from the assignment to them respectively of the octave, fifth, and fourth in Ptolemy's *Harmonics* IV, 5 (see no. 5), which would make their numbers 12, 9, and 8.

7. An epigrammatic statement of the Pythagorean as against the Aristoxenian view.

8. Possibly because the soul, in Plato's *Timaeus* 54ff., is made of plane geometrical figures.

9. Possibly because the stage of creation corresponding to the gods in the *Timaeus* is the division of the World-Soul by arithmetical units.

10. Two of the *Conclusiones de mathematicis secundum opinionem propriam* (ed. cit., p. 74) are also relevant:

> 7. As medicine moves the spirits principally in order to regulate the body, so music moves the spirits in order to serve the soul.
> 8. Medicine heals the soul through the body, but music heals the body through the soul.

Did Oscar Wilde knowingly echo this in *The Picture of Dorian Gray,* when he wrote: ". . . to cure the soul by means of the senses, and the senses by means of the soul"?

29 · FRANCHINO GAFORI

1. On this emblematic woodcut, see James Haar, "The Frontispiece of Gafori's *Practica Musica* (1496)" in *Renaissance Quarterly* 27 (1974), pp. 7–22; Edgar Wind, "Gafurius on the Harmony of the Spheres" in his *Pagan Mysteries in the Renaissance*; S. K. Heninger in *The Cosmographical Glass*, pp. 136–138. Athanasius Kircher had the illustrations redrawn in more baroque taste for his own *Arithmologia* of 1665.

2. *Metam*. V, 280: *Mnemonides (cognorat enim) consistite dixit.* [CAM]

3. Referring to the Homeric Hymn to the Muses.

4. This verse, which is the heading of [Gafori's] diagram, . . . is attributed to Ausonius. See Haar, "The Frontispiece of Gafori's *Practica Musica* (1496)," p. 11. [CAM]

5. See note 6 to the twelfth-century anonymous poem (no. 20).

6. Twenty-eight is both 7 + 6 + 5 + 4 + 3 + 2 + 1, and the sum of its factors 14 + 7 + 4 + 2 + 1, the latter making it a "perfect number."

7. The following passage, up to the mention of Callimachus, is based entirely on Fulgentius, *Mythologia* I, 5, "The Fable of the Nine Muses" (see Bibliography for a translation with useful notes, pp. 55–57). Gafori allows the reader to assume that he, rather than Fulgentius, is familiar with the obscure writers whom he cites.

8. A salutary warning to scholars!

9. "When he doesn't see the shoots appear he is consumed by hunger." Fulgentius, ed. cit., p. 56.

10. This quotation (not, in fact, from *Poimandres*), is also given by Fulgentius.

11. The attributions of zodiacal signs are the traditional planetary rulerships.

12. See Aristeides (no. 9).

13. *De Musica* II, 16 (not in our extract).

30 · FRANCESCO GIORGI

1. See Frances Yates, *The Occult Philosophy in the Elizabethan Age*, pp. 29–36 and *passim* on Giorgi's situation among the intellectual currents of his time.

2. By Giorgi's time it had become commonplace to style Plato a Pythagorean, partly because of the long period during which the only Platonic dialogue known in Latin was the explicitly Pythagorean *Timaeus*,

and partly also because of the idea of an ancient theology whose last representatives, according to Ficino, were Pythagoras, Philolaus, and Plato. Ernest McClain's researches show that this may have been closer to the truth than the modern scholarship that has been at pains to separate the two (see his *The Pythagorean Plato*).

3. The diatonic scale in Pythagorean tuning; see note 12 to Plato (no. 1).

4. This entire passage is a paraphrase from Proclus on the *Timaeus* (see no. 12).

5. See note 26 to Proclus.

6. Giorgi borrowed his scheme of novenaries or enneads from Pico della Mirandola (see no. 28), who in his *Heptaplus* assigns the three worlds (angelic, celestial, and sublunary) nine levels each:

1. Nine orders of angels
2. Primum mobile, firmament, plus 7 planetary spheres
3. Animals, plants, zoophytes, mixtures, metals, plus 4 elements.

The number 27 is interesting as 3^3 and the limit of the World-soul numbers in the *Timaeus*, but it becomes even more so with the addition of 1, representing the unity of God above the chain of being: for 28 is a perfect number, the sum of all its factors $(14 + 7 + 4 + 2 + 1)$; also the sum of the first seven digits; and the sum-total of the notes of the Greater Perfect System in all three genera (diatonic, chromatic, enharmonic), as Giorgi observes in Chapter 16, below.

Another feature of Pico's numerology is his use of simple or root numbers, squares, and cubes to represent degrees of being, either those of the three worlds (which are successive degrees of "solidification," culminating in the cubical and solid world of the four elements), or within each one of them. The *Timaeus* had contained this idea in its use of the plain, square, and cubic numbers, but Pico leans more toward Kabbalah, in which the Hebrew letters are numerically equivalent to the first nine digits, then to multiples of ten, then to hundreds. All the tens are counted as "squares," all the hundreds as "cubes." Therefore the "novenary of hundreds" referred to by Giorgi at the beginning of this chapter is the ninefold sublunary world whose numbers are 100, 200 . . . 900. To it corresponds, in man, the ninefold body described here as made from the four material elements, four subtle ones, plus the vital spirit. See Heninger's *The Cosmographical Glass*, pp. 92–94, for a diagram and further explanation of this numerical system.

While Giorgi never set out a complete table of correspondences, it is possible, and instructive, to compile one from various parts of his work. In Book I, 4 (La Boderie's translation, pp. 103ff.) he demonstrates how he has worked out the correspondences of angelic orders with spheres on the basis of the classical writers, the Fathers and Doctors of the Church. In I, 8, 21 (ibid., pp. 329ff.) he explains that the spheres receive their planetary powers from the Sephiroth and the Names of God. These correspondences are already present in Hebrew sources (e.g., *Sifra di-Zeni'uta* IV, 32—Mathers' translation, p. 104), while some neo-Kabbalists also incorporate the trans-Saturnian planets, substituting Uranus for the fixed stars and Neptune for the primum mobile (e.g., Regardie, *A Garden of Pomegranates*, p. 93). Here is a reconstruction of Giorgi's table of correspondences:

Names of God	Sephiroth	Angelic orders	Spheres
Ehieh asher ehieh	Kether	Seraphim	P.M.
Sem mephoras	Chokhmah	Cherubim	stars
Hihvh	Binah	Thrones	Saturn
El	Chesed	Dominations	Jupiter
Elohim	Gevurah	Virtues	Mars
Ihvh	Tiphereth	Powers	Sun
Elohim Zervaoth	Netzach	Principalities	Venus
Ihvh Zevaoth	Hod	Archangels	Mercury
Sadai, Elhai,			
Adonai, Melec	Yesod	Angels	Moon
	[Malkuth]		[Elements]

It will be noticed that the angel-sphere correspondences are different from those of Anselmi (see no. 24) and his followers. Giorgi's, however, are the same (except for the common exchange of Powers with Virtues) as those of the *Magical Calendar* popularly attributed to Tycho Brahe but more likely deriving from the Abbot Trithemius' circle (see Adam McLean's edition). Giorgi's Sephiroth-sphere correspondences are also the same, and his names of God similar to those of the Calendar.

Following the suggestion near the end of our extract, one might place Saturn at low A and the other planets ascending the scale. But Giorgi is far from believing it possible, or useful, to assign actual tones to any of these entities; questions such as planetary sounds and distances are foreign to his contemplative character.

7. These two bodies would seem to be what the Neoplatonists of Iran knew as the two *jasads*. See H. Corbin, *Spiritual Body and Celestial Earth*, pp. 91–96.

8. The inner articulations and operations of the twenty-sevenfold being are now likened to numerical progressions, hence to musical harmonies.

9. Chapters 11–14 are omitted here. Ch. 11 is on the three types of proportions and what they signify, drawing again on Proclus. Ch. 12 tells why the Pythagoreans and Academics enclosed the discourse of the

world and the soul within seven limiting numbers (1, 2, 3, 4, 9, 8, 27). It explains the symbolism of numbers from both Hebrew and Christian sources. Ch. 13 shows how the World-soul's intervals are filled out, giving Proclus' scale that runs from 384 to 10,368. Ch. 14 is on the agreement of what is said of the intelligences of the spheres with the scriptures.

10. Proportions according to the formula x : x + 1, such as the *Timaeus* scale also uses. See note 38 to Proclus.

11. The quotation used by Saint Paul in his speech on the Areopagus of Athens (Acts 17:28) is sometimes attributed to Epimenides; Paul continues "For we are indeed his offspring," quoting from Aratus' *Phaenomena*.

12. The Kabbalists speak of a mystical experience which removes the consciousness from the body in such a way that physical death ensues (cf. Lemech's Vision and Death, in *MM&M*). Pico della Mirandola was greatly preoccupied with this experience, according to Frances Yates, *Giordano Bruno and the Hermetic Tradition*, p. 99, with references to his works.

13. Cf. Jacques de Liège (no. 22), for whom the types of knowledge are meridional and vespertine. The "knowledge of things in themselves" is not to be taken in a Kantian sense: it is the inferior of the two types of knowledge.

14. The *Sepher Yetzirah*, the oldest book of the Hebrew Kabbalah (see Bibliography).

15. I.e., Solomon. The reference is to Wisdom 11:20.

16. I.e. twelve as limit of the tuning matrix 6:8:9:12, also known as the four-stringed Lyre of Hermes.

17. The following account of the enlargement of the lyre is compiled from Boethius and Nicomachus.

18. Giorgio's vision anticipates the "Pythagorean Table" of Albert von Thimus (see no. 49), which fills in the intervals between the limbs of the lambda. Hans Kayser (*Lehrbuch der Harmonik*, pp. 167ff.) would later develop the idea into the third dimension, just as is suggested here.

31 · HEINRICH GLAREAN

1. Hymn to Hermes, *The Homeric Hymns* II, 51. [CAM]

2. Vergil, *Eclogues* II, 36. [CAM]

3. See note 6 to the Twelfth-Century Anonymous (no. 20).

4. *Saturnalia* I, 19, 1 [CAM]

5. *Republic* X, 617b.

6. Diodorus Siculus, *History* I, makes Hermes' lyre trichord, but Nicomachus says that it had seven strings (*Excerpta* Ch. 1; v. Jan, p. 266). Boethius *De Inst. Mus.* I, 20 makes it tetrachord and aligns its strings with the four elements. The tetrachord tuning naturally followed the 6:8:9:12 Tetraktys,—see note 10 to Plato (no. 1)—and made the lyre an archetype of cosmic harmony.

7. The diagram is an elaboration of one in Giorgio Valla, *De Expetendis et fugiendis rebus opus* (Venice, 1501), vol. I, f.e8'; reproduced in Heninger, *The Cosmographical Glass*, p. 139. Glarean's version is an ingenious presentation of the moon-low scale of Cicero and the moon-high one of Boethius. On the left the eight strings are given the Greek note-names, from *mese* (stars) to *proslambanomenos* (moon). The arcs mark off the various intervals, with, at the bottom, the tetrachordal matrix A D E A' (see note 6 above). On the right, the strings go down from the high A of the moon to the low A of the stars.

8. The reference is to the passage cited in note 7 above. This is certainly a novel solution of Pliny's scale (see note 4 to Pliny, no. 2).

9. Aeneid IX, 525: *"Vos, o Calliope, precor aspirate canenti."*

10. Aeneid VII, 37: *"Nunc age, qui reges, Erato, quae tempora rerum."*

11. Glarean gives the same order of muses as Hesiod, *Theog.* 77–79, thus differing from the Martianus-derived correspondences of Ramis (no. 27) and Gafori (no. 29). Glarean's diagram shows the following, moon-high system, using the Aeolian mode transposed up a fourth:

D'	Moon	Clio
C	Mercury	Euterpe
B♭	Venus	Thalia
A	Sun	Melpomene
G	Mars	Terpsichore
F	Jupiter	Erato
E	Saturn	Polyhymnia
D	Fixed Stars	Urania

32 · GIOSEFFO ZARLINO

1. St. Ambrose, *Super Lucam* VIII, 6. [GZ]

2. Job 38:137; see note 1 to Aurelian, no. 15.

3. *Asclepius* 10 may be intended, where the cosmos is said to be God's *ornamentum*. Zarlino's reference reads "Pimandro Ser. 10."

4. The four numbers 8:12:18:27 make a continuous geometrical progression, multiplying at each step by 3/2.

5. Boethius *De Cons. Phil.* II, metr. 9. [GZ] I use the 1609 translation of "I.T.," revised by H. F. Stewart in the Loeb edition (London, 1918), p. 265. Stewart remarks that "this poem is a masterly abridgement of the first part of the *Timaeus*, and was eagerly fastened on by commentators of the early Middle Ages, whose direct knowledge of Plato was confined to the translation of that dialogue by Chalcidius" (p. 262n).

6. *De Cons. Phil.* IV, metr. 6. [GZ]

7. Ovid, *Metamorphoses* I, 25–32, in Arthur Golding's 1567 translation.

8. Aristotle, *De Gen. et Corr.* II, 5–6. Cf. the Kabbalistic "squares" and "cubes" mentioned in note 6 to Giorgi (no. 30).

9. *De Remed.* I. [GZ]

10. *Symposium*, 187.

33 · JEAN BODIN

1. Queen Christina of Sweden and John Milton are among those known to have read the work in manuscript. On Bodin, see Frances Yates, *The Occult Philosophy in the Elizabethan Age*, pp. 67–71, and D. P. Walker, *Spiritual and Demonic Magic from Ficino to Campanella*, pp. 171–177.

2. Octavius is the Muslim, Coronaeus (the host) is a Catholic, and the narrator is employed by Coronaeus as a reader and secretary, who takes the conversations down in shorthand.

3. Fridericus is the Lutheran, and Curtius, who replies to him, the Calvinist.

4. Senamus is the pagan.

5. Toralba is the philosophic naturalist. The seventh speaker, who does not appear in this extract, is Solomon, the Jew.

6. Seneca, *Quaest. nat.* IV A. 19–20; VI, 2. [MDLK]

7. H. Diels, *Fragmente der Vorsokratiker*, 6th ed (Berlin, 1954), II.97.49. [MDLK]

8. *Metaphys.* V.8.10.[MDLK]

9. Diels, op. cit., II.114.134. [MDLK]

10. The notes intended appear to be respectively *lichanos hypaton* (d), *nete hyperbolaion* (a'), and *mese* (a). The principle is the Platonic one of the necessity of means to join opposites (*Timaeus* 31b–c).

11. Curtius' poem invokes a Trinity with terms deliberately reminiscent of Hermes Trismegistus. The "ladling" comes from *Timaeus* 35–36; the whole is much indebted to Boethius' poems on universal harmony, already quoted by Zarlino (no. 32).

34 · JOHANNES KEPLER

1. Sun: one sign a month; Moon: whole Zodiac each month; conjunctions of Saturn and Jupiter: every fourth sign.

2. In the theoretical terms of Kepler's time, "perfect" consonances were the octave, fifth, and fourth; "imperfect" ones the thirds and sixths.

3. The five "Platonic" or regular solids: tetrahedron or "pyramid", cube, octahedron, icosahedron, and dodecahecron.

4. These are the plane figures needed to generate the respective solids geometrically.

5. The aspects and intervals under discussion are:

Opposition	180°	Octave 2:1	G-g
Quincunx	150°	no interval	
Biquintile	144°	Ma. sixth 5:3	G-e♮
Sesquiquadrate	135°	Mi. sixth 8:5	G-e♭
Trine	120°	Fifth 3:2	G-d
Square	90°	Fourth 4:3	G-c
Quintile	72°	Ma. third 5:4	G-B♮
Sextile	60°	Mi. third 6:5	G-B♭
Semisquare	45°	no interval	
Semisextile	30°	no interval	
Conjunction	0°	Unison 1:1	G-G

6. One cannot add intervals as one can aspects: two squares, or a trine plus a sextile, make an opposition, but two fourths do not make an octave, neither do a fifth and a minor third. Herein lies the snag in trying to compare the arithmetically divided circle of the Zodiac with the logarithmically divided string. Kepler recognizes this, in a way, at the end of this chapter, and wisely leaves it to others to push the analogies to absurdity.

7. Marin Mersenne, in *Harmonie Universelle*, gave an even fuller table of possible interval-aspect correspondence, but rejected the one as explaining the other.

8. He refers to the exposition of the harmonic division of strings, and the way in which the possible harmonic divisions within the octave form an almost complete scale. The term "harmonic division" is taken not in its strict mathematical sense, but to mean the division already used in our extract from *Mysterium Cosmographicum* in which the two resulting parts of the string are consonant both with the whole string and with each other. Such divisions are those into 1 + 1, 2 + 1, 3 + 1, 4 + 1, 5 + 1, 3 + 2, and 5 + 3, giving (with the whole string tuned to G) the "scale" G, B♭, B, C, E♭, E, G.

9. He refers to the forthcoming discussion in his chapters 7–9 (not translated here) of the possible harmonies between two or more planets, in which intervals too small for melodic use become important in order to make pure tunings of one kind or another.

10. An important principle of harmonics, which assumes that because of octave equivalences any number representing a tone can be halved or doubled any number of times for purposes of comparison. Therefore every note/number whatsoever can theoretically be brought within a single octave space.

11.

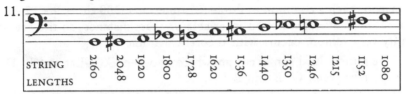

STRING LENGTHS: 2160 2048 1920 1800 1728 1620 1536 1440 1350 1246 1215 1152 1080

FIGURE 36: *The Chromatic Scale and Its Numbers*

12. I.e, the slowest motion, that of Saturn at aphelion, will correspond to the longest string length, that of low G.

13. See note 8. The harmonic division of a string tuned to G at A, giving the proportions 8 + 1, produces discords. So does the division at F (see below). There is nothing surprising in this: A and F are each a dissonant whole tone away from G. But Kepler was struck by the coincidence when he came to construct his planetary scales.

14. Just as he had given two different scales starting from the two extreme positions of Saturn, now he shows how the two scales tuned from Venus' extreme positions each involve consonances with different planets.

15. This is the chapter in which the exact intervals between planets are calculated at length.

16. "From aphelion to perihelion" would be more accurate, since it is now not the extreme motions/notes that are in question but the differences between them.

17. Kepler is describing what is now called a glissando.

18. The moon does not enter into the planetary harmony. Its music is presumably calculated from the point of view of the earth, not from that of the sun, as the others are.

19. The 7th and 8th modes have their final on G and begin with a major third, G A B; the 1st and 2nd have their final on D and begin with a minor third, D E F. Hence their assignment to the planets which cover respectively a major and a minor third. But on this basis Saturn could just as well be given the 5th and 6th (F G A, etc.) and Jupiter the 3rd and 4th (E F G, etc.).

20. Kepler is referring to the invention, unknown to the ancients, of polyphonic music which best reflects the situation in the heavens.

21. If Saturn and Jupiter both start from 1° Aries, they will not coincide again until they meet at about 9° Capricorn, approximately twenty years later, when Saturn has gone most of the way round the Zodiac once, Jupiter nearly twice. The intervals between successive grand conjunctions (as these conjunctions between the two slowest Chaldean planets are called) are what Kepler describes as "leaps." 800 years, i.e., 40 leaps, are needed for the 81° separating conjunctions and the 360° of the Zodiac to reach a common multiple (81 times 40 = 360 times 9 = 3240). But 81° is an average, not an invariable or exact distance, so the eventuality very seldom occurs with theoretical tidiness.

22. The minor third can occur anywhere between G, B-flat and B, D; the major between G, B and B-flat, D.

23. Later in the chapter Kepler describes the relationship of Earth and Venus in picturesque terms as a marital one, oscillating from the "masculine" G#–E to the "feminine" G–E. This is particularly audible on the recording *"The Harmony of the World"—A Realization for the Ear* (LP 1571) by Willie Ruff and John Rogers, which synthesizes these planet songs in various combinations.

24. The remainder of this chapter, not translated here, sets out in columns the various consonant combinations that can be made from the whole range of pitches available to each planet, e.g., Saturn–G, Jupiter–B, Mars–G, Earth–G, Venus–E, Mercury–E. Kepler finds four possible six-part chords, each in various spacings, and four other five-part ones if Venus is omitted.

35 · ROBERT FLUDD

1. See Huffman, *Robert Fludd and the End of the Renaissance*, p. 234, for a bibliographic outline of this complicated work.

2. This most basic version of the pyramids is from Fludd's *Philosophia Sacra et vere Christiana* (1626), p. 212. The intersecting light and dark pyramids shown in the illustration are one of Fludd's symbols for the cosmic interplay of form and matter. They stretch from the Earth, which is total matter, to God, who is pure form. Everything in between is compounded of both, in different proportions. In order to save space, Fludd represents only a sector of the universe, with the Earth at the bottom and a part of the Sun's orbit in the middle.

3. The vibrations of music are regarded here as making the affected air denser or subtler, in conformity with Fludd's belief in music as a for-

mative force. This is of course the case, but only on the smallest scale, as the air is alternately compressed and rarefied by each vibration.

4. These are the three groupings of the nine orders of angels. The *Epiphania* or "apparitions" comprise the Seraphim, Cherubim, and Thrones.

5. Psalm 19:4. The Sun, although only halfway up the cosmic ladder, plays a special part in Fludd's system as God's tabernacle within the created universe. Macrobius, whom Fludd cites at the outset of the work, mentions the Sun as being to the universe what the heart is to the creature (*De Somn. Scip.* I, 22). This parallel, a most important one for Fludd, is made explicit in the illustration on page 248. It would be profitable to consider Fludd's ambiguity—being astronomically a geocentrist but metaphysically a heliocentrist—in the light of what von Thimus (no. 49) says about exoteric and esoteric cosmologies.

6. From *Utriusque Cosmi Historia* I, a (1617), p. 90.

7. For several more illustrations of monochords and pyramids with musical notes, with explanations, see Godwin, *Robert Fludd . . .* , pp. 42–53.

8. From *Utriusque Cosmi Historia* II, a, 1 (1619), p. 275.

9. The lower half of the plate should be interpreted in a temporal as well as a spatial sense: as the darkness of death, nonbeing, and precosmic chaos from which a world, or human consciousness, arises like the sun rising above the horizon, and becomes extinct again as the sun sets.

36 · MARIN MERSENNE

1. See Frances A. Yates, *The French Academies of the Sixteenth Century*, pp. 284–290 *et passim*, on Mersenne's place in the intellectual history of his time; also her *Giordano Bruno and the Hermetic Tradition*, pp. 434–440 on Mersenne's controversy with Fludd.

2. See Mersenne's *Questions Harmoniques*, p. 5; *Quaestiones celeberrimae in Genesim*, pp. 1554–1555. In the latter work, pp. 1555ff., Mersenne mentions the Fludd-Kepler controversy and initiates an acrimonious exchange with Fludd himself.

3. Book 19 of the *Problems* attributed to Aristotle concerns harmony.

4. "For there is a point at which a state may attain such a degree of unity as to be no longer a state, or at which, without actually ceasing to exist, it will become an inferior state, like harmony passing into unison, or rhythm which has been reduced to a single foot" (*Politica* 1263b, 33–36, trans. Benjamin Jowett).

5. I.e., the *Speculum Musicae* of Jacques de Liège (see no. 22).

6. Whereas in the Middle Ages, Aristotle was admitted to the canon of spiritual philosophers through the attribution to him of works actually by Plotinus and Proclus, the Renaissance scholars, with their clearer sense of what was and was not his philosophy, often took up a defiantly anti-Aristotelian stance, contrasting him unfavorably with Plato.

7. A reference to I Corinthians 15:28.

8. Mersenne was one of the first theorists to explain consonances via the vibration theory. See Walker, *Studies in Musical Science . . .*, pp. 13, 30–33.

9. The mythographers usually attribute the power of building by means of music to Amphion, who with his brother Zethus raised the walls of Thebes. Orpheus is generally held only to have moved stones.

10. The Dark Aleph is the inconceivable Absolute enclosed within itself. It is the same as the En Soph ("no end"), whereas the Light Aleph corresponds to the En Soph Aur ("limitless light"), the *fiat lux* which initiates creation. The Light Aleph is the goal of positive, the Dark that of negative theology.

11. On this odd form of Kabbalistic arithmetic, see note 6 to Giorgi (no. 30).

12. This refers to the vibration theory of consonance, the "degrees of unison" being the coincident vibrations.

13. Proposition XI, on pp. 41–46 of *Harmonie Universelle* in the English translation by Roger E. Chapman (see Bibliography).

37 · ATHANASIUS KIRCHER

1. This is the subtitle of Book 6 of *Musurgia Universalis*.

2. Subtitle of Book 6, part 1.

3. Subtitle of Book 3, chapter 4.

4. For Kircher the soul is a *numerus numerans*, hence able to judge number-relationships, but its own numerosity is fixed from the beginning. See Ulf Scharlau, *A. K. als Musikschriftsteller*, pp. 132–133.

5. This experiment, also described by Galileo Galilei, is discussed in D. P. Walker's *Studies in Musical Science*, pp. 28–29. It seems to originate with Theon of Smyrna, who describes two Pythagoreans "taking several similar vessels of the same capacity: one was left empty and the other filled half way with a liquid, then they were both struck, thus obtaining the consonance of the octave. Again leaving one vessel empty and filling the other up to one quarter, then striking them, the consonance of the fourth was obtained. For the accord of the fifth, a third

of the vase was filled" (Theon II, 12; Lawlors' translation, p. 39). Actually a glass, whether struck or rubbed, sounds lower as it is filled with water; a bottle, blown across the top, sounds higher, but that is another experiment. I have not managed to obtain the ripples which Kircher describes. The "experiment" was criticized by G. A. Beer in *Schediasma Physicum*, f.A4.

6. The following passage is adapted from Robert Fludd, *Utriusque cosmi historia* I, a (1617), pp. 71–72.

7. Kircher now cites Fludd explicitly; cf. our no. 35 for Fludd on the World-Monochord.

8. Just as Francesco Giorgi's book (see no. 30) was divided into "songs," etc., so the chapters of *Musurgia*, Book 10, are called "Registers," referring to the World-Organ, which Kircher depicts as an allegory of creation in Vol. II, p. 366 (see illustration in Godwin, *Athanasius Kircher*, p. 1).

9. The treatise on music by the fourteenth-century Byzantine writer Manuel Briennius was one of Kircher's main sources on ancient Greek music. Briennius, in turn, relies heavily on Aristeides Quintilianus (see no. 9).

10. "Pythagoras" here is the Pythagorean philosopher Timaeus, as reported in Plato's dialogue of that name.

11. Kircher's journey to Sicily and Malta as companion of Friedrich, Landgraf of Hesse-Darmstadt, coincided with the eruption of Etna and the destruction of Saint Euphemia Island. The reflections occasioned by the journey gave rise to Kircher's most popular work, *Mundus Subterraneus* (Amsterdam, 1665 and 1678), alluded to at the end of the paragraph.

12. This instrument without a player, on which Kircher was the first expert, became an image of supernatural harmony for the Romantics. Andrew Brown's study (see Bibliography) also includes an entire essay on the Aeolian Harp by Dalberg (see no. 44).

13. The simile comes from Zarlino's chapter (see no. 32) on *musica humana*; see *MM&M* for a translation.

14. A Greek term for a Spartan provincial governor, probably favored for its assonance with "harmony."

15. Part of Kircher's *Ars Magna Lucis et Umbrae* (Rome, 1646).

16. Kircher adhered, for public purposes, to Tycho Brahe's cosmological system, the one officially approved by the Jesuits. This makes the stars and planets circle the sun, which, with the moon, revolves around the earth.

17. Kircher's planetary music does not pretend to astronomical accuracy, as Kepler's does. In the preceding sections (vol. II, pp. 365ff.) he has considered, and rejected, Kepler's planetary harmonies. The Greek

terms in the example are those of the high, middle, and low tetrachords and the lowest note of the system. Here again, the connection is evocative rather than precise.

18. The Accademia dei Linceii, founded 1603, to which Galileo Galilei belonged. The first impressions of Saturn's rings, seen through the primitive telescope, seemed like two neighboring bodies such as Kircher describes.

19. Kircher's *Ars Magna Lucis et Umbrae.*

20. Like Anselmi (no. 24) and Gafori (no. 29) before him, Kircher regards the spheres of the planets as reflecting the nine orders of angels. To reproduce the whole of his presentation here would be impossible, because of his extreme verbosity which the reader has doubtless noticed. The reference is to the ninefold diagram in Vol. II, p. 392; see *MM&M* and also Godwin, "Athanasius Kircher and the Occult," p. 21.

21. Presumably a working title for Kircher's *Oedipus Aegyptiacus* (Rome, 1652–54).

22. See note 12 to Giorgio (no. 30).

23. There follows, concluding the second volume of *Musurgia Universalis,* the passage on the Ultimate Music which I have already given in *MM&M.*

38 · ANGELO BERARDI

1. The observation is Friedrich Blume's, in *MGG,* vol. I, col. 1673, s.v. "Berardi."

2. This table seems to have been compiled by Berardi from various sources. His Sephiroth-angel-sphere correspondences are the same as Giorgi's (see no. 30), but he applies the scale in the contrary direction. I have added the Hebrew names for the ten Sephiroth.

3. What follows is an alternative Kabbalistic scale: one of ten notes, corresponding to the ten Sephiroth and believed to be the subject of Psalm 33:2: "Praise the Lord with the lyre; make melody to him upon a harp of ten strings!" The Sephiroth are an archetypal pattern which manifests at every level of existence; in particular, Kabbalists consider them as ruling the four worlds of Atziluth (World of Emanation), Briah (World of Creation), Yetzirah (World of Formation), and Assiah (World of Action). Berardi gives the first of these with a triangular diagram (reminiscent of a symbolic triangular harp in Robert Fludd's *Utriusque Cosmi Historia,* I, a, p. 92) and accompanied by the conventional and not altogether accurate translations of their Hebrew names (see the preceding chart). The Sephiroth are here in reverse direction, com-

pared to the other orderings in this essay. The long tabulation that follows consists of three ten-note scales from f to a', each representing one of the three lower worlds and the things in them that reflect the ten Sephiroth.

4. Saint Thomas of Villanova (1486–1555), Spanish Augustinian and philanthropist, canonized 1658.

39 · ANDREAS WERCKMEISTER

1. Werckmeister comes closer than any previous writer to the musical metaphysics of Hans Kayser, (1891–1964), in whose philosophy the first members of the harmonic series were so richly symbolic of creation, in every sense. See especially Kayser's posthumous *Orphikon*. Here Werckmeister may have been inspired by Kircher's image of the days of creation as an organ with six registers (see note 8 to Kircher, no. 37).

2. An ingenious use of the parallel of discord with necessary evil, to explain the vexing problem of the out-of-tune seventh harmonic. Werckmeister invokes Kepler to support his omission of 7, on the grounds that since a circle cannot be divided geometrically into seven equal parts [but see John Michell], seven is not a harmonic number in music nor an effective angle in astrology.

3. I.e., 7:1 is a discord, whereas 6:1 is not.

4. Daniel Schwenter (1585–1636), *Deliciae physico-mathematicae* (Nuremberg, 1636), reissued 1651–53 with two additional books by Georg Philipp Harsdorffer.

5. As a practically-minded organist, Werckmeister was much concerned with problems of tuning and temperament, and invented a system which, though it is not exactly equal temperament, made possible transposition into all twelve keys. This is one of the temperaments that J. S. Bach may have had in mind for his *Well-tempered Clavier*.

6. The idea of the triad as an image of the Holy Trinity first appears in the works of Johannes Lippius, collected in his *Synopsis musicae novae* (Strasburg, 1612). See translation by Benito V. Rivera (Colorado Springs, 1977), p. 41.

7. Werckmeister is probably alluding to the fact that the music of the Jewish synagogue is unharmonized, whereas in their secular music, Jews have often adopted the style of their surroundings—in this case, popular German drinking-songs.

8. Mid-sixteenth-century compiler of medical and theological works, particularly those of Melancthon. The work cited here does not appear in the standard catalogues.

9. The Mercy Seat (Exodus 25:17) was a cover to the Ark of the Cov-

enant, 2½ cubits long by 1½ broad (i.e., the proportion 5:3, a major sixth). No third dimension is given. The Ark itself is 2½ by 1½ by 1½. Noah's Ark had the dimensions 300 x 50 x 30 cubits (Genesis 6:13), which gives the ratios 10:1 (major third plus two octaves), 6:1 (fifth plus two octaves), and 5:3. Ernest McClain shows in *Meditations through the Quran*, p. 79, that the Ka'aba at Mecca is in the proportions of a perfect triad (5:6:8).

10. On the correlation of musical proportions with the forms of government, see Kircher, *Musurgia Universalis*, Vol. II, pp. 435–438.

11. Despite Werckmeister's old-fashioned ideas, he fully accepted the Baroque system of general-bass composition, as opposed to the Renaissance counterpoint still being taught and practiced (e.g., by Berardi, see no. 38). Perhaps because Lippius (see note 6) had, earlier in the century, given the general bass a symbolic meaning, Werckmeister was willing to see it as a symbol of the foundation of all things in the divine order.

40 · ISAAC NEWTON

1. See our extract from Pliny, no. 2.

2. These words do not appear in the chapter of Macrobius cited, but seem to be compiled from that chapter and Book II, Ch. 3 (see our extract from Macrobius, no. 11).

3. See our extract from Proclus, no. 12.

4. Newton here explains the fallacy latent in the classical account of Pythagoras' experiment; see our extract from Nicomachus, no. 3.

5. I.e., the heliocentric system, which Newton believed Pythagoras to have known and secretly taught. The main evidence is the cosmology of the Pythagorean Philolaus, who taught that the earth moves around a "central fire." Von Thimus (see no. 49) will present further evidence for this.

6. See the Orphic Hymn to Apollo (no. 7), especially note 3.

7. From Orphic Hymn no. 11 to Pan.

8. A reference to the *Krater*, a lost Orphic work.

9. I.e., Hermes.

10. I.e., Asclepius.

11. Newton also cites "on the same subject Hermesianax [poet of Colophon, third century B.C.E.]: Pluto, Persephone, Ceres and kindly Venus and the Loves, the Tritons, Nereus, Thetis, Neptune and Mercury, Juno, Vulcan, Jupiter and Pan, Diana and Phoebus the dartsman are one God." [JEMcG and PMR]

41 · JEAN-PHILIPPE RAMEAU

1. Rameau is referring to the very small intervals, such as commas, which distinguish the different tunings off the larger intervals.

2. Charles Etienne Briseux (1680–1754), author of *Traité du beau essentiel dans les arts . . . avec un traité des proportions harmoniques* (Paris, 1752); see Godwin, *L'Esotérisme musical en France, 1750–1950*, pp. 27–29. Brian Juden, in *Traditions Orphiques et Tendances Mystiques dans le romantisme français (1800–1855)*, p. 149, says that this association of music with architecture makes one think of Amphion, who built the walls of Thebes with his lyre. Perhaps the Amphion legend did refer, in one respect, to the harmonic principles of ancient architecture. Mersenne also argued the theorem the "Architecture and its proportions are similar to the musical consonances and concerts" (*Harmonie Universelle*, p. 404, quoted in S. K. Heninger, *Touches of Sweet Harmony*, p. 387).

3. Presumably an allusion to the division of the World-Soul in the *Timaeus*.

4. Rameau here seems to be taking the Aristoxenian position, that the ear is the best judge of what is significant, hence dismissing the subtler shades of tuning that have obsessed theorists for so long. This is only because he has substituted for the ancient tunings the mathematical and equally dogmatic principle of the fundamental bass. The "geometrical progression" is around the circle of fifths, attained by successive multiplication by three, e.g. 1 = Bb, 3 = F, 9 = C, 27 = G; Rameau will allude to it in the second part of our extract. The Abbé Roussier and his followers would make much of this "triple progression"; see Godwin, *L'Esotérisme*, pp. 38–45.

5. We omit a passage in which Rameau explains the need for temperament as it arises on various instruments.

6. The eleventh harmonic from C falls between f"-natural and f"-sharp.

7. Rameau's thought was so inveterately harmonic that he could not imagine the Greek tetrachord as arising from anything but an implied harmony, which he gives as the bass notes proper to the melodic progression B C D E. The numbers refer to the "geometric progression" mentioned above.

8. When Newton was experimenting with the spectrum, he directed his assistant to divide it into seven colors analogous to the diatonic scale. He was therefore obliged to invent the color "indigo," never before used in color theory and unknown to painters except as a vegetable dye, in addition to the three primary and three secondary colors. After Newton, color theorists tend to follow either his sevenfold system, which for the Hermetic temperament leads naturally to parallels and corre-

spondences with the seven planets, or else the sixfold system of the painters, which is readily doubled to provide analogies with the zodiac, the chromatic scale, etc. There is much on this in Godwin, *L'Esotérisme*, pp. 17–27.

42 · GIUSEPPE TARTINI

1. An Ariadne's thread through the labyrinth of Tartini's thought is provided by D. P. Walker in *Studies in Musical Science in the Late Renaissance*, pp. 123–170, on "The Musical Theory of Tartini." But this is only of limited help when it comes to the *Scienza Platonica*, which was not published in time for Walker to discuss it.

2. The formula is $AH = G^2$.

3. Tartini assigns the treatment of whole numbers to arithmetic, that of fractions to harmonics, because a string is always divided, never multiplied. Geometry alone can deal with irrationals such as π and the square root of 2.

4. Tartini refers to his discovery of the harmonics sounded by a string, which he heard as only three in number: the fundamental, third, and fifth harmonics. This is the subject of Ch. 1 of the Trattato. (Cf. Peter Singer, no. 48.)

5. The ground for this assertion is that the progression $1 : \frac{1}{3} : \frac{1}{5}$ answers to the formula of the harmonic mean, not the arithmetic or geometric (see note 8 to Plato, no. 1). Tartini's discovery of the harmonic progression both in the physical or acoustical realm (the string) and in the mathematical or geometrical realm (the circle) convinced him that it must be at the foundation of all phenomena.

6. As demonstrated by the harmonics of a string.

7. In the following passage Tartini attempts to find both straight and curved elements in the vibrating string, in order to bring together the two realms, geometrical and acoustical.

8. The reference is to the phenomenon of difference tones, which Tartini discovered.

9. The argument seems to be that the square, being prior, should be the one to contain the secondary figure.

10. There follows the demonstration that the circle is "intrinsically harmonic," too involved to include or summarize here but admirably explained in D. P. Walker's *Studies in Musical Science in the Late Renaissance*, pp. 145ff.

11. For a suggestion of what this extravagant claim might mean, see Alfred Rubeli's commentary on Ch. 2 of Tartini's *Trattato*, ed. cit.,

pp. 80–85. Rubeli cites Hans Kayser, for whom the science of harmonics provided the much sought-after bridge between the material world of vibrations and the psychological or soul-world of tone: another attempt on the link between matter and mind that has so troubled Western philosophy. Rubeli considers the *Trattato* a step in this direction, although, as he says, Tartini's "enthusiam for the idea that only the synthesis of the harmonic thought of Antiquity with the latest discoveries of physics could lead to the true understanding of all musical questions, led him to develop his theory on insufficiently grounded and partially absurd hypotheses" (p. 84). I have not studied *Scienza Platonica* sufficiently to judge whether Tartini really got any further in that work; it demands the long-term commitment of a specialist.

12. "Hence, you profane!" (Vergil, *Aeneid* VI, 258).

13. "They have become futile in their thinking" (Romans 1:21).

43 · LOUIS CLAUDE DE SAINT-MARTIN

1. Ideas of a universal, primordial language were much in vogue during this period. Athanasius Kircher's work on the Egyptian hieroglyphs (*Oedipus Aegyptiacus*, 1652-54) was not yet discredited by the discoveries of Champollion, and Kircher's claim that they enshrined the original language of mankind was taken up at even greater length by a friend of Saint-Martin, the Protestant savant Antoine Court de Gébelin, in *Le Monde primitif* (9 vols., Paris, 1775-84). Court de Gébelin posited as the source of all languages and alphabets a primitive language and a primitive alphabet. The latter was "taken from nature; each one of its elements painted a particular object; its pronunciation instantly caused the idea to be born in the one who heard it" (quoted in Gwendolyn Bays, *The Orphic Vision*, p. 33.) With the rediscovery of this language, he believed, a golden age would ensue. Fabre d'Olivet was to endow Hebrew with a similar virtue in *La Langue hebraïque restituée* (1815). Saint-Martin was not a philologist like these, but this very deficiency enabled him to pierce through the babel of tongues to a stage anterior even to the spoken word. In a later work, *De l'esprit des choses* (1800), he would write: "It is an injustice to man to wish to deny the legitimate source of languages in order to substitute one of his own invention" (vol. II, p. 143), and explain that when Adam was created, "the names of all the things which surrounded him must have been infused in him by his Principle simultaneously, as those of objects today are, progressively, in children" (p. 212). "True language" is therefore the knowledge of all things instilled in the human spirit from the beginning and still supposedly accessible to us.

2. One cannot count as an alteration the transposition of any of its

notes, from which result the differently named chords, seeing that this transposition introduces no new note into the chord, and consequently cannot change its true essence. [LCSM, transferred from text] Saint-Martin here transposes to the metaphysical level the theory of Rameau concerning the archetypal primacy of the triad.

3. It was Tartini's discovery (see no. 42) that every note we hear, unless extremely pure in timbre, contains as harmonics the common chord.

4. The use of numbers in *Des Erreurs et de la Vérité* is extremely obscure. The meaning here relates to the Tetraktys (1, 2, 3, 4), whose numbers added together make ten: the number of completion for the Pythagoreans. With ten, another order of being is entered (cf. note 6 to Giorgi, no. 30, on Kabbalistic numeration). In Hebrew gematria, One is the number of Aleph, but ten is another beginning: it is the number of Yod, the initial letter of God's revealed name YHVH. Similarly, in the scale the attainment of the octave completes one level and begins the unfolding of a higher one. As Saint-Martin says in the next sentence, these are analogies, not exact images.

5. The intervals of the common chord are likened to the first, spiritual, or good Principle (the fourth) presiding over the world of duality (symbolized by the two different thirds). The number of the Principle is 4, suggesting another reference to the immutable Tetraktys.

6. The two primordial Principles are Good and Evil.

7. A reference to acoustics would have strengthened the argument, for the flat seventh does in fact intervene in the harmonic series before the octave is reached. One might say that the resolution of the dominant seventh to the common chord is like an idealized purification of the harmonic series.

8. It may seem far-fetched of Saint-Martin to derive the diatonic scale from the dissonance of the seventh and eighth harmonics, but this is not so different from the procedure of the Demiurge in the *Timaeus* (see no. 1), who takes the dissonant 8:9 whole-tone brought about by dividing the octave, and then uses it to create a scale.

9. Nine is the cube of three, and as such has been interpreted as the number of solidity ever since Plato's *Timaeus*, as many of our extracts bear witness. Saint-Martin turns the musical seventh into the number nine by taking its three thirds separately, e.g., GAB, BCD, DEF. The octave G would then be counted as tenth: another solution to the remark which is treated in note 4.

10. Cf. Kepler (no. 34) in *Mysterium Cosmographicum*, when he starts his parallel by defining the consonances within the octave.

11. Saint-Martin is referring to the conventional eighteenth-century cadence:

FIGURE 37: *The Cadence of Human Life*

12. The A to which tuning standards are related, now fixed arbitrarily at 440 Herz but more variable in Saint-Martin's time. This passage is directed against those who try to build speculative systems on absolute pitches, as opposed to relative ones.

13. The reference again seems to be to the Principle—though of a lower, temporal grade—above the duality of Nature; cf. note 4.

14. This error is the seeking of principles in matter rather than in the Intellect.

15. Note Saint-Martin's acceptance of the aesthetic of his time, which made of music an art of imitation.

16. Inasmuch as one can summarize the preceding argument, it would be to say that Saint-Martin is combining a Platonic doctrine of the arts with a Christian idea of fallen man, and the potential for human regeneration as taught by Martinez de Pasqually. More particularly, he is persuading us to look for rules and models in the Intellectual, not the temporal world, but he does not give any precise instructions on how this might be achieved in music.

44 · JOHANN FRIEDRICH HUGO VON DALBERG

1. For the source of this myth, see Plato, *Symposium* 187d.

45 · ARTHUR SCHOPENHAUER

1. "An occult exercise of arithmetic on the part of the soul, acting unconsciously." G. W. von Leibniz, *Letters*, ed. Kortholt, no. 154.

46 · FABRE D'OLIVET

1. *Timaeus* 55c; Timaeus of Locri, *On the Nature of the World and the Soul*, sect. 35.

2. Founder of Manicheism, mid-third century.

3. It is surprising that Fabre d'Olivet does not cite Jakob Boehme on the Septenary, which is far more important to Boehme's theosophy. Boehme is cited in almost identical terms, quoting from the *Aurora*, in Fabre d'Olivet's *Les Vers dorés de Pythagore*, p. 360.

4. Presumably J. S. Bailly, author of *Histoire de l'astronomie ancienne* (Paris, 1775).

5. Boethius, *De Inst. Mus.* I, 20; P. J. Roussier, *Mémoire sur la musique des anciens*, pp. 11ff.

6. Macrobius, *Saturnalia* I, 21–22.

7. Julian in his *Oration to the Sovereign Sun* does not actually mention seven rays; however, Thomas Taylor, in his translation of the work (London, 1793), prefaces it with a poem of his own in which the Sun is addressed as "All-beauteous, *seven-rayed*, Supermundane god!" (his italics).

8. Martianus Capella, *De Nupt.* II, 183–183.

9. The *LiuTzu*, by Liu Chou (sixth century C.E.·), included in the *Tao Tsang* (Taoist Patrology).

10. "Tu septiformis munere," from the Whitsunday hymn *Veni Creator Spiritus*.

11. This refers to Louis-Claude de Saint-Martin (see no. 43), in particular to the chapter on music in his *L'Esprit des choses* (Paris, 1800), vol. I, pp. 170ff.

12. Starting on any note B and rising by successive fourths, obtained on the monochord by shortening the string by a quarter of its length each time (4:3), one finds the series B, E, A, D, G, C, F, B♭, E♭, etc. Starting on the note F and rising by successive fifths, obtained by shortening the string by a third of its length each time (3:2), one finds the series F, C, G, D, A, E, B, F#, C#, etc. Only the first seven tones of each series are identical; these form the diatonic scale.

13. Fabre d'Olivet adds in a note: "It is important to observe that the Ancients called 'Venus' or 'Juno' the planet that we nowadays call Mercury; and 'Mercury,' 'Hermes,' or 'Stilbon' the resplendent, the one we now call Venus. It is a change of name made at the epoch when vulgar opinions took the upper hand over philosophical ones. It is indifferent in itself, but one must be aware of it in order to understand several passages in ancient writings. Thus when I say 'Venus,' I mean the planet [now] called Mercury, and when I say 'Mercury,' I mean the one called Venus. In ordinary astronomy one can keep the accepted names, but in musical astronomy this is impossible and one must reinstate the true names."

14. The F series reaches B after three and a half octaves; the B series reaches F after two and a half.

15. This passage would have pleased Johannes Kepler, discoverer of the

planets' elliptical orbits (see no. 34). The parallel with the World-Egg of the Orphics was suggested by Nicholas Fréret in a memoir to the Académie des Inscriptions (vol. 25), and taken up by C. F. Volney in his very popular *Les Ruines, ou méditations sur les révolutions des empires* (1791), where Fabre d'Olivet probably found it.

16. Fabre d'Olivet defines the Eumolpids as "perfect ones," the disciples of Orpheus, in his *Histoire philosophique du genre humain* (1979), vol. I, p. 297, and gives his etymology of their name in *Les Vers dorés de Pythagore*, p. 73n., as coming from *eumolpos*, *"la voix accomplie."* Fabre d'Olivet called the French verse-form which he invented *vers eumolpiques*.

17. This statement is a summary of the cosmology of the Pythagorean Philolaus.

18. A reference again to the doctrines of the emperor Julian concerning the physical and the intelligible sun, here conflated with the Philolaic scheme of the Sun as distinct from the Central Fire or Hearth of the Universe.

19. These figures give the two intervals of Pythagorean diatonic tuning: all whole-tones are in the ratio 9:8, all leimmas in the ratio 256:243.

20. That is, the order working outward from the Earth considered as center: presumably different from the order of the physical solar system, which is heliocentric. Cf. Albert von Thimus (no. 49).

21. That is, if one continues the progression from F beyond B, the next term is F#, a nondiatonic tone.

47 · ALPHONSE TOUSSENEL

1. He explained it in his work of 1822, at first called *Traité d'association domestique agricole*, then *Théorie de l'unité universelle*. See especially Fourier's *Oeuvres complètes*, vol. IV, pp. 241ff., and Godwin, *L'Esotérisme musical en France*, pp. 87–109.

2. See Fourier's introduction to the second edition of his *Théorie des quatre mouvements*, in *Oeuvres complètes*, vol. I.

3. Fourier constantly coined his own terms, here reproduced as faithfully as the English language allows.

4. A "series," in Fourier's system, is a certain number of things put together with a certain interior organization. In the third degree, there are 32 things, plus a pivot, organized as 12 + 12 + 4 + 4; another example is found in the teeth, pivoted on the lower jaw. See Fourier, "Des Séries mesurées."

5. These are all terms with a specialized meaning in Fourier's system.

The best introduction currently available is in Simone Debout's edition of *Théorie des quatre mouvements*.

6. The five moons are to be the asteroids Phoebina, Juno, Ceres, Pallas, and the planet Mercury.

7. As far as one can be precise about Fourier's symbolism, he seems to be imagining a two-manual instrument with a range of two octaves, with one manual or rank tuned to a major, the other to a minor scale.

8. This designates the upper octave of the manual in question; *hypo-* refers to the lower one.

9. It is impossible to exaggerate the lengths to which Fourier goes in tracing "passional movement" through every domain of life, but especially in his fantasies of what "harmonian" existence will someday be like. In a typical example, he applies these same numbers and arrangements to an exhibition of prize pears, put on by the inhabitants of an idealized commune.

10. These correspond to the seven diatonic and five chromatic tones. In minor, the diatonic scale has eight different notes, because the seventh degree can be either flat or natural.

11. The early name for the planet Uranus.

12. "Aromas" are the supersensible exhalations of planets, etc.

13. Fourier predicts that telescope lenses made from fusible diamond and fixed mercury will one day teach us to read the "unitary harmonic language spoken in the Sun and the harmonized planets, and in all the suns and vortices of the celestial vault." (*Traité d'association domestique agricole* (1822), vol. I, p. 534.

14. An early name for the planet Neptune.

48 · PETER SINGER

1. The acute ear can observe this in all perfect tones, but it is easiest with deep ones, especially with the low tones of a harmonium [*Physharmonika*], because at low pitch the vibrations are slower, and consequently the perception of the substratum that makes up a perfect tone is easier. Abbé Vogler has made the same observation with deep organ pipes. This would also explain why on large organs every tone is accompanied by a complete triad, i.e. a third and a fifth beside the predominant tone, and these in the relationships shown above. In cases where the natural pipe tone does not entirely possess the three components needed for a

full sound, this lack is made good through artificially added partials, and only thus does the full organ achieve a round and perfect tone. However, since these artificial partials are calculated by every intelligent organ builder on the basis of all the registers sounding together, they are only effective when this is the case; if one draws too few or weak fundamental registers, their [the mixtures'] predominance makes the effect quite unpleasant, because the unity so essential to each tone is lost: it does not serve purely for the manifestation of the primordial harmony, but is used for countless other disordered purposes. [PS]

2. Beside the evidence of the musical ear, large organs likewise give proof of this, where every tone sounds accompanied by its whole triad, both in each pipe and with the [mixtures of the] full organ, without the unity of any tone being thereby disturbed or destroyed. Indeed, if the three tones did not appear as a unity, the full organ would be quite unusable, for the actual playing of a triad would be unbearable to the ear, since it would hear in these three components the most jarring dissonances. For example, with the triad CEG sound CEG, EG#B, and GBD; yet for all that, the ear hears only the tones CEG, evident proof that the complete, perfectly tuned triad sounds like a single perfect tone. [PS]

3. Our music achieves a true nobility from this sublime analogy, and from the disposition inherent in it: that the Godhead manifests not only in the harmonious government of the whole universe, but also, and especially, shows forth its work of redemption in multifarious ways to the sensitive human heart. For although the universe is so great and wonderful—more to be wondered at, indeed, than understood—this work of God's omnipotence, wisdom, and goodness is incomparably greater than the whole work of creation; so that if one were to compare the latter to single drops from the infinite ocean of the Godhead, then the miracle of incarnation would be the entire sea—for here humanity truly receives the infinite Godhead itself, clothed in our image. And whereas creating, sustaining, and ruling the whole universe cost God only a word—that is, only a simple act of will—the redemption cost him his life, as a divine person in human form! Through the descent and humiliation of the Son of God, the whole universe has now come into a new and closer relationship with the Godhead, and through him is sanctified and as it were regenerated. [PS]

4. Thus also the third person of the Trinity plays a part in the Incarnation, while in the new realm of grace, the deeds are mostly attributed to the Holy Spirit. [PS]

5. Singer goes on in the next section to explain how the $\frac{6}{5}$ chord on the

subdominant is complementary to the dominant seventh, but he has no further theological meanings to offer, and the remainder of the treatise is devoted to plain harmonic theory.

49 · ALBERT VON THIMUS

1. The musicologist Siegmund Levarie and the composer Ernst Levy have done much to publicize a musical philosophy that grows out of their study of von Thimus and Kayser; see their works cited in the Bibliography. In Vienna, Professor Rudolf Haase has founded the "Hans-Kayser Institut für harmonikale Forschung"; see his *Geschichte des harmonikalen Pythagoreismus* for a comprehensive survey and bibliography of the subject.

2. The author has explained (Vol. I, pp. 132ff.) that in Iamblichus' Commentary on the *Principles of Arithmetic* of Nicomachus of Gerasa there is a diagram called the *lambdoma* (from the Greek letter lambda), which places the monad at the peak, and along the two arms respectively the increasing series of whole numbers (2, 3, 4 . . .) and the decreasing series of their reciprocal fractions (½, ⅓, ¼ . . .). Von Thimus fills it out (see illustration, taken from Levarie and Levy, *Musical Morphology*) with its "pleroma" of intermediate fractions, each of which can be interpreted as a musical pitch in relation to the monad. The "Pythagorean Table" thus developed is theoretically expandable ad infinitum.

3. Reference is to the first six terms of the horizontal series (overtones, major triad C E G) and of the vertical series (undertones, minor triad C A♭ F). Von Thimus consistently omits the seventh term as musically unusable.

4. Arithmetic progression: 1, 2, 3, 4 . . . Harmonic progression (analogous to the division of a string of length 1 into its harmonics): 1, ½, ⅓, ¼ . . .

5. The prime numbers 1, 3, and 5; the others (7, 11, 13, etc.) do not give harmonically viable pitches.

6. Von Thimus' version of the Pythagorean Table reflects vibration-ratios rather than string-lengths, as do those in later sources (Levy and Levarie, Kayser, etc.).

7. *Perissos* = increasing series; *artios* = fractional series. Von Thimus gives the Greek words these "esoteric" meanings rather than the usual "exoteric" ones of even and odd numbers (Vol. I, pp. 71ff.).

8. In his preceding discussion of the Taoist scripture *I Ching* (Vol. I, pp. 79ff.), von Thimus has determined the number 720 (= 6!) as that of the all Yang hexagram *Kien*, Heaven. Its opposite, *Kouen*, is Earth, hence the reciprocal value 1/720 is given to it.

9. Pitches to which a caret is added are a comma (80:81) higher than those of pure tuning. E.g., F is the note reached from A♭ by the ascent of a major sixth (3:5), whereas F^ is reached via four steps of the circle of fifths, A♭–E♭–B♭–F, to give the ratio 16:27.

10. That is, the circle of fifths taken as developed through powers of three: the familiar *Timaeus* sequence 1, 3, 9, 27.

11. The text has B♭ here.

12. Exoterically, "oddly-odd," i.e., multiples of prime numbers (ibid., pp. 74ff., citing Euclid VII, 11). Here they are interpreted as the harmonic series arising from the notes on the circle of fifths.

13. Exoterically, "evenly-even," i.e., powers of 2. Here they are the reciprocals of the previous class of progressions.

14. Valentinian Gnosis is in question here; von Thimus has treated it in Vol. I, pp. 165ff.

15. The tones for this diagram are set out on p. 175 of von Thimus's text; they form the central line of our illustration. The crossed arrows, a fundamental element in the Thimean symbology, are described by him on pp. 94ff.: "In the cosmological-harmonic semiotics of the *Sepher Yezirah* there appears for the *Belimah* ["nothingness"] the symbol [crossed arrows], derived from the form X of the letter Tau. The crossing of the two world axes is symbolized here . . . i.e., the earth's axis and the axis of the sphere of the ecliptic. . . ." Von Thimus uses the symbol for the "inexpressible" juncture of the tritone at the center of the octave: an irrational value, the square root of 2, which obsessed ancient theorists, as demonstrated in Ernest McClain's works. The development of the tone-sequence into a circular diagram, following his instructions, is the Editor's.

16. "The so-called Oth Aleph ‎ב ‎לא‎ת‎ו‎א, i.e., the ancient hieratic Tau-sign T, meaning both Aleph and Tau, adorned with the circular . . . *coronula* represented on ancient Asiatic scuptures in manifold versions [specimens follow]." Ibid., p. 93.

17. In Greek theory, *diazeuxis* (disjunction) is the whole-tone between two disjunct tetrachords, here D E F G ˙ A B C D.

18. The author explains on p. 94 of his text: "The number-harmonic types of this ideal root-unity [in the *Sepher Yezirah*] are ten numbers. . . : on the left, the ratios of the five diatonic steps, with their enharmonic neighbors, a comma away: CC^, D DD^, EE^, FF^, GG^—which in this context are representative of the masculine, major principle; and on the right the ratios of the following five upward steps AA^, BB^, cc^, d^v dd^, dd^, which represent the Shibboleth or the female, minor principle." Later (pp. 174f.) the tetrachord D EE^ F G is given as the lower

half of the scale of the Dorian-Hypomixolydian octave, the tetrachord A^ B cc^ d as the upper half. The two sets together form the "dekas-scale" of ten different notes, symmetrical about D, if regarded thus:

FIGURE 38: *The Dekas Scale*

The intervals C-D and D-E^ are the large 8:9 whole-tone; those between c^-D and D-E the small, 9:10 whole-tone.

19. Von Thimus's investigations into the origin of the scale have led him to the same parallel as at the end of our extract from Fabre d'Olivet (no. 46).

20. A page-long digression on ancient and modern science is omitted here.

21. A term from the *Sepher Yezirah* for the "crossing." (Ibid., p. 95)

22. Evidently referring to Job 26:7; but no warrant for this "crossing" in the Vulgate or any English version.

23. The symmetry of reciprocals around D is referred to: see note 18, above.

24. Von Thimus now refers again to the first diagram, giving its esoteric interpretation. Ring 3 in the diagram is now assigned to the Sun; around it are the circles of the solar system (Kosmos); inside, or in fact beyond it, is the universe (Ouranos), with the divine Unity at the center.

50 · ISAAC RICE

1. Siegmund Levarie is the first modern musicologist to have rediscovered Isaac Rice and his work. See his forthcoming article "Isaac Rice: *What Is Music*" in the Festschrift for Wiley Hitchcock.

2. Karl Friedrich Julius Sondhauss (born 1815) and Camillo Hajech, acousticians and physicists.

3. August Seebeck (1805–1849), mineralogist and acoustician.

4. The second edition of Rice's book (1879) contained an appendix, "How the geometrical lines have their counterparts in music," in which this analogy was developed much further. One could make an interesting study of the few attempts there have been to link music with geometry.

5. Now called ultraviolet light.

6. Compare the end of our last excerpt from Fabre d'Olivet (no. 46) for another approach to this perennial principle of Hermetic cosmology, that of the twin forces of expansion and contraction.

7. Cf. our excerpt from Albert von Thimus (no. 49) for an elaboration of this idea of ancient heliocentricity.

8. Rice is referring to the *Dream of Scipio*, where Cicero's cosmology is explicit, and his planet-tone system implicit. See Macrobius (no. 11).

9. I do not know what source Rice used for this account, but it attributes far more to Cicero than is present in his text.

51 · SAINT-YVES D'ALVEYDRE

1. In two concluding paragraphs that follow our excerpt, Saint-Yves suggests one of several new series of names for the intervals. This, and much else on Saint-Yves' music, is treated in Godwin, *L'Esotérisme musical en France*, Chapter 8, pp. 202–220.

2. For an analysis of the history and constitution of the Archéomètre, see Godwin, "The Genesis of a Universal System."

3. These are the Hebrew and cognate alphabets of twenty-two letters, which Saint-Yves derive from a more ancient alphabet called Vattan. See Godwin, "Saint-Yves d'Alveydre and the Agarthian Connection," part I, for information on this otherwise unknown alphabet.

4. One of the strongest statements of the Pythagorean, as opposed to the Aristoxenian position. However, when it comes to tuning the scale, Saint-Yves rejects the Pythagorean system for archeometric reasons, and prefers the Ptolemaic one.

5. Saint-Yves' language here betrays two of his influences: Fourier, in speaking of the "harmonic pivot," and Fabre d'Olivet, author of *La Musique expliquée comme science et comme art* (see no. 46).

6. This refers to the placement on the Archéomètre, which is essentially a zodiac-based diagram, of the twelve houses and the planets in their respective houses of rulership.

7. Pythagorean tuning, which is that of the *Timaeus* scale, is excellent for unaccompanied melody (or melody with a drone), but unusable for harmony since the major thirds are excessively large. This problem is solved in the Ptolemaic tuning, with its two different sizes of whole-tones (8:9 and 9:10).

8. Saint-Yves' vision of the "universities" of antiquity, in which the true social order of Synarchy was taught along with correct arts and sci-

ences, was revealed in his *Mission des Juifs* (Paris: Dorbon,1884) and supplemented by the astral voyage of the *Mission de l'Inde* (Paris: Dorbon, 1885, withdrawn, reissued 1910).

9. The reason is that the three letters in the Archéomètre that constitute the name of Jesus add up to 96, a number that enables one to construct a diatonic scale (between the octave 96:48) in Ptolemaic, but not in Pythagorean intonation. The name of Maria adds up to 240, another such number.

10. The essay is unfinished. It was probably intended as the introduction to the musical part of his great work on the Archéomètre, of which the volume as published contains only fragments.

Bibliography
of Works Cited in the
Introductions and Notes

Abert, Hermann. *Die Lehre vom Ethos in der griechischen Musik.* Leipzig, 1899; 2nd ed. Tutzing: Schneider, 1968.

Barbara, C. André. "The Persistence of Pythagorean Mathematics in Ancient Musical Thought." Ph.D. diss., University of North Carolina at Chapel Hill, 1980.

Bays, Gwendolyn. *The Orphic Vision: Seer Poets from Novalis to Rimbaud.* Lincoln: University of Nebraska Press, 1964.

Beer, Georg Alexander. *Schediasma Physicum de Viribus Mirandis Toni Consoni.* Wittenberg, 1672.

Bragard, Roger. "L'harmonie des sphères selon Boèce." *Speculum* 4 (1929): 206–213.

Brown, Andrew. *The Aeolian Harp in European Literature 1591–1892.* Cambridge: Bois de Boulogne, 1970.

Burkert, Walter. *Lore and Science in Ancient Pythagoreanism.* Trans. Edwin L. Minar, Jr. Cambridge, Mass.: Harvard University Press, 1972.

Corbin, Henry. *Spiritual Body and Celestial Earth.* Trans. Nancy Pearson. Princeton, N.J.: Princeton University Press, 1977.

Cornford, Francis M. *Plato's Cosmology.* New York: Routledge, 1937.

Dhorme, E. *A Commentary on the Book of Job.* Trans. Harold Knight. London: Nelson, 1967.

Fabre d'Olivet. *Music Explained as Science and Art.* Trans. J. Godwin. Rochester, Vt.: Inner Traditions International, 1987.

Festugière, A. J. "L'âme et la musique d'après Aristide Quintilien." *Transactions of the American Philological Association* 85 (1954): 55–78.

Ficino, Marsilio. *The Letters of Marsilio Ficino.* Trans. by members of the Language Dept. of the School of Economic Science. 4 vols. London: Shepheard–Walwyn, 1975, 1978, 1981, 1988 or 1989.

Fludd, Robert. *Escritos sobre Música.* Trans. Luis Robledo. Madrid: Editora Nacional, 1979.

Fourier, Charles. "Des Séries Mesurées." *La Phalange* 2 (1845): 352–384; 3 (1846): 5–19.

———. *Oeuvres Complètes.* 6 vols. Paris, 1841–1845.

———. *Théorie des Quatre Mouvements et des Destinées Générales.* Ed. Simone Debout. Paris: Pauvert, 1967.

———. *Traité d'association domestique agricole.* Paris & London, 1822. 2nd ed. (1841–1843) called *Théorie de l'Unité universelle.*

Fulgentius. *Mythologia.* Trans. Leslie George Whitbread. Columbus, Ohio: Ohio State University Press, 1971.

Gerbert, Martin. *Scriptores ecclesiastici de musica sacra.* St. Blasius, 1784. Repr. Broude.

Godwin, Joscelyn. *Athanasius Kircher: A Renaissance Man and the Quest for Lost Knowledge.* London: Thames & Hudson, 1979.

———. "Athanasius Kircher and the Occult." *Athanasius Kircher und seine Beziehungen zum gelehrten Europa seiner Zeit.* Ed. John Fletcher. Wiesbaden: Harrassowitz, 1988, pp. 17–36.

———. *L'Esotérisme musical en France.* Paris: Albin Michel, 1991.

———. "The Genesis of a Universal System: The Archéomètre of Saint-Yves d'Alveydre." *Alexandria* 1 (1991): 229–249.

———. *Harmonies of Heaven and Earth. The Spiritual Dimension of Music from Antiquity to the Avant-Garde.* Rochester, Vt.: Inner Traditions International/London: Thames & Hudson, 1987.

———. *Music, Mysticism and Magic. A Sourcebook.* London: Routledge & Kegan Paul, 1986.

———. *Robert Fludd, Hermetic Philosopher and Surveyor of Two Worlds.* London: Thames & Hudson, 1979. Reprint, Grand Rapids: Phanes Press, 1991.

———. "Saint-Yves d'Alveydre and the Agarthian Connection." *Hermetic Journal* 32 (1986): 24–34; 33 (1986): 31–38.

Guthrie, Kenneth Sylvan, ed. *The Pythagorean Sourcebook and Library.* Grand Rapids: Phanes Press, 1987.

Haar, James. "The Frontispiece of Gafori's *Practica Musica* (1496)" in *Renaissance Quarterly* 27 (1974): 7–22.

———. "Musica Mundana: Variations on a Pythagorean Theme." Ph.D. dissertation, Harvard University, 1960.

———. "Roger Caperon and Ramos de Pareja." *Acta Musicologica* 4 (1969): 26–36.

Haase, Rudolf. *Geschichte des harmonikalen Pythagoreismus.* Vienna, 1969.

Handschin, Jacques. "Die Musikanschauung des Johannes Scotus

(Erigena)." *Deutsche Vierteljahrschrift für Literaturwissenschaft und Geisteswissenschaft* 5 (1927): 316–341.

Heath, Sir Thomas. *Greek Mathematics.* 2 vols. London: Oxford University Press, 1921.

Helmholtz, Hermann L. F. *On the Sensations of Tone.* 2nd English ed., trans. and rev. A. J. Ellis. London, 1885. Repr. Dover, 1954.

Heninger, S. K. *The Cosmographical Glass. Renaissance Diagrams of the Universe.* San Marino, Calif.: Huntington Library, 1977.

———. *Touches of Sweet Harmony: Pythagorean Cosmology and Renaissance Poetics.* San Marino, Calif.: Huntington Library, 1974.

Huffman, William H. *Robert Fludd and the End of the Renaissance.* London: Routledge & Kegan Paul, 1988.

Hutton, James. "Some English Poems in Praise of Music." *English Miscellany* 2 (1951): 13–26.

Jan, Carl von. "Die Harmonie der Sphären." *Philologus* NF 6 (1894), 13–37.

———. *Musici Scriptores Graeci.* Leipzig, 1845.

Juden, Brian. *Traditions orphiques et tendances mystiques dans le romantisme français (1800–1855).* Paris: Klincksieck, 1971. Repr. Slatkine.

Kayser, Hans. *Lehrbuch der Harmonik.* Zurich: Occident Verlag, 1950.

———. *Orphikon. Eine harmonikale Symbolik.* Ed. Julius Schwabe. Basel: Schwabe, 1973.

Levarie, Siegmund, and Ernst Levy. *Musical Morphology: a Discourse and a Dictionary.* Kent, Ohio: Kent State University Press, 1983.

———. *Tone: a Study in Musical Acoustics.* Rev. ed., Kent, Ohio: Kent State University Press, 1980.

Levin, Flora R. *The Harmonics of Nicomachus and the Pythagorean Tradition.* University Park, Pa.: American Philological Assoc., 1975.

MGG. Die Musik in Geschichte und Gegenwart. Ed. Friedrich Blume. 17 vols. Cassel: Bärenreiter, 1949–1986.

The Magical Calendar of Tycho Brahe. Ed. Adam McLean. Edinburgh: Magnum Opus Hermetic Sourceworks, 1980.

Marinus. *The Life of Proclus.* Trans. Kenneth Sylvan Guthrie. Grand Rapids: Phanes Press, 1986.

Mathers, S. L. McGregor. *The Kabbalah Unveiled.* London: Redway, 1887.

McClain, Ernest G. *Meditations through the Quran.* York Beach, Me.: Nicolas-Hays, 1981.

———. *The Pythagorean Plato: Prelude to the Song Itself.* Stony Brook, N.Y.: Nicholas Hays, 1978.

Mead, G. R. S. *Orpheus.* London: Theosophical Publishing Soc., 1896. Repr. Watkins, 1965.

Meijering, E.P. *Orthodoxy and Platonism in Athanasius.* Leiden: Brill, 1968.

————. *Questions Harmoniques*. Paris, 1634.

Michell, John. *Ancient Metrology. The Dimensions of Stonehenge and of the Whole World as Therein Symbolised*. Bristol: Pentacle Books, 1981.

————. *The Dimensions of Paradise: The Proportions and Sacred Numbers of Ancient Cosmology*. London: Thames & Hudson, 1988.

Moutsopoulos, Evanghélos. "Mousike kinesis kai psuchologia en tois eschatois Platonikois dialogois." *Athena* 64 (1960): 194–208.

————. "Sur la 'participation' musicale chez Plotin." *Philosophia* 1 (1971): 379–389.

Münxelhaus, Barbara. "Aspekte der Musica Disciplina bei Eriugena." *Jean Scot Erigène et l'histoire de la philosophie*. Colloque International du C.N.R.S. (Paris, 1977): 253–262.

Nasr, Seyyed Hossein. *An Introduction to Islamic Cosmological Doctrines*. Boulder, Col.: Shambhala / London: Thames & Hudson, 1978.

Needham, Joseph. *A History of Embryology*. 2nd ed. Cambridge: Cambridge University Press, 1959.

Neugebauer, Otto. *A History of Ancient Mathematical Astronomy*. New York: Springer, 1975.

The New Grove Dictionary of Music and Musicians. Ed. Stanley Sadie. 20 vols. London: Macmillan, 1980.

Pietzsch, Gerhard. *Die Klassifikation der Musik von Boetius bis Ugolino con Orvieto*. Halle: Niermayer, 1929.

Reese, Gustav. *Music in the Middle Ages*. London: Dent, 1941.

Regardie, Israel. *A Garden of Pomegranates*. St. Paul, Minn.: Llewellyn, 1975.

Reinach, Théodore. "La Musique des sphères." *Revue des études grecques* 13 (1900): 432–449.

Roussier, Pierre Joseph. *Mémoire sur la musique des anciens*. Paris, 1770. Repr. Broude.

Ruelle, Charles-Emile. "Le chant gnostico-magique des sept voyelles grecques." With "Analyse musicale" by Elie Poirée. *Congrès international d'histoire de la Musique* (Paris, 1900): 15–38.

Schäfke, Rudolf. *Geschichte der Musikästhetik in Umrissen*. 3rd ed. Tutzing: Hans Schneider, 1982.

Scharlau, Ulf. *Athanasius Kircher (1601–1680) als Musikschriftsteller*. Marburg: Studien zur hessischen Musikgeschichte, 1969.

Seay, Albert. "Ugolino of Orvieto, Theorist and Composer." *Musica Disciplina* 9 (1955): 111–166.

Sepher Yetzirah: The Book of Creation. Trans. Irving Friedman. New York: Samuel Weiser, 1977.

Shiloah, Amnon. "Un ancien traité sur le `ud d'Abu Yusuf al Kindi." *Israel Oriental Studies* 4 (1974): 179–205.

————, trans. "L'épître sur la musique des Ikhwan al-Safa." *Revue des Etudes islamiques* 32 (1965): 125–162; 34 (1967): 159–193.

Shiloah, Amnon. "Un ancien traité sur le `ud d'Abu Yusuf al Kindi." *Israel Oriental Studies* 4 (1974): 179–205.

———, trans. "L'épître sur la musique des Ikhwan al-Safa." *Revue des Etudes islamiques* 32 (1965): 125–162; 34 (1967): 159–193.

Taylor, A. E. *A Commentary on Plato's Timaeus.* Oxford: Oxford University Press, 1928.

Taylor, Thomas. *The Mystical Hymns of Orpheus.* 2nd ed. London, 1824.

———. *Selected Writings.* Ed. George Mills Harper and Kathleen Raine. Princeton, N.J.: Princeton University Press, 1969.

———. *The Theoretic Arithmetic of the Pythagoreans.* London, 1816. Repr. Weiser, 1978.

Thorndike, Lynn. *A History of Magic and Experimental Science.* 8 vols. New York: Columbia University Press, 1923–1958.

Timaios of Locri. *On the Nature of the World and the Soul.* Trans. Thomas H. Tobin. Chico, Calif.: Scholars Press, 1985.

Toussenel, Alphonse. *Passional Zoology; or, Spirit of the Beasts of France.* Trans. M. E. Lazarus. New York, 1852.

Walker, D. P. *Spiritual and Demonic Magic from Ficino to Campanella.* London: Warburg Institute, 1958.

———. *Studies in Musical Science in the Late Renaissance.* London: Warburg Institute, 1978.

Waszink, J. H. *Studien zum Timaioskommentar des Calcidius.* 2 vols. Leiden: Brill, 1964.

Wetherbee, Winthrop. *Platonism and Poetry in the Twelfth Century.* Princeton, N.J.: Princeton University Press, 1972.

Westerink, L. G., ed. *Anonymous Prolegomena to Platonic Philosophy.* Amsterdam: North-Holland Publishing Co., 1962.

Wind, Edgar. *Pagan Mysteries in the Renaissance.* New and enlarged ed. London: Faber, 1968.

Wolfson, H. A. "The Internal Senses in Latin, Arabic, and Hebrew Philosophic Texts." *Harvard Theological Review* 28 (1935): 69–133.

———. "The Problem of the Souls of the Spheres, from the Byzantine Commentaries on Aristotle through the Arabs and St. Thomas to Kepler." *Dumbarton Oaks Papers* 16 (1962): 67–93; also in his *Studies. . .* II: 250–314.

———. *Studies in the History of Philosophy and Religion.* Cambridge, Mass.: Harvard University Press, 1973.

Yates, Frances A. *The French Academies of the Sixteenth Century.* London: Warburg Institute, 1947, repr. Kraus, 1973.

———. *Giordano Bruno and the Hermetic Tradition.* London: Routledge & Kegan Paul, 1964.

———. *The Occult Philosophy in the Elizabethan Age.* London: Routledge & Kegan Paul, 1979.

Index Nominum

Achilles Tatios, 411
Alexander of Aetolia, quoted by
    Theon, 18–19
Alexander the Great, 94–6
Al-Hasan al-Katib, 119–22, 440–1
Al-Kindi, 120, 438
Ammonius, quoted by Hunayn, 92–3,
    435
Amphion, 67, 458, 463
Anonymous of the Twelfth Century,
    123–5, 146, 441
Anselmi, G., 145–51, 170, 443–4
Apollo
    and Bacchus or Dionysus, 77, 188
    as decad, 434
    his lyre, 48–9
    and Marsyas, 55, 56, 423
    and Muses, 66, 85, 178–9
    Orphic hymn to, 46–7
    his plectrum, 268
    as sun, 306, 420, 433
    in various realms, 431
Archytas, quoted by Hunayn, 97
Aristeides Quintilianus, 51–9, 181,
    411, 422–5, 435, 459

Aristotle,
    cited or quoted by Bodin, 217
    Glarean, 201
    Hunayn, 95, 435
    Jacques de Liège, 132, 135, 137,
        140
    Mersenne, 252
    Rice, 392
    Schopenhauer, 343
    Ugolino, 143–4
    Zarlino, 212, 453
    on embryology, 418
    in Middle Ages, 458
    on pitch, 427
Aristoxenus, 419–20, 463, 475
Athanassakis, A. N., translator of
    Orphic hymn, 46–7
Athena, 423
Aurelian of Réôme, 99–103, 436
Azbel (E. A. Chizat), 399–401

Bacchus, nine different, 433–4
Bach, J. S., 291, 461
Bailly, J. S., 468
Barbara, A., 425

Beer, G. A., 459
Beethoven, L. van, 383, 394
Benivieni, D., 163–4
Berardi, A., 286–91, 460–1
Bodin, J., 214–18, 453
Boehme, J., 323–4, 346, 467–8
Boethius, 9, 86–8, 434–5
  anonymous addition to, 123–5
  on brass spheres, 410
  cited or quoted by Aurelian, 102
    Giorgi, 186
    Glarean, 199, 201
    Kircher, 265
    Regino, 111, 438
    Ugolino, 144
    Zarlino, 208, 210, 453
  on Hermes' lyre, 452
  ignored by Anselmi, 145
  his work used by Bodin, 453
    Fabre d'Olivet, 467
    Giorgi, 451
    Jacques de Liège, 128
    Ramis, 170
Bower, C., translator of Boethius,
  87–8
Bragard, R., 408
Brahe, T., 264, 408, 450
Brahmins, 347
Brethren of Purity, 112–18, 438–40
Briennius, M., 459
Briseux, C-E., 309, 310, 463
Bruckner, A., 363
Bry, J. T. de, 236

Calcidius, 4, 60–3, 426
Caperon, R., 446
Caspar, M., translator of Kepler, 222
Castel, L. B., 362
Censorinus, 40–45, 411, 418–20
Chaldaeans, embryology of, 41–2
Chinese, system of, 347, 472
Chizat, E. See Azbel
Cicero
  cited or quoted by Glarean, 199,
    200–1
    Regino, 111
    Rice, 475
    Twelfth-century Anonymous, 124
    Zarlino, 206
  Macrobius' commentary on, 64–70
  planetary system of, 87–8, 416
    contrasted with Pliny's, 7

Cornford, F. M., 3
Court de Gébelin, A., 465
Creuzer, G. F., 370
Cylennius, quoted by Newton, 308–9

Dalberg, J. F. H. von, 335–7, 459,
  467
Dante, source for Anselmi, 146, 444
David, King. See Psalms
Debout, S., 470
Demiurge, Plato's, 3–6, 57, 62, 66,
  404
  like Apollo, 46
  Proclus on work of, 77, 85
Diodorus Siculus, 178, 452
Dion Cassius, 353
Dionysius the Areopagite, 72, 187,
  442
Dionysos, modal theorist, 121
Dupuis, J., translator of Theon,
  17–20
Düring, I., translator of Ptolemy,
  22–39, 411–12

Empedocles, quoted by Calcidius, 61
Ephorus, quoted by Hunayn, 98
Eratosthenes, 19–20, 407, 411
Eriugena, John Scotus, 104–8, 436–8
Etruscans, on the Muses, 67
Euclid, 98, 314
Eumolpids, 348–50, 469
Eusebius, cited by Newton, 306

Fabre d'Olivet, 336, 344–55, 446,
  465, 467, 475
Farndell, A., translator of Ficino, 164–9
Favonius Eulogius, 437
Festugière, A.-J., translator of
  Proclus, 73–5
Ficino, M., 163–9, 175, 177, 433,
  445–6, 449
Fludd, R., 236–49, 456–7
  cited by Kircher, 264, 267–8, 459
  compared with Berardi, 460
    Fabre d'Olivet, 344
    Mersenne, 250
Fourier, C., 356–61, 469–70, 475
Fréret, N., 469
Fulgentius, 448

Gafori, F., 177–84, 440
Galen, system of, 216, 443

Galileo, 460
Giorgi, F., 185–195, 448–51
  and Berardi, 287
  compared with Berardi, 460
    Mersenne, 250
  Kircher's use of, 264
Glarean, H., 196–204, 451–2
Godwin, J., translator of Al-Hasan,
    120–2
  Anselmi, 146–51
  Azbel, 400–1
  Berardi, 287–91
  Calcidius, 61–3
  Censorinus, 41–5
  Dalberg, 336–7
  Eriugena, 105–8
  Fabre d'Olivet, 345–55
  Fludd, 237–49
  Giorgi, 186–95
  Isaac ben Haim, 153–9
  Jacques de Liège, 129–40
  Kepler, 222–35
  Kircher, 264–85
  Mersenne, 251–62
  Nicomachus, 10–15
  Pico, 176
  Pliny, 8
  Ptolemy, 22–39
  Rameau, 310–13
  Ramis, 171–4
  Saint-Martin, 323–34
  Saint-Yves, 396–8
  Singer, 363–9
  Tartini, 315–21
  Thimus, 371–81
  Toussenel, 356–61
  Twelfth-century Anonymous, 124–5
  Werckmeister, 293–301
  Ugolino, 142–4
  Zarlino, 206–13
Gushee, L., 436

Haar, J., 411
Haase, R., 222, 472
Haldane, R. B., translator of
    Schopenhauer, 339–43
Handschin, J., editor of Twelfth-
    century Anonymous, 123
Heracleitus, 421
Hermes, 423, 452. *See also* Mercury,
    Hermeticism
Hermesianax, 462

Hesiod, on Muses, quoted by Glarean,
    203, 452
  Macrobius, 66
  Ramis, 173–4
  Zarlino, 207
Homer
  Aristeides' interpretation of, 52, 54
  on Sirens, 74
Hucbald, 392–3
Huffman, W., 456
Hunayn, 91–8, 435

Iamblichus, 175, 409, 410, 472
Ibn Latif, 126–7, 441–2
Ikhwan al-Safa'. *See* Brethren of
    Purity
Isaac ben Abraham. *See* Ibn Latif
Isaac ben Haim, 152–9, 444–5
Isidore of Seville, 99, 131, 135, 442

Jacques de Liège, 128–140, 442, 451
Jan, C. von, 417, 439
Jean de Muris, cited by Mersenne,
    252
Jesus, name of, 476
Juden, B., 463
Julian, Emperor, 347, 468, 469

Kayser, H., 371, 412, 451, 461, 465
Kemp, J., translator of Schopenhauer,
    339–43
Kepler, J., 221–35, 454–6
  and Anselmi, 145
  and Eriugena, 105
  and Kircher, 264
  and Mersenne, 250
  and Ptolemy, 21, 416
  and Werckmeister, 461
Kircher, A., 263–85, 362, 458–60
  on hieroglyphs, 465
  uses Gafori's illustration, 448
  and Werckmeister, 461–2
Kuntz, M. L. D., translator of Bodin,
    215–18

Lawlor, R. and D., translators of
    Theon, 17–20
Le Fèvre de la Boderie, G., 186
Leibniz, 467
Levarie, S., 472, 474
Levin, F., 408–9
Levy, E., 472

Lippius, J., 461–2

Macrobius, 64–70, 426–7
    cited or quoted by Fabre d'Olivet,
            346
        Glarean, 197
        Newton, 306, 307
        Ramis, 174
        Regino, 111
    criticized by Glarean, 201
    known to Twelfth-century
            Anonymous, 123
    used by Fludd, 457
        Pico, 447
        Ramis, 174
    Saturnalia, 443
Mani, 345
Maria, name of, 476
Marinus, biographer of Proclus, 71
Martianus Capella, 347, 407, 436–7
    quoted by Regino, 110–11
    used by Ramis, 170, 174, 446
Martinez de Pasqually, 323, 324, 467
Mary Protase Le Roux, Sister,
            translator of Regino, 110–11
Mathiesen, T., translator of
            Aristeides, 51–9
    quoted, 51, 422–5
McClain, E. G., 3–4, 405, 449, 462,
            473
McGuire, J. E., quoted, 305
Mead, G. R. S., 428, 434
Medici, Cosimo de', 163–4
Mercurius Trismegistus, cited by
            Zarlino, 209. *See also*
            Hermeticism
Mercury, inventor of the lyre, 198,
            213. *See also* Hermes
Mersenne, M., 250–62, 454, 457–8
Michell, J., 4, 8, 405, 407, 461
Miller, C. A., translator of Gafori,
            177–84
        Glarean, 197–204
Moses, interpreted by Isaac ben
            Haim, 154–5
Moutsopoulos, E., 423
Muses. *See also* Correspondences
    distinguished from Sirens, 73–5, 85
    etymology, 67
    in Hesiod, 66
    led by Apollo, 47

theories of Anselmi, 151
    Aristeides, 56
    Aurelian, 102
    Dalberg, 336–7
    Gafori, 178–81
    Glarean, 202–4, 452
    Proclus, 427–8, 433–4
    Ramis, 174
    Varro, 178, 202
    Zarlino, 207

Nasr, S. H., 444
Neoplatonists, 52, 167–8
Neugebauer, O., 417
Newton, I., 305–8, 410, 462
    criticized by Rameau, 313, 463–4
Nicolaus of Capua, 442
Nicomachus of Gerasa, 9–15, 408–10
    Boethius based on, 86–7
    cited by Al-Hasan, 119
        Giorgi, 194, 451
    on harmonic mean, 440
    on Hermes' lyre, 452
    used by Pico, 175
        Thimus, 472

Orpheus, Hymns of, 46–7, 420–1,
            430, 433–4. *See also* Orphic
            doctrines
    cited by Newton, 308
    described by Vergil, 441
    Ficino on his theory, 165
    invented four-stringed lyre, 198
    powers of, 67, 254, 458
    Proclus on, 72
Ovid, quoted by Zarlino, 211–12, 213

Pan
    his crook, 58
    god of harmony, 307–8
    a name of Apollo, 47
    of the universe, 346, 420
Paolini, F., 441
Papus (G. Encausse), 396
Phanes, 434
Philolaus, 378, 462, 469
Pico della Mirandola, 175–6, 447,
            449, 451
    quoted by Berardi, 288
Plato, 3–6, 403–6, 448–9
    Cratylus, cited by Proclus, 75

Laws
  cited by Proclus, 77–8
  quoted by Schopenhauer, 343
opinions of Anselmi, 148
  Gafori, 183
  Mersenne, 253
Parmenides, cited by Giorgi, 186
Phaedrus, cited by Giorgi, 192
Politicus, quoted, 457
quoted by Hunayn, 93–4, 95–7
Republic
  cited by Anselmi, 150
  Glarean, 202
  Pico, 447
  Tartini, 315
  list of sciences in, 16, 17, 410
  Macrobius on, 66
  Proclus' commentary, 71–5
Symposium, 213, 423, 467
Timaeus
  Aristeides' commentary, 51–9, 425
  cited by Fabre d'Olivet, 344
  Tartini, 320
  Zarlino, 208
  Porphyry's lost commentary, 426
  commentary by Giorgi, 185–8
  Proclus' commentary, 74, 75–85
  translated by Calcidius, 60–3
  and Twelfth-century Anonymous, 441
  used by Bodin, 453
  Nicomachus, 9, 15, 410
  two principles of planetary music, 437
Plethon, G. Gemistos, 163
Pliny the Elder, 7–8, 407
  cited by Newton, 306
  quoted by Glarean, 199, 452
  Ugolino, 143
  on zones of the earth, 420
Plotinus, 52, 421
Plutarch, 411, 424
Ponte, J., translator of Aurelian, 100–3
Porphyry, 52, 426
  quoted by Macrobius, 68
Proclus Diadochus, 4, 71–85, 406, 427–34
  cited by Giorgi, 186–8, 449
  former life as Nicomachus, 9
  quoted by Newton, 306

used by Pico, 175
Ptolemy, 21–39, 411–18
  cited by Zarlino, 208–9
  criticized by Kepler, 227
  known to Al-Hasan, 119, 440
  Hunayn, 435
  Ibn Latif, 442
  scale of, 397
Pythagoras and Pythagoreans
  astronomical theories of, 18, 44, 305–8
  "Grand Unified System" of, 9
  instruments used by, 11
  Nicomachus records myths of, 9, 12–14
  silence of, 8, 319
  teachings reported by Censorinus, 42–3, 419
  Fabre d'Olivet, 344
  Ficino, 165–6
  Gafori, 184
  Giorgi, 190
  Jacques de Liège, 137–8
  Kircher, 269
  Mersenne, 250
  Newton, 305–8
  Pliny, 7–8
  Ptolemy, 29
  Ramis, 171
  Regino, 110
  Rice, 392
  Thimus, 376–81
  Varro, 419
  Zarlino, 208
  Theon an inheritor of, 16

Rameau, J.-P., 309–13, 362, 463–4, 466
Ramis de Pareja, 170–4, 178, 446–7
Rattansi, P. M., quoted, 305
Regino of Prüm, 109–11, 438
Reinach, T., 411
Rheita, G. astronomer, 277–8
Rice, I. L., 382–94, 474–5
Robledo, L., translator of Fludd, 237
Rousseau, J.-J., 323, 336
Roussier, P.-J., 362, 463
Rubeli, A., translator of Tartini, 315, 464–5
Ruelle, C.-E., translator of Nicomachus, 10–15, 409

St. Athanasius, 48–50, 421
St. Augustine cited or quoted by
    Gafori, 178
  Giorgi, 193
  Jacques de Liège, 133, 139
  Mersenne, 257, 259
  Ramis, 172
  Ugolino, 143
  Werckmeister, 298
Saint-Martin, L.-C. de, 323–34, 347,
    465–7, 468
St. Paul, quoted by Giorgi, 191, 451
  Jacques de Liège, 133–4
  Kircher, 271
  Mersenne, 258, 261–2
St. Thomas of Villanova, quoted by
    Berardi, 291
Saint-Yves d'Alveydre, 395–8, 475
Schopenhauer, A., 338–43, 467
Seneca, cited by Bodin, 216
Shakespeare, W., quoted by Rice, 387
Shiloah, A., translator of the Brethren
    of Purity, 113–18, 435,
    438–40
  Isaac ben Haim, 153–9, 444–5
Singer, P., 362–9, 470–1
Sirens
  in Anselmi, 150
  in Aristeides, 56
  in Glarean, 202
  in Proclus, 73–5, 85, 428, 434
  in Zarlino, 206–7
Socrates, 411
Sonne, I., translator of Hunayn, 92–8
  Ibn Latif, 126–7
Stewart, H. F., 453
Sufis, 92, 146

Tartini, G., 314–21, 370, 464–5, 466
Taylor, T., cited, 410, 421, 428, 430–1,
    433, 468

translator of Plato, 4
  Proclus, 73, 75–85
Terpander, 213
Thales, cited by Newton, 308
Thebit ben Corat, 443
Theon of Smyrna, 16–20, 410–11,
    458–9
Thimus, A. von, 10, 370–81, 451,
    472–4
Thrasyllus, astrologer, 16
Timaeus of Locris, 3–6, 467. See also
    under Plato, Timaeus
Toussenel, A., 356–61, 469–70
Trithemius, 450
Tullius. See Cicero

Ugolino of Orvieto, 141–4, 443

Valla, G., cited by Glarean, 200, 452
Varro, 419
Vergil, quoted, 179, 197, 441
Vincent, A.-J.-H., 409
Vogler, Abbé, 362, 363, 470
Volney, C. F., 469

Wagner, R., 338, 362
Walker, D. P., 464
Wallis, J., translator of Ptolemy, 22
Waszink, J. H., 426
Werckmeister, A., 292–301, 461–2
Werner, E., translator of Hebrew and
    Arabic writers, 438
Wilde, O., quoted, 447
Wind, E., 434

Yates, F. A., 448, 451, 457

Zamorensis, A., 442
Zarlino, G. 205–18, 255, 405, 452

Index Rerum

Academy of Athens, 435
Accademia dei Lincei, 460
Acoustics
  Aristotle, 427
  Fludd, 237, 456–7
  Kircher, 265–6
  Macrobius, 68–9
  Nicomachus, 11–12, 408
  Rice, 388
  Singer, 470–1
Aeolian harp, 271, 459
Aleph, Dark and Light, 257, 458
Ancients and Moderns, 299–300, 344
Angels, nine orders of. *See also*
      Spheres, souls of
  Anselmi, 150–1
  Berardi, 288, 290
  Dionysius, 442
  Fludd, 457
  Giorgi, 189
  Jacques de Liège, 134
  Kircher, 281–2, 460
  planetary, 281–2
  Twelfth-century Anonymous, 125
Arabic alphabet, 117

Arabic music theory, 97, 121–2
Archéomètre of Saint-Yves, 395–8, 475–6
Architecture, 309, 310–11, 463
Ark, dimensions of, 461–2
Asteroids, 400–1
Auditions voilées of Azbel, 399
Aulos, 56, 423–4

Beauty, 23–4, 384–6
Bode's Law, 399
Body, harmony with soul, 244–9
Brass spheres, 409–10

Cabala. *See* Kabbalah
Cadences, symbolism of, 466–7
Canobic Inscription, 417–18
Cataract of the Nile, 70, 207, 270
Central Fire, 374, 378
Chinese scriptures, 468, 472
Christ, two natures of, 297
Color, 255–6, 313, 388–9, 390, 463–4
Common chord, symbolism of, 294–8,
      323–9, 363–9
Consonance and dissonance, 43, 166–7,
      217–18, 254, 258, 264–6

between planets, 38–9, 209, 240–1,
    273–5
like good and evil, 461
Correspondences
cube with earth, Brethren of
    Purity, 115
cube with elements, Giorgi, 194
four voices with kingdoms of
    nature, Schopenhauer, 341–3
Greek tetrachords with aspects of
    sun and moon
    Ptolemy, 36–7
    Zarlino, 208
harmonic series with the creation,
    Werckmeister, 294–5
harmonic series with planets, Azbel,
    400–1
intervals or tones with elements
    Berardi, 290
    Ficino, 167–8
    Fludd, 239–40
    Giorgi, 194
    Glarean, 198
    Jacques de Liège, 139
    Kircher, 267–9
    Zarlino, 209–13
intervals or tones with powers or
    parts of the soul
    Al-Hasan, 120–1
    Berardi, 287–8
    Fludd, 248–9
    Pico, 176
    Ptolemy, 26–8
    Toussenel, 357–9
intervals with astrological aspects
    Ficino, 168–9
    Kepler, 227–8, 454
    Mersenne, 261
    problematic, 418
    Ptolemy, 31–3, 37–8
    Zarlino, 208–9
intervals with embryology
    Brethren of Purity, 117
    Censorinus, 42–3, 419
intervals with geometrical prin-
    ciples, Proclus, 431
intervals with pagan gods, Ficino,
    165
intervals with planetary distances
    Azbel, 400–1
    Boethius, 87

Brethren of Purity, 115–16
Calcidius, 62–3
Censorinus, 44–5
Eriugena, 106–8, 437
Fludd, 241
Glarean, 199–201
Macrobius, 68
Newton, 306–7
Pliny, 8
Regino, 110–111
Theon, 18
Thimus, 375–7, 379–81
Zarlino, 208
keyboard with planets, moons,
    Fourier, 470
major and minor scales with
    planetary motions, Kepler,
    230–2
means with daughters of Themis,
    Pico, 176
means with number, measure, and
    weight, Giorgi, 193
modes with humors, emotions,
    planets, Muses, Ramis,
    171–4
modes with journey of the sun,
    Orpheus, 420
modes with Muses
    Aurelian, 102
    Gafori, 180–2
modes with planetary motions
    Aurelian, 101
    Isaac ben Abraham, 127
    Kepler, 233–4
modes with states of the soul,
    Ptolemy, 28–9
modes with zodiac, Ptolemy, 415
modes with zones of the earth,
    Ptolemy, 35
Muses with parts of the mouth,
    Gafori, 179
musical ratios with divisions of
    world-soul
    Anselmi, 147
    Aristeides, 57–8
    Giorgi, 191
    Kircher, 284–5
    Macrobius, 65–6, 68
    Nicomachus, 15
    Plato, 5
    Proclus, 78–9, 81, 84–5

names of God with Sephiroth,
angels, spheres, deduced
from Giorgi, 450
oval shape of Muses' chorus with
ear and instruments, Ficino,
165–6
pitch changes with movements of
stars
Ptolemy, 33–4, 414–15
Zarlino, 208
pitch changes with three genera
Anselmi, 149
Ptolemy, 34
Zarlino, 208
planets with divine names, Giorgi,
194
rhythms with heavens and hells,
Brethren of Purity, 117
scale with zodiac
Al-Hasan, 122
Ficino, 168–9
Kayser, 412–13;
Ptolemy, 29–31, 412–13
senses with elements, Aristeides,
412
strings with temperaments, humors,
seasons, and elements
Al-Kindi, 438
Brethren of Purity, 114–15
Hunayn, 97
Isaac ben Haim, 154, 158
tones with angelic orders
Anselmi, 150–1
Berardi, 288
Fludd, 242
Giorgi, 189
Kircher, 281–2
Twelfth-century Anonymous, 125
tones with colors, Rice, 388–90
tones with epicycles, Ptolemy, 415
tones with Muses, Sirens, planets
Glarean, 202–3, 452
Zarlino, 207–8
tones with passions, planets, moons,
tones, colors, Toussenel,
359–60
tones with planetary motions
Anselmi, 149
Berardi, 290
Boethius, 87–8
Eriugena, 105–6, 437

Gafori, 181–2
Giorgi, 194
Glarean, 199–200
Isaac ben Haim, 157
Jacques de Liège, 136–7, 138
Kepler, 228–35
Kircher, 274
Macrobius, 66, 69
Nicomachus, 10
Proclus, 73–4
Regino, 111
Theon, 18–19
Twelfth-century Anonymous,
124–5
Ugolino, 143
Zarlino, 206
tones with planets, days of the
week, and hours of the day,
Fabre d'Olivet, 352–4
tones with Platonic solids
Brethren of Purity, 115
Kepler, 225–7
Kircher, 269
tones with Sephiroth, angels,
planets, powers of the soul,
elements, Berardi, 288–90,
460
tones with twelve tribes, Apostles,
Singer, 368
tones with zodiac, Hebrew letters,
colors, planets, Saint-Yves,
395–6
two-octave system with planets,
Eriugena and Favonius, 437
Cosmological system, Kircher, 264
Counter-Earth, 374, 378
Creation, musical parallel, 293–5

Dance of Muses, 47
Days of the week, 352
Death of the kiss, 192, 285, 451
Diatonic scale, origin of, 347–54
Distances of planets
Censorinus, 44–5
Eriugena, 107–8, 438
Macrobius, 68
Pliny, 7–8, 407
Porphyry, 426–7
Theon, 18
Dodecahedron, 115, 345. *See also*
Platonic solids

Ear, symbolism of, 165–6
Earth
  bass to the planets, 275
  dimensions of, 407
  not silent, 174
  silent, 87, 108, 174
  sounds produced by, 269–70
Ecliptic, 29
Elements. *See* Correspondences
Ellipse. *See* Oval
Embryology, 40–5
Emmeleis, defined, 411
Enneachords, 285
Etymology of "music," 131
Experiments
  Kircher, 266–7
  Pythagoras, 12–14, 43–4, 307
  Theon, reported by Galileo, 458–9
Eye and ear, 388–9

Funeral customs, 207

Genera, use of in Ancient Greece,
    407, 427
Geometry and music, 316–17, 474
Gnosis, 374
Gnostic powers, 433–4
God
  author of heavenly harmony, 151,
    218, 228
  coupled with mud, 245
  and creation, 471
  as lyre-player, 48–50, 421
  musical preferences of, 153–7
  musician, 283
  self-revelation of, 296–7
  of seven rays, 346–7
Gospels, quoted, 192, 193, 262
Greater Perfect System, outlined,
    412–13
Greek music theory, 10–14, 19, 28,
    34–7, 70, 87, 111, 149
  rediscovery of, 177
  system, 412–13

Harmonic division of string, 454–5
Harmonic series, 292, 294–5, 372,
    400–1
Harmonics, science of, 315–18, 465
Harmony of universe, Athanasius,
    49–50

Harmony of the Spheres, explained.
    See also under Correspon-
    dences
  Aurelian, 101
  Calcidius, 60–1
  Cicero, 66, 70
  Eriugena, 107
  Gafori, 183
  Glarean, 201–2
  Isaac ben Abraham, 127
  Jacques de Liège, 136–9
  Proclus, 73–5
  Regino, 110
  Theon, 17
  Ugolino, 143–4
  Zarlino, 206–7
Harmony, three types of, Proclus, 75
Healing powers of music, 114
Heliocentricity
  ancient, 348–9, 374–81, 392, 462
  Fludd's, 457
Hermeticism
  of Brethren of Purity, 112–13
  doctrines and principles, 146, 345,
    435, 475
  and Florentine Platonism, 163, 164
Hermetic writings quoted
  Bodin, 453
  Fludd, 246
  Zarlino, 209, 452
Hours ruled by planets, 352–3
Humors, defined, 443. *See also under*
    Correspondences

Ideas, Platonic, 341
Instruments, 55–6, 153–5, 362, 421–2
Intervals, musical. *See* Ratios,
    Correspondences
Inverse-square law, 305, 307

Jewish theory and philosophy, 126–7,
    152–9
Job, Book of
  passage on concert of the heavens,
    436
  quoted, 101, 137, 139, 207, 291,
    474
Jupiter, moons of, 277–9, 359, 400–1

Kabbalah
  Berardi's use of, 287–91

correspondences in, 450
doctrine of four worlds, 460
doctrines of, 257, 259, 375, 451,
    458
Giorgi an expert, 185, 250
known to Saint-Martin, 323
numerology of, 192–3, 440, 449,
    466
suggested by Ibn Latif, 442

Lambda, 404, 451, 472
Lambdoma, defined, 10, 472. *See also*
    Pythagorean Table
Language, original, of mankind, 465
Leimma, defined, 420
Liberal Arts, seven, 127, 157, 179,
    436
Lost system of music, 397
Love and Chaos, 355
Lute, Arab. *See* Oud
Lyncean School, 275, 278
Lyre. *See also* Correspondences, Oud
    of Apollo, 47, 308
    of David, 287–91
    eight-stringed, 14, 19–20
    four-stringed, 105, 194, 213
    Greek, 408
    of Hermes, 423, 451, 452
    meaning of different numbers of
        strings, Glarean, 197–9
    of Orpheus, 268
    preferable to aulos, 56–7
    seven-stringed, 14, 213
Macrocosm and microcosm, 239, 247–9.
    See also Correspondences
Mathematics, 16–18, 23–4, 176
Means, the three, 76, 78–9, 176, 193,
    314, 404–5, 430, 440, 464.
    See also Correspondences
Medieval music theory, 100, 171–4,
    196–7
Melody, symbolism of, 343
Mercy Seat, harmonic numbers of,
    298
Meter, symbolism of, 331–3
Modes. *See also* Correspondences
    compared to zones of earth, 34–5
    defined and distinguished, 446,
        455
    effects of, 28–9, 47, 171–2
    Glarean's expansion of, 196–7

to Muses, 173–4, 180–1, 183
origin of seven, 353
planetary, 8
to planets, 172–3, 196–7
Monochord, Fludd's symbolic, 236,
    239, 243–4, 247–9, 267–8
Moon, phases of, 36
Mouth, parts of, and Muses, 179
Music, effect on body and soul, 447
Musica anamastica, 205
Musica angelica, 442
Musica caelestis (celestis, coelestis), 130–6,
    141, 145
Musica cantabilis, 145
Musica humana, 86, 119, 128–30, 171–2,
    286, 443
Musica instrumentalis, 86, 119–20,
    128–30, 145, 171–2
Musica mundana, defined, 86, 119,
    128–9, 131, 135–6, 141–4,
    286, 376
    essays on, 135–40, 205–13
Musica organica, 205
Musical numbers, 396–7

Nature, harmony in, 311–13
Neoplatonic doctrines, 429, 433–4
Neoplatonists, Iranian, 450
Noeane, etc. 102–3
Numbers
    male and female, 404
    perfect, 439

Opposites, harmony of, 216–17
Order of planets, 411
Organ, acoustics of, 470–1
Origin of music, 336–8
Orphic doctrines, 348, 433–4, 446,
    469
Oud
    and lyre, 435–6
    symbolism of, 97, 113, 115, 121–2,
        152
Oval (elliptical) shape of cosmos,
    165–6, 348
    orbits of planets, 221

Pallas, 447
Passions, 358–9
Perennial Philosophy, 112, 163
Pipe of Pan, 307–8, 346

Planetary
  aphelia and perihelia compared,
    229–33
  gender, 181, 183
  good and evil influxes, 273, 280
  intelligences, 194
  order, 19–20, 408
  scales
    Boethius, 434–5
    Brethren of Purity, 439
    Canobic Inscription, 417
    Censorinus, 411
    Naples manuscript, 439
Planets, periods of rotation, 444, 456
Planet-tone theories. See Correspondences
Platonic solids, 225–7, 269, 433–4,
  454. See also correspondences
Poles, North and South, 35, 47
Power of music, 67, 92–8, 147–8,
  154–5, 254, 337, 426
Principles, two, in universe, 326
Prisci theologi, prisca theologia, 163, 175
Proportion. See Ratio
Psalms, cited, 127, 133, 136, 139,
  157–9, 188–9, 257, 262,
  283–4, 442
Ptolemaic tuning, 396
Pyramids, Fludd's symbolic, 236, 237–41,
  267–8, 456
Pythagorean Table, 371–4, 383, 451,
  472
Pythagorean tuning, 345, 397, 475

Quaternary. See Tetraktys

Ratios
  and astrological aspects, 31–3, 37–9
  classified by Proclus, 430
  consonant, 102
  of distribution of life in the world,
    191
  of elements, 239
  and gestation, 43
  of intervals, 12–15
  more important to the Ancients,
    184
  in natural sounds, 270
  in nature and architecture, 311
  between planets, 240–3, 350
  of planets and tones, 106

pleasing, 215–16
  of soul, 58–9, 80–2, 84, 120, 248–9
  in string, 316
  symbolism of, 164–5
  of universe, 284–5
Rays, seven, 468

Saturn, rings of, 275–7
Scale. See also Common chord, Triad,
    Correspondence
  derivation of diatonic, 466, 468
  directions of, 407–8
  meaning of its degrees, 391–4
  Timaeus, explained, 403–6, 410
Sciences, division of , 101–2
Sea, sounds of, 270
Seasons, motions of, 47, 139–40,
    212–13, 268, 386. See also
    correspondences
Sepher Yetzirah (Book of Formation),
    192, 378, 381, 473
Sephiroth, 442, 450, 460
Septenary, 22, 69, 77, 124, 187–9,
    197–9, 306, 328–9, 347, 441,
    447
Series, measured, 469
Soul. See also Correspondences,
    World-soul
  ascent of, 192
  and body, 423
  descent through spheres, 52, 53–5,
    421, 422, 435
  effects of music on, 52–3
  made from numbers, 62
  nature and parts of, 76, 82–3, 403,
    412, 427–9, 440
  powers or parts of, 25–7
Space and time, 383–7, 391
Spheres,
  beyond planetary, 444
  souls of, 442, 444
Sun. See also Heliocentricity
  central, 243, 377–8, 457
  triple, 469
Symphony of the world, 284
Synarchy of Saint-Yves, 475–6

Talmud, 445
Temperaments, four, 97, 113–14, 154,
    171–2. See also Correspon-
    dences
Tempered scale, 311–12, 345, 397

Tetraktys, 9, 58, 353, 414, 425
Timaeus scale, 427–33
Trees, harmonies of, 270
Triad
  acoustics of, 471
  relation to circle, 316
  symbolism of, 294–8, 323–9, 363–9,
    466
Trinity, Holy, 296–7, 364–5, 461
Triple progression, 463
Trumpet scale, 292

Unison, 166–7, 251–62

Vattan, ancient alphabet, 475
Vespertine knowledge, 134, 451
Vibrations, universal, 387–9
Virtues, 26–8

Winds, whistling, 47, 53, 421

Wisdom, Book of, quoted, 193, 209
World-soul. *See also* Correspondences
  Anselmi, 147
  Aristeides, 57–9
  Calcidius, 60–3
  Giorgi, 187–9, 449–50
  Macrobius, 65–6
  Nicomachus, 15
  Pico, 447, 449
  Plato, 3–6
  Proclus, 75–6, 403–6
World-Egg, 446, 469
Worlds, three, 449

Zodiac. *See also* Correspondences
  musical interpretations of, 30–2,
    101, 122, 208, 222–5, 227–8
  origin of, 346
Zodiacal houses, symbolism of, 168–9
Zones of the earth, defined, 420